Study Guide

Volume 1

Sears and
Zemansky's

UNIVERSITY
PHYSICS

with Modern Physics

11th Edition

Young & Freedman

JAMES R. GAINES
University of Hawaii

WILLIAM F. PALMER
Ohio State University

D1501110

PEARSON

Addison
Wesley

San Francisco • Boston • New York
Capetown • Hong Kong • London • Madrid • Mexico City
Montreal • Munich • Paris • Singapore • Sydney • Tokyo • Toronto

ISBN 0-8053-8778-1

3 4 5 6 7 8 9 10–PBT–07 06 05 04
www.aw.com/bc

CONTENTS

PREFACE

The emphasis in any calculus-based physics course for scientists and engineers is on problem solving. This Guide as been written to help you learn how to solve the kind of problems you will encounter in homework assignments and examinations.

Each chapter in the Guide corresponds to a chapter in the parent text, *University Physics*, Eleventh Edition, by Young and Freedman (Addison Wesley, 2004). The organization of the chapters is as follows:

Objectives. All chapters begin with a statement of learning objectives, stressing application of basic physical principles to problems and calculations.

Review. Primary concepts, definitions, formulas, units, and physical constants are then reviewed and highlighted.

Supplement. Some chapters have supplementary material which builds the basis for or expands upon materials of the text. These supplements are primarily in the early chapters, where they are concerned with applying basic skills of vector manipulation and calculus to physical problems.

Problem-Solving Strategies. Many chapters have a section on problem solving strategies, including step-by-step guidance on how to set up and complete a problem.

Questions and Answers. This section poses questions that are conceptual rather than mathematical in nature. Occasionally, a short calculation is required to answer a question.

Examples and Solutions. In this section examples similar to the problems at the end of the chapter in the text are worked out, step-by-step, with commentary on organization of given information, principles applied, set-up procedure, pitfalls, and alternative solutions.

Quiz. Each chapter has a short quiz, with answers. In addition, any of the solved examples may be used as a self-quiz.

The Guide is thus an organizer of study, review, and self-diagnostic activities, as well as a ready reference of solved problems. If difficulty is encountered with a homework problem, a similar one can usually be found in the Guide, fully worked out. Rather than stare at the homework problem, with no useful forward progress, you can refer to the similar one for guidance. This is one of the chief uses to which the Guide is put. Solved exercises as a pedagogical tool are a tradition in physics.

We have attempted to make the solutions clear and detailed. We have emphasized problems which demonstrate the main concepts, avoiding those which require only substitution into a formula of the text.

After reading the text and notes you may wish to read the objective and review sections, and then work through the examples, referring back to the text and supplement sections when an unfamiliar concept or method is met. This process may occur before, after, or during the process of attempting the homework.

A word of caution is in order when reading through solved problems. The acid test of understanding comes only when you challenge yourself by taking a self-administered examination. This occurs each time you sit down to do a homework problem, or you can make things more realistic by testing yourself with an example in the Guide under examination conditions: Do not peek at the solution. Use no books or notes. Then score yourself according to correct method and correct answer.

The results of such an exercise may be enlightening. Remember your instructor regards the statement "I understand the material but don't know how to solve the problems" as self-contradictory. In a sense, the problems **are** the material. A complete mastery of the material consists in the ability to (1) <u>identify</u> the appropriate physics concepts underlying the problem, (2) <u>set up</u> the problem, applying these concepts, (3) <u>execute</u> a solution with confidence and care to a final result, and (4) <u>evaluate</u> your result to ensure that it makes sense and that there are no computational errors.

To that end we hope this Study Guide will make your way a little less rough. Good luck!

James R. Gaines
William F. Palmer
Honolulu and Columbus, 2003

1
UNITS, PHYSICAL QUANTITIES, AND VECTORS

OBJECTIVES

In the first part of this chapter you are introduced to idealized physical models and to physical quantities, the result of measurements, which are related to each other by the laws and formulas of physics. Your objectives are to:

Recognize the standards and units for time, length, and mass.

Write down physical quantities with the correct number of *significant figures* and the correct *units*.

Calculate *absolute* and *percent* uncertainties.

Treat units *algebraically* and make unit *conversions*.

Make *estimates* and find the *order of magnitude* of quantities.

In the second part of this chapter you are introduced to *vectors* and vector operations. Your objectives are to:

Plot a vector.

Add and subtract two vectors.

Resolve a vector into its *rectangular components*.

Add and subtract vectors using the *component method*.

Write a vector and sums and differences of vectors in terms of *unit vectors*.

Multiply vectors using the *scalar product*.

Multiply vectors using the *vector product*.

REVIEW

Models of physical systems relate measurements of physical quantities. All physical quantities in the early parts of the text are given in terms of the basic mechanical *units* of measurement: meters (m) for length, kilograms (kg) for mass, and seconds (s) for time, in the SI system. For example a volume is given in terms of m^3; a density or mass per unit volume is given in terms of kg/m^3.

All quantities derived from measurement have *uncertainties or errors*. For example, if a length is known with some certainty to be between 10.1 m and 9.9 m, it is written as

$$L = 10.0 \pm 0.1 \text{ m.}$$

The absolute uncertainty is 0.1 m. The percent uncertainty is the absolute uncertainty divided by the measurement itself (the relative uncertainty) and then multiplied by 100. In this case:

$$\text{Percent uncertainty} = \frac{(0.1 \text{ m})}{(10.0 \text{ m})} \times 100\% = 1\%.$$

In the last example the measurement is known to three *significant figures*. The mass (26.24 ± 0.02) kg is known to four significant figures. If it were written (26.243 ± 0.02) kg, the last digit in the measurement would not be significant. If a quantity is written down without an uncertainty, the last digit on the right must be a significant figure: 26.24 kg is the correct way of writing the mass in the last example.

The units on each side of an equation must be the same or multiples of each other: this affords a computational check of the algebra. In the equation m = ρV (mass = density x volume), if ρ = 8 g/cm^3 and V = 10 cm^3, we have:

$$m = (8 \text{ g/cm}^3)(10 \text{ cm}^3) = 80 \text{ g.}$$

The result automatically comes out in the correct units of mass (g) when the units are cancelled as algebraic quantities.

To convert units from one unit of mass to another, use the conversion relations as algebraic substitutions: for example, since

$$1 \text{ g} = 10^{-3} \text{ kg (conversion relation),}$$

we know that

$$80 \text{ g} = 80 \times 10^{-3} \text{ kg} = 0.080 \text{ kg.}$$

Similarly

$$1 \text{ m} = 10^{-3} \text{ km and } 1 \text{ s} = (1/3600) \text{ hr,}$$

so that

$$20 \text{ m/s} = (20 \times 10^{-3} \text{ km})(3600)/\text{hr}$$

$$20 \text{ m/s} = 72 \text{ km/hr.}$$

An *order of magnitude* estimate is the result of a rough calculation accurate to factors of about two or even ten.

A vector $\vec{A}$ is a quantity with a direction and a magnitude. The magnitude is written $|\vec{A}| = A$. The sum or resultant $\vec{R}$ of the vectors $\vec{A}$ and $\vec{B}$,

$$\vec{A} + \vec{B} = \vec{R},$$

may be obtained by graphical construction or by other methods discussed in the supplementary material below. Now is a good time to review the vector addition problem solving strategy in the main text.

A vector may be resolved into component vectors: $\vec{A} = \vec{A}_x + \vec{A}_y$. In terms of *unit vectors* $\hat{\imath}, \hat{\jmath}$, we may write the component vectors as

$$\vec{A}_x = A_x \hat{\imath}, \quad \vec{A}_y = A_y \hat{\jmath},$$

and

$$\vec{A} = A_x \hat{\imath} + A_y \hat{\jmath} = \vec{A}_x + \vec{A}_y$$

The components A_x and A_y may be positive or negative.

The scalar product of two vectors $\vec{A}$ and $\vec{B}$ is the scalar quantity

$$\vec{A} \cdot \vec{B} = AB \cos \theta = A_x B_x + A_y B_y$$

where θ is the angle between $\vec{A}$ and $\vec{B}$.

The vector product of two vectors $\vec{A}$ and $\vec{B}$ is the vector $\vec{C}$,

$$\vec{C} = \vec{A} \times \vec{B},$$

whose magnitude is $AB|\sin \theta|$ and whose direction is perpendicular to $\vec{A}$ and $\vec{B}$, with sense given by the right hand rule.

SUPPLEMENTARY MATERIAL

All quantities in the first portion of the text may be expressed algebraically in terms of quantities with dimension mass, length and time.

Some quantities are pure numbers, without dimension, such as the ratio of two lengths. These quantities do not have units.

A dimensional physical quantity is not specified unless its units are given. The equation s(length) = 10 is meaningless because it could mean 10 meters or 10 miles. In the table below the three *basic* quantities of length, time and mass and one *derived* quantity, speed, are analyzed according to their dimension and units in the S.I. and British systems.

QUANTITY	SYMBOL	DIMENSION	UNIT(S.I.)	UNIT(BRI.)
distance	s (basic)	L	meter	foot
time	t (basic)	t	second	second
mass	m (basic)	m	kilogram	pound-mass
speed	v (derived)	L/t	meter/second	feet/second

Note the dimension is independent of the system of units.

Equations involving physical quantities must be dimensionally consistent. For example in the equation

$$x = x_0 + vt$$

if x and x_0 have dimension L, and v has dimension L/t, then each side has the dimension L and the equation is dimensionally correct. This is a useful check in any result you derive.

The *units* of an equation, on the other hand, need not be identical on right and left:

$$1 \text{ inch} = 2.54 \text{ cm} = 2.54 \times 10^{-2} \text{ m}$$

is a correct relation expressing the dimensionless ratio

$$\frac{(2.54 \text{ cm})}{(1 \text{ inch})} = 1 = \frac{(2.54 \times 10^{-2} \text{ m})}{(1 \text{ inch})}$$

Such a relation is useful in unit conversions when it is noted that units are algebraic quantities which combine and cancel, like numbers, according to the rules of algebra: Suppose it is known that v = s/t and s = 10 inch, t = 5 s. What is the speed v in cm/s?

$$v = \frac{(10\ \text{inch})}{(5\ \text{s})} = 2\ \text{inch/s} = (2\ \text{inch/s})\frac{(2.54\ \text{cm})}{(1\ \text{inch})}$$

$$= 5.08\ \text{cm/s}.$$

Note we have simply inserted the factor 1 = 2.54 cm/inch and cancelled the units.

When two numbers are combined algebraically the number of significant figures of the result is the number of significant figures of the least significant of the two:

2.145 m + 3.0 m = 5.1 m,

(2.3 m)(1.234 m) = 2.8 m².

Two vectors $\vec{A}$ and $\vec{B}$ are equal, $\vec{A} = \vec{B}$, if they have the same magnitude and direction, as shown in Fig. 1-1.

Figure 1-1

Note they need *not* be coincident. *For purposes of adding and subtracting, vectors may be moved about in space as long as the magnitude and direction are not changed.*

The negative of the vector $\vec{A}$, denoted $-\vec{A}$, is defined as a vector equal in magnitude but opposite in direction to $\vec{A}$, as shown in Fig. 1-1.

Two vectors may be added according to the rule of placing the tail of one to the head of the other; the sum or resultant $\vec{R}$ is shown in Fig. 1-2. $\vec{R}$ is the vector drawn from the tail of the first to the head of the second, completing the triangle. The construction in Fig. 1-2 illustrates that the addition operation does *not* depend on the order, $\vec{A} + \vec{B} = \vec{B} + \vec{A}$.

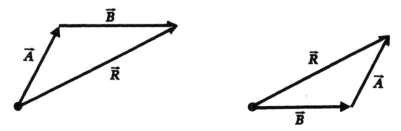

Figure 1-2

Vectors may be multiplied by numbers or scalars. $a\vec{A}$ is a vector with the same direction as $\vec{A}$ and a magnitude $a|\vec{A}|$, if a is positive. If a is negative, $a\vec{A}$ is opposite in direction to $\vec{A}$.

The sum $\vec{A} + \vec{B} + \vec{C} = (\vec{A} + \vec{B}) + \vec{C} = \vec{A} + (\vec{B} + \vec{C})$, is shown in Fig. 1-3. Note the parentheses are unnecessary.

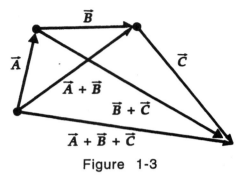

Figure 1-3

Subtraction of vectors is defined as follows

$$\vec{A} - \vec{B} = \vec{A} + (-\vec{B}),$$

as shown in Fig. 1-4, where the vector $-\vec{B}$ is found from the given vector $\vec{B}$ by reversing its direction. The vector $-\vec{B}$ is then added to $\vec{A}$ by the usual procedure.

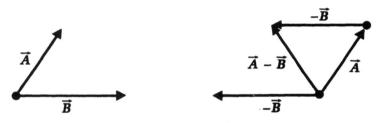

Figure 1-4

The rectangular components of a vector, as indicated in Fig. 1-5,

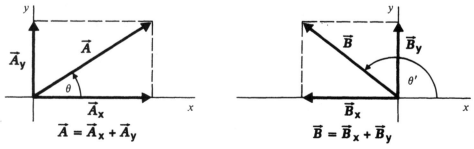

$$\vec{A} = \vec{A}_x + \vec{A}_y \qquad \vec{B} = \vec{B}_x + \vec{B}_y$$

are the projections of $\vec{A}$ on the x and y axis. The magnitudes of the component vectors $\vec{A}_x$ and $\vec{A}_y$, A_x and A_y, are the called the *components* of $\vec{A}$, *with agreement that an x component is positive if it points in the positive axis direction*, with a similar rule for the y-component sign.

From Fig. 1-5 we see

$$\frac{A_x}{A} = \cos \theta \qquad\qquad \frac{A_y}{A} = \sin \theta$$

$$\frac{B_x}{B} = \cos \theta' \qquad\qquad \frac{B_y}{B} = \sin \theta'$$

Note the magnitudes of $\vec{A}$, $\vec{B}$ are always positive but the components may be positive (A_x, A_y and B_y) or negative (B_x) as shown in Fig. 1-5.

You can use various angles to describe the orientation of a vector, as shown in Fig. 1-6.

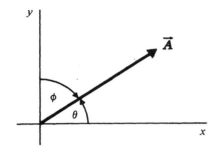
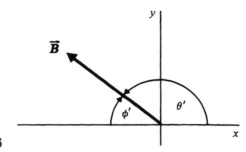

Figure 1-6

In terms of the angles ϕ and ϕ', complementary to θ and θ', we have

$$A_x = A \sin \phi, \qquad\qquad B_x = - B \cos \phi',$$

$$A_y = A \cos \phi, \qquad\qquad B_y = B \sin \phi',$$

and the formulas look different but yield the same numerical answer. For example, take $A = B = 2$, $\theta = 30°$, $\phi = 60°$, $\theta' = 150°$, $\phi' = 30°$.

Then by the first method

$$A_x = A \cos \theta = 2 \cos 30° = 1.7,$$

$$A_y = A \sin \theta = 2 \sin 30° = 1.0,$$

$$B_x = B \cos \theta' = 2 \cos (150°) = - 1.7,$$

$$B_y = B \sin \theta' = 2 \sin (150°) = 1.0,$$

7

and by the second method

$$A_x = A \sin \phi = 2 \sin 60° = 1.7,$$

$$A_y = A \cos \phi = 2 \cos 60° = 1.0,$$

$$B_x = -B \cos \phi' = -2 \cos 30° = -1.7,$$

$$B_y = B \sin \phi' = 2 \sin 30° = 1.0.$$

Given the components of a vector you can reverse the process and calculate its magnitude and direction. From Fig. 1-5, the magnitude $\vec{A}$ and orientational angle θ are:

$$A = [A_x^2 + A_y^2]^{1/2},$$

$$\theta = \arctan \left(\frac{A_y}{A_x} \right)$$

In the above numerical example

$$\theta = \arctan \left(\frac{A_y}{A_x} \right) = \arctan \left(\frac{1.0}{1.7} \right) = 30°,$$

$$\theta' = \arctan \left(\frac{B_y}{B_x} \right) = \arctan \left(\frac{-1.0}{1.7} \right) = 150° \text{ or } 30°\,?$$

The last example illustrates the pitfalls of blind calculation. 150° and $-30°$ have the same tangent; to decide which angle is correct you must refer to the diagram or examine the signs B_x and B_y to see which quadrant $\vec{B}$ lies in. By reference to Fig. 1-5 we see that $\theta = 150°$.

To master the material in this text you must become efficient and accurate at component resolution. In later chapters you will often be required to sum forces. The component method is often the most convenient:

If $\vec{C} = \vec{A} + \vec{B}$, then

$$\vec{C}_x = \vec{A}_x + \vec{B}_x, \qquad C_x = A_x + B_x,$$

$$\vec{C}_y = \vec{A}_y + \vec{B}_y, \qquad C_y = A_y + B_y.$$

Another useful way to write vectors is in terms of *unit vectors*, which have magnitude unity, are dimensionless, and point along the positive axes in a rectangular system, as shown in Fig. 1-7.

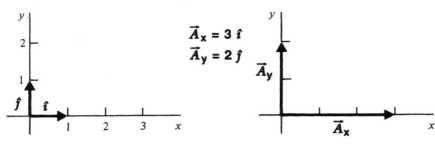

$$\vec{A}_x = 3\,\hat{\imath}$$
$$\vec{A}_y = 2\,\hat{\jmath}$$

Figure 1-7

The component vectors are

$$\vec{A}_x = A_x\hat{\imath} \qquad \vec{A}_y = A_y\hat{\jmath}$$

and the vector $\vec{A}$ is reconstructed as

$$\vec{A} = \vec{A}_x + \vec{A}_y = A_x\hat{\imath} + A_y\hat{\jmath}$$

Extending this to three dimensions, we have

$$\vec{A} = A_x\hat{\imath} + A_y\hat{\jmath} + A_z\hat{k},$$

where $\hat{k}$ points along the positive z axis. The usual formulas for vector addition and subtraction can be rewritten to utilize this new notation, e.g.

$$\vec{A} + \vec{B} = (A_x + B_x)\hat{\imath} + (A_y + B_y)\hat{\jmath} + (A_z + B_z)\hat{k}$$

Vectors may be 'multiplied' with each other but the operation is different from the simple multiplication of two scalars and must be carefully defined and distinguished. There are two useful products which will be used in the text, the *scalar product* and the *vector product*.

The scalar product $\vec{A}\cdot\vec{B}$ is a scalar found from $\vec{A}$ and $\vec{B}$ according to the rule

$$\vec{A}\cdot\vec{B} = AB\cos\theta$$

where θ is the angle between $\vec{A}$ and $\vec{B}$, measured from $\vec{A}$ to $\vec{B}$, as shown in the Fig. 1-8a and 1-8b.

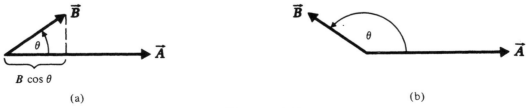

(a)   (b)

Figure 1-8

9

Since cos θ changes sign at θ = 90°, the scalar product is positive for 0 < θ < 90° and negative for 90° < θ < 180° and vanishes at θ = 90°. As shown, the scalar product is the projection of $\vec{B}$ on $\vec{A}$ multiplied by the magnitude of A. In the example of Fig. 1-8 with A = 2, B = 1, and θ = 30°,

$$\vec{A} \cdot \vec{B} = AB \cos \theta = 2 \cdot 1 \cdot \cos 30° = 1.73$$

If $\vec{A}$ and $\vec{B}$ are parallel, $\vec{A} \cdot \vec{B}$ = AB; if they are anti-parallel, θ = 180° and $\vec{A} \cdot \vec{B}$ = − AB. If they are perpendicular $\vec{A} \cdot \vec{B}$ = 0.

The scalar product obeys the law

$$(\vec{A} + \vec{B}) \cdot \vec{C} = \vec{A} \cdot \vec{C} + \vec{B} \cdot \vec{C}$$

Thus with the rectangular resolution we can show

$$\vec{A} \cdot \vec{B} = (A_x \hat{\imath} + A_y \hat{\jmath}) \cdot (B_x \hat{\imath} + B_y \hat{\jmath})$$

$$= A_x B_x + A_y B_y$$

because $\hat{\imath} \cdot \hat{\imath} = \hat{\jmath} \cdot \hat{\jmath} = 1$ and $\hat{\imath} \cdot \hat{\jmath} = 0$.

The *vector product* $\vec{C}$ of two vectors $\vec{A}$ and $\vec{B}$ is a vector obtained from $\vec{A}$ and $\vec{B}$ by the rule

$$\vec{C} = \vec{A} \times \vec{B}$$

Magnitude: C = |AB sin θ|

Direction: perpendicular to $\vec{A}$ and $\vec{B}$ with sense given by the right hand rule.

According to the right hand rule, if you curl the fingers of your right hand around the perpendicular to $\vec{A}$ and $\vec{B}$ in the direction of rotation of $\vec{A}$ into $\vec{B}$, your thumb points toward $\vec{C}$. As a consequence, the scalar product changes sign if the order of the factors is reversed:

$$\vec{B} \times \vec{A} = - \vec{A} \times \vec{B}$$

An important expression for $\vec{A} \times \vec{B}$ in terms of unit vectors is given by

$$\vec{A} \times \vec{B} = (A_y B_z - A_z B_y)\hat{\imath} + (A_z B_x - A_x B_z)\hat{\jmath} + (A_x B_y - A_y B_x)\hat{k}$$

$$\vec{A} \times \vec{B} = \det \begin{vmatrix} \hat{\imath} & \hat{\jmath} & \hat{k} \\ A_x & A_y & A_z \\ B_x & B_y & B_z \end{vmatrix}$$

If $\vec{A}$ and $\vec{B}$ are parallel, $\vec{A} \times \vec{B}$ = 0 because sin θ = 0. If $\vec{A}$ and $\vec{B}$ are perpendicular, C = AB.

QUESTIONS AND ANSWERS

Question. Do absolute uncertainty and relative uncertainty have the same units?

Answer. In general, no. Absolute uncertainty has the same units as the physical quantity measured. For instance, if the measured quantity is "time", then absolute uncertainty has dimensions of time and the units of seconds. Relative uncertainty is dimensionless and has no units.

Question. Does a vector quantity depend on the choice of origin of a coordinate system?

Answer. No. A vector is defined by a magnitude (including units) and a direction, the particular coordinate system and its origin being unimportant. We have used this fact when we added and subtracted vectors by the polygon method.

EXAMPLES AND SOLUTIONS

Example 1

Suppose we know that $x_0 = 2.0$ m, $v_0 = 3.0$ m/s, $a = -9.8$ m/s^2, in the equation,

$$x = x_0 + v_0 t + \frac{1}{2} at^2,$$

Find the value of x when t = 4.0 s.

Solution:

$$x = 2.0 \text{ m} + (3.0 \text{ m/s})(4 \text{ s}) + (1/2)(-9.8 \text{ m/s}^2)(4.0 \text{ s})^2 = -64 \text{ m}$$

This is accurate to 2 significant figures. Note how the units cancel to produce the unit of meters in each term and that the correct answer has a sign and a unit.

Example 2

Convert 60.0 mph to a speed in km/hr, given 1 inch = 2.54 cm.

Solution:

$$\frac{(60.0 \text{ mi})}{(1 \text{ hr})} = \frac{(60.0 \text{ mi})}{(1 \text{ hr})} \left[\frac{(5280 \text{ ft})}{(\text{mi})}\right]\left[\frac{(12 \text{ in})}{(\text{ft})}\right]\left[\frac{(2.54 \text{ cm})}{(\text{in})}\right]\left[\frac{(10^{-5} \text{ km})}{(\text{cm})}\right] = \frac{(96.6 \text{ km})}{(1 \text{ hr})}$$

(The square bracketed expressions, [ ], are all of magnitude unity and may be inserted anywhere in the expression. Note how the units cancel.)

Example 3

Consider the density of water to be exactly 1.00 g/cm³. What is its value in pound-mass/ft³ to three significant figures? (The "pound-mass", a common household unit, is equal to 0.454 kg.)

Solution:

For this conversion you must know the relation between a gram (g) and a pound-mass, and the relation between a cubic centimeter (cm³) and a cubic foot (ft³). Since there are 12 inches in a foot and 10^3g in a kg,

$$\frac{1}{12} \text{ foot} = 1 \text{ inch} = 2.54 \text{ cm},$$

$$1 \text{ pound-mass} = 0.454 \times 10^3 \text{g}.$$

In the last expression we have kept only three significant figures.

To convert units, multiply the original expression by the factors

$$1 = \frac{\text{pound mass}}{454 \text{ g}} \qquad \text{and } 1 = \left[\frac{2.54 \text{ cm}}{(1/12) \text{ ft}}\right]^3$$

Then we have

$$1 \text{ g}/\text{cm}^3 = \left(1 \frac{\text{g}}{\text{cm}^3}\right)\left[\frac{1 \text{ pound mass}}{454 \text{ g}}\right]\left[\frac{2.54 \text{ cm}}{(1/12) \text{ ft}}\right]^3$$

$$= 62.4 \text{ pound-mass/ft}^3.$$

Alternatively you may substitute for a gram in the original expression its value from the conversion equation and the same for the centimeter (cm):

$$1 \text{ gram (g)} = \left[\frac{1 \text{ pound} - \text{mass}}{454}\right] \qquad \text{and } 1 \text{ cm} = \left(\frac{1}{2.54}\right) \cdot \left(\frac{1}{12}\right) \text{ ft},$$

Completing these substitutions yields:

$$1 \text{ g}/\text{cm}^3 = \left[\frac{1 \text{ pound- mass}}{454}\right] \cdot \left[\left(\frac{1}{2.54}\right) \cdot \left(\frac{1 \text{ ft}}{12}\right)\right]^{-3}$$

$$= 62.4 \text{ pound-mass/ft}^3.$$

Example 4

Compute the number of km in a mile to three significant figures.

Solution:

$$1 \text{ mi} = (1 \text{ mi}) \left[\frac{5280 \text{ ft}}{\text{mi}} \right] \left[\frac{12 \text{ in}}{\text{ft}} \right] \left[\frac{2.54 \text{ cm}}{\text{in}} \right] \left[\frac{10^{-5} \text{ km}}{\text{cm}} \right]$$

$$= 1.61 \text{ km}$$

Example 5

Suppose $\pi = 3.142...$ is approximated by $(10)^{1/2}$. Compute the percent error made.

Solution:

$$\text{absolute error} = \sqrt{10} - \pi = 0.02;$$

$$\text{percent error} = \left(\frac{\sqrt{10} - \pi}{\pi} \right) \times 100\%$$

$$= 0.6\%.$$

In this case both π and $(10)^{1/2}$ are known to any desired accuracy, as is the percent error, and it is a matter of judgement and convenience as to how many significant figures to present.

Example 6

Estimate the number of times you blink in a day.

Solution:

Counting your own or someone else's blinks, you might time the frequency of blinks to be *about* one per second (probably no greater than 3 per second nor less than one every ten seconds.) An order of magnitude estimate is thus

1 blink per second = 60 blinks per minute

$$= (60)(60) \text{ blinks per hour}$$

$$= (60)(60)(24) \text{ blinks per day}$$

$$= \text{about } 10^5 \text{ blinks per day}$$

For part of the day your eyes are closed in sleep and that reduces the result by about 1/3, but doesn't change the order of magnitude.

Example 7

A "cloudburst" drops 2 cm of rain in one hour. Estimate the number of drops of rain that fall on your head if you are in the storm for one minute.

Solution:

A raindrop leaves a spot about one quarter of a centimeter in diameter. (You may argue that a typical drop is larger or smaller by a factor of two or three, and that uncertainty is part of our order-of-magnitude estimate.) The volume of the drop is

$$(4/3)\pi(\text{radius})^3 \approx 4(\text{cm}/8)^3 = (1/128) \text{ cm}^3$$

In 10 minutes, <u>256</u> such drops must fall on 1 cm^2 to reach a height of 2 cm. If the top of your head is a square about 10 cm on a side, its area is 100 cm^2, and thus

$$(100 \text{ cm}^2)(256 \text{ drops/cm}^2) \approx 25600$$

drops fall on your head in an hour, or 427 drops in one minute.

Example 8

An example of a vector is a force with a magnitude measured in the unit N = newtons.
(a) Find the vector sum or resultant of the two forces in Fig. 1-9.
(b) Find their difference $\vec{D} = \vec{A} - \vec{B}$.

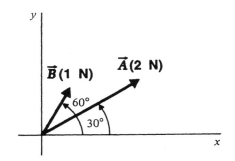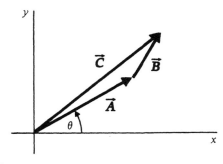

Figure 1-9

Solution:

(a) First resolve the individual vectors into their rectangular coordinates, following the vector addition problem solving strategy of the main text:

$$A_x = A \cos 30° = 2 \cos 30° = 1.7 \text{ N},$$

$$A_y = A \sin 30° = 2 \sin 30° = 1.0 \text{ N},$$

$$B_x = B \cos 60° = 0.50 \text{ N},$$

$$B_y = B \sin 60° = 0.87 \text{ N},$$

Then find the components, magnitude, and direction of $\vec{C}$:

$$C_x = A_x + B_x = 2.2 \text{ N}$$

$$C_y = A_y + B_y = 1.87 \text{ N}$$

$$C = (C_x^2 + C_y^2)^{1/2} = [(2.2)^2 + (1.87)^2)]^{1/2} = 2.9 \text{ N}$$

$$\theta = \arctan \frac{C_y}{C_x} = \arctan \frac{1.87}{2.2} = 40°$$

(For a check, compare these results with the graphical solution in Fig. 1-9. Even a rough sketch serves to check your analytical solutions for gross errors.)

We similarly find the vector difference $\vec{D} = \vec{A} - \vec{B}$:

15

$$D_x = A_x - B_x = 1.7 \text{ N} - 0.5 \text{ N} = 1.2 \text{ N}$$

$$D_y = A_y - B_y = 1.0 \text{ N} - .87 \text{ N} = 0.03 \text{ N}$$

$$D = (D_x^2 + D_y^2)^{1/2} = [(1.2)^2 + (.03)^2]^{1/2} = 1.2 \text{ N}$$

$\vec{D}$ is a vector in the first quadrant ($D_x > 0$, $D_y > 0$) with

$$\theta = \arctan \frac{D_y}{D_x} = \arctan \frac{0.3}{1.2} = 1.43°,$$

where θ is measured counter-clockwise from the positive x axis.

Example 9

Consider two vectors $\vec{A}$ and $\vec{B}$ in the x-y plane of Fig. 1-10. Find their cross product $\vec{A} \times \vec{B}$.

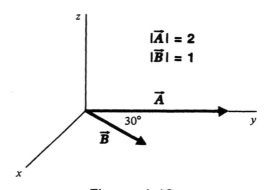

Figure 1-10

Solution:

The direction of $\vec{A} \times \vec{B}$ is along the negative z axis. (If your thumb points down along that axis your fingers indeed curl around the z axis from $\vec{A}$ to $\vec{B}$.)

Alternatively, by the component method:

$$A_x = 0 \qquad A_y = 2 \qquad A_z = 0$$

$$B_x = B \sin \theta = 1 \sin 30° = 0.50$$

$$B_y = B \cos \theta = 1 \cos 30° = 0.87$$

$$B_z = 0$$

and

$$C_x = A_yB_z - A_zB_y = 2(0) - 0(0.87) - 0$$

$$C_y = A_zB_x - A_xB_z = 0(0.5) - 0(0) = 0$$

$$C_z = A_xB_y - A_yB_x = 0(0.87) - 2(0.5) = -1.0.$$

$$C = 1$$

Example 10

A car drives 3 km east and then 4 km north. Find the resultant displacement vectors:

Solution:

$\vec{A}$: magnitude 3 km, direction east; $\vec{B}$: magnitude 4 km, direction north;

$\vec{R} = \vec{A} + \vec{B}$ = vector sum of $\vec{A} + \vec{B}$.

This sum may be found in several ways.

(a) Graphical Solution

This method requires graph paper, a protractor for measuring angles and a ruler for measuring lengths. You must also choose a scale, e.g., one box = 1 km. Then your plot will look like Fig. 1-11.

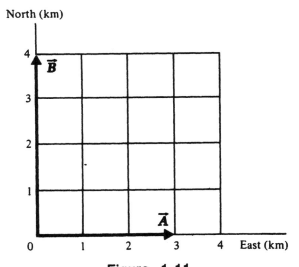

Figure 1-11

To add $\vec{A}$ and $\vec{B}$, move A or B so that the head of one vector is coincident with the tail of the other, as shown in Fig. 1-12.

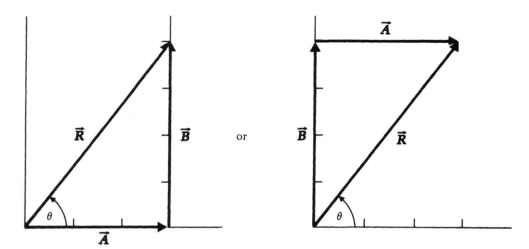

Figure 1-12

$\vec{R}$ is the vector from the open tail to the open head. Now measure the length of $\vec{R}$ in km, according to your scale. R ≈ 5 km. Measure the angle θ with a protractor: θ ≈ 53°. This graphical method inevitably suffers from errors of measurement.

(b) Geometric or Triangle Method.

Here you need only draw a single sketch to guide the eye, as shown in Fig. 1-13.

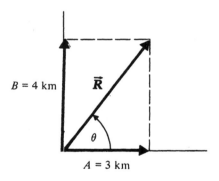

Figure 1-13

By the Pythagorean Theorem, the magnitude of the resultant is

$$R = (A^2 + B^2)^{1/2} = (3^2 + 4^2)^{1/2} = 5 \text{ km}$$

The angle between $\vec{R}$ and $\vec{A}$ is

18

$$\theta = \arctan \frac{B}{A} = \arctan \frac{4}{3} = 53.1°$$

(Note these last two relations are true only when $\vec{A}$ and $\vec{B}$ are perpendicular.)

(c) Component Method--Rectangular Resolution.

To add $\vec{A}$ and $\vec{B}$ by this method, first resolve $\vec{A}$ and $\vec{B}$ into their x and y components:

$A_x = 3$ km, $\qquad\qquad A_y = 0,$

$B_x = 0,$ $\qquad\qquad B_y = 4$ km.

Then

$C_x = A_x + B_x = 3 + 0 = 3$ km

$C_y = A_y + B_y = 0 + 4 = 4$ km

$C = (C_x{}^2 + C_y{}^2)^{1/2} = (3^2 + 4^2)^{1/2} = 5$ km

$$\theta = \arctan \frac{C_y}{C_x} = \arctan \frac{4}{3} = 53.1°$$

QUIZ

1. How many people standing hand-to-outstretched-hand are needed to span the width of the United States, coast to coast?

Answer: About 2.5 million

2. Find the sum ($\vec{C}$) and difference ($\vec{D}$) of vectors $\vec{A}$ and $\vec{B}$ shown in Fig. 1-14.

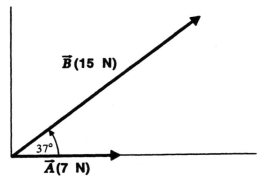

$\vec{B}$(15 N)

37°

$\vec{A}$(7 N)

Figure 1-14

Answer: $\vec{C}$ = 21 N, 25° measured counter-clockwise from the x-axis.

$\vec{D}$ = 10 N, 241° measured counter-clockwise from the x-axis.

3. Two forces $\vec{F}_1$ and $\vec{F}_2$ act on a body at the origin as shown in Fig. 1-15. Find a third force $\vec{F}_3$ acting at the origin such that the sum of all forces acting on the body is zero.

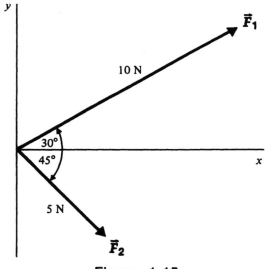

y

$\vec{F}_1$

10 N

30°

45°

x

5 N

$\vec{F}_2$

Figure 1-15

Answer: F_3 = 12 N, θ = 187° measured counter-clockwise from the x-axis.

4. Find the scalar product of the vectors in Fig. 1-14

Answer: 84

5. What are the magnitude and direction of the vector product $\vec{F}_1 \times \vec{F}_2$ in Fig. 1-15?

Answer: magnitude: 48; direction: down into page.

6. On your right hand, arrange your thumb, index finger, and middle finger so they are mutually perpendicular. Label the direction in which these fingers point **t**, **i**, and **m** respectively, and let **t**, **i**, and **m** have unit length.

What is **t** x **m**?; **t** • **i**?; (**i** x **m**) • **t**?

Answer: − **i**, 0, 1.

2
MOTION ALONG A STRAIGHT LINE

OBJECTIVES

In this chapter you will describe the motion of an object on a straight line by its *position*, *velocity*, and *acceleration*. A particularly important special case is when the acceleration is *constant*. Your objectives are to:

Calculate *average* and *instantaneous* velocity.

Calculate average and instantaneous acceleration.

Calculate instantaneous velocity when the position is given as a function of time by *differentiation*.

Calculate the position as a function of time when the instantaneous velocity is given as a function of time by *integration*.

Find the *equations of motion* for the case of constant acceleration.

Apply these equations of motion to a variety of problems involving *constant* acceleration, such as motion of a body falling in the earth's gravitational field.

Comments on the Objectives

The calculus techniques of differentiation and integration may be new concepts to many of you. Indeed, some of you will be seeing an integral for the first time. But don't panic: the entire chapter can be read without serious attention to the calculus techniques of differentiation and integration. Most of the problems at the end of the chapter can be solved without use of calculus. You will not be blown out of the water if you can't integrate or differentiate, at least not at this stage.

However, this is a calculus-based text and the introduction to these techniques comes most naturally here in connection with velocity and acceleration. You will be missing an important part of this chapter if you do not meet all of the above objectives. Eventually, sometime during the course, these objectives must be met.

REVIEW and SUPPLEMENT

The *average velocity*, between time t_1 and time t_2, of a particle moving on a straight line from coordinate x_1 to coordinate x_2 is

$$v_{av} = \frac{x_2 - x_1}{t_2 - t_1} = \frac{\Delta x}{\Delta t} = \text{slope of the chord shown in Fig. 2 - 1}$$

During the interval between t_1 and t_2 the particle may speed up, slow down, even reverse its motion: the average velocity depends only on the initial and final positions. In a displacement x versus time t graph, Fig. 2-1a, the average velocity is the slope of the chord joining the points (x_1,t_1) and (x_2,t_2).

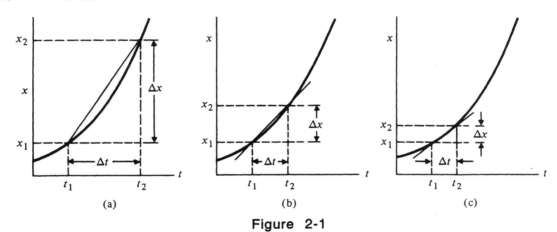

Figure 2-1

The last form may also be written

$$x_2 - x_1 = v_{av}(t_2 - t_1)$$

When the initial time is $t_1 = 0$, the initial displacement is $x_1 = x_0$, and the final displacement is x,

$$x = x_0 + v_{av}t$$

If the initial point is the origin,

$$x_0 = 0, \quad x = v_{av}t.$$

The *instantaneous velocity* is the limit of v as the increments Δx and Δt tend toward zero, that is, the limit of the *slope of the chord* in Fig. 2-1 as t_2 approaches t_1; in Fig. 2-1a, 2-1b, and 2-1c we have a sequence of successively smaller choices of $t_2 - t_1 = \Delta t$ and correspondingly smaller increments Δx. The smaller Δt, the more nearly is the chord tangent to the displacement curve as t_2 approaches t_1. The limit of the sequence of average velocities v_{av} for successively smaller Δx and Δt is called the derivative of x with respect to t and written dx/dt. The value of the derivative is the slope of the tangent curve. The instantaneous velocity is defined in terms of the limit as

$$v(t) = \lim_{\Delta t \to 0}\left(\frac{\Delta x}{\Delta t}\right) = \frac{dx}{dt} = \text{slope of the tangent line}$$

In Fig. 2-1 the coordinate x increases with increasing t, so the body is moving in the positive direction and has a positive velocity, corresponding to the positive slope and positive increment $\Delta x = x_2 - x_1$. In Fig. 2-2a the displacement x decreases with increasing time, the body is moving in the negative x direction and has a negative velocity corresponding to a negative slope and negative increment $\Delta x = x_2 - x_1$.

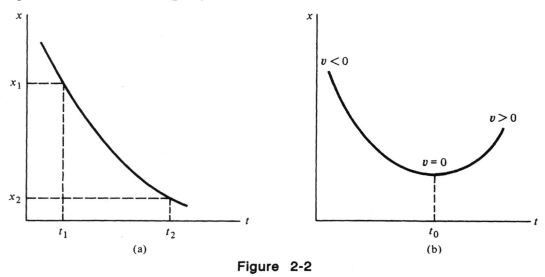

Figure 2-2

(By convention we take $t_2 > t_1$ so Δt is always positive.) In Fig. 2-2b the body starts out with negative velocity, which gets less negative, until it is zero at t_0 where the tangent is horizontal and has zero slope. For $t > t_0$, v is positive.

In practice you can find the instantaneous velocity at a time t graphically by measuring the slope of the tangent to the x vs t graph at time t, or by differentiation. If $x = ct^n$ where c is a constant, the derivative of x is

$$v(t) = \frac{dx}{dt} = \frac{d}{dt}(ct^n) = nct^{n-1}$$

Note the derivative of a constant is zero.

Together with the fact that the derivative of a sum is the sum of the derivatives of the summands,

$$\frac{d}{dt}[f(t) + g(t)] = \frac{df}{dt} + \frac{dg}{dt}$$

24

this rule will enable you to differentiate any polynomial. For example, if $x = a + bt^2$, then $v = 2bt$.

Just as velocity is the rate of change of displacement with time, acceleration is the rate of change of velocity with time. The average acceleration a_{av} is defined by

$$a_{av} = \frac{v_2 - v_1}{t_2 - t_1} = \frac{v(t_2) - v(t_1)}{t_2 - t_1} = \frac{\Delta v}{\Delta t}$$

and the instantaneous acceleration, a, by

$$a = \text{Lim}_{\Delta t \to 0} \left(\frac{\Delta v}{\Delta t} \right) = \frac{dv}{dt}$$

Returning to the example $x = a + bt^2$, we found

$$v(t) = 2bt$$

If we wish to find the acceleration at t,

$$a = \frac{dv}{dt} = \frac{d}{dt}(2bt) = 2b$$

In Fig. 2-3a, b, c we show the coordinate $x = a + bt^2$ for $a > 0$, $b > 0$, the corresponding velocity $v = dx/dt = 2bt$, and the acceleration $a = dv/dt = 2b$.

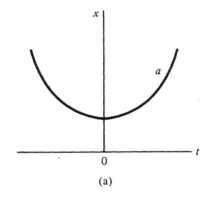
(a)

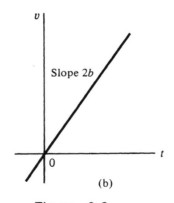
(b)

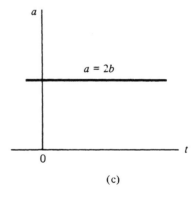
(c)

Figure 2-3

When the acceleration is constant, the displacement and velocity are given as functions of time by

$$x = x_0 + v_0 t + \frac{1}{2} at^2$$

$$v = v_0 + at.$$

25

The time may be eliminated in the two equations yielding a third equation, which although not independent of the other two, is very useful:

$$v^2 = v_0^2 + 2a(x - x_0)$$

Another way of deriving these and other relations is to use the method of integration and integral calculus. Consider the velocity versus time graph (v vrs. t) of Fig. 2-4.

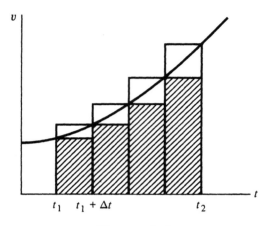

Figure 2-4

The time axis has been broken into small increments of width Δt. In each time increment, $\Delta x = v\, \Delta t$. If we sum all the increments Δx between $x_2 = x(t_2)$ and $x_1 = x(t_1)$ we have the interval length,

$$\sum (\Delta x) = x_2 - x_1 = \sum (v_{av} \Delta t).$$

In each interval the average velocity v_{av} lies somewhere between the highest and lowest instantaneous velocity in that interval. Thus, referring to Fig. 2-4, $v_{av}\, \Delta t$ for each slice is an area somewhere between the smaller shaded slice and the larger slice, which includes the unshaded part. The sum $\sum v_{av}\, \Delta t$ is thus bounded from above by the sum S_2 of the large slices and from below by the sum S_1 of the smaller slices,

$$S_1 < \sum v_{av}\, \Delta t < S_2$$

As Δt is made smaller and smaller the common limit of S_1 and S_2 is the area under the v curve between t_2 and t_1, called the definite integral of v from t_1 to t_2, and denoted

$$\int_{t_1}^{t_2} v\, dt.$$

Thus we have

$$x_2 - x_1 = \int_{t_1}^{t_2} v\, dt = \int_{t_1}^{t_2} \left(v_0 + at \right) dt$$

If $t_1 > t_2$ the area is counted negative when v is positive.

When $v = v_0 + at$, as in Fig. 2-5,

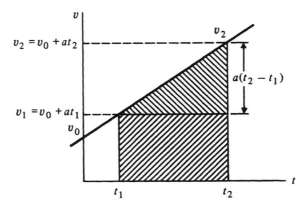

Figure 2-5

the area under the curve is the area of the box = $v_1(t_2 - t_1)$ plus the area of the triangle $(t_2 - t_1)\, a(t_2 - t_1)/2$. Thus

$$x_2 - x_1 = v_1\left(t_2 - t_1 \right) + \frac{1}{2} a \left(t_2 - t_1 \right)^2$$

When $t_1 = 0$, $v_1 = v_0$ and $x_1 = x_0$ we regain our earlier result.

$$x = x_0 + v_0 t + \frac{1}{2} at^2$$

For another demonstration of the power of these methods, consider the limit of

$$a_{av} = \frac{\Delta v}{\Delta t} = \frac{\Delta v}{\Delta x}\frac{\Delta x}{\Delta t} = \frac{\Delta v}{\Delta x} v$$

as all small increments tend to zero, when we obtain the instantaneous acceleration

$$a = \frac{dv}{dx} v$$

27

This is an example of the "chain" rule. Then we have

$$\sum a_{av}\left(\Delta x\right) = \sum v\left(\Delta v\right)$$

In the limit of small increments, these expressions define the integrals

$$\int_{x_1}^{x_2} a\, dx = \int_{v_1}^{v_2} v\, dv$$

These integrals respectively, are the shaded areas in Fig. 2-6a and Fig. 2-6b when a = constant,

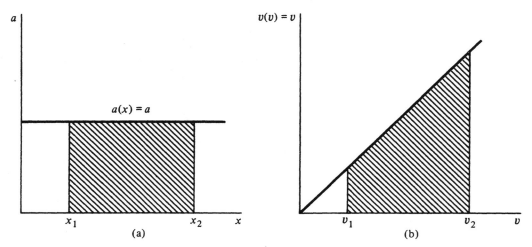

Figure 2-6

$$\int_{x_1}^{x_2} a\, dx = a\left(x_2 - x_1\right) \quad \text{and} \quad \int_{v_1}^{v_2} v\, dv = \frac{1}{2}\left(v_2^2 - v_1^2\right)$$

Thus we have

$$a\left(x_2 - x_1\right) = \frac{1}{2}\left(v_2^2 - v_1^2\right)$$

or, with $x = x_2$, $x_1 = x_0$, $v_2 = v$ and $v_1 = v_0$,

$$v^2 = v_0^2 + 2a(x - x_0).$$

You can not always easily calculate the areas under curves as in the simple examples above. Fortunately there is a simple theorem, true for sufficiently smooth but otherwise arbitrary curves f(x), that does the job:

28

$$\int_a^b f(x)\ dx = F(b) - F(a) \qquad \text{if } \frac{dF(x)}{dx} = f(x)$$

Thus one must find the "anti-derivative" of f, that function which when differentiated yields f. The anti-derivative of v is $v^2/2$ because

$$\frac{d}{dv}\left(\frac{v^2}{2}\right) = v$$

Thus we find

$$\int_{v_1}^{v_2} v\ dv = \left[\frac{v^2}{2}\ \text{at } v_2\right] - \left[\frac{v^2}{2}\ \text{at } v_1\right]$$

$$\int_{v_1}^{v_2} v\ dv = \left[\frac{v^2}{2}\right]_{v_1}^{v_2} = \frac{v_2^2}{2} - \frac{v_1^2}{2}$$

yielding the area of Fig. 2-6b.

The acceleration of a body may be positive or negative depending on whether its positive velocity is increasing or decreasing. The constant acceleration formula applies to freely falling bodies in a constant gravitational field. The sign of displacements, velocities, and accelerations depends on the orientation of the coordinate system, a matter of choice. In Fig. 2-7a, up is positive; in Fig 2-7b, down is positive.

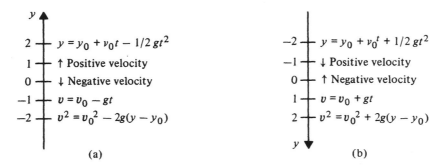

(a) (b)

Figure 2-7

In Fig. 2-7a a particle moving upward (wherever it is, above or below the axis) has positive velocity; moving downward, negative velocity. The acceleration is negative: bodies moving upward are slowed down and bodies moving downward have velocities which are negative and are becoming more negative. The same physics is described in Fig. 2-6b, although the formulas look different because the axis is oriented to make downward velocities and accelerations positive.

In all of the above, the motion has been referred to a single reference frame. If there are two reference frames in motion <u>with respect to</u> (WRT) each other (say the earth E and a moving observer B) then using the definitions:

V_{AE} = velocity of A WRT E

V_{AB} = velocity of A WRT B

V_{BE} = velocity of B WRT E

we have (the addition law of relative velocities)

$V_{AE} = V_{AB} + V_{BE}$

Note adjacent subscripts on the right are the ones which do not appear on the left. The first subscript refers to the body and the second to the reference frame.

HINTS and PROBLEM-SOLVING STRATEGIES

A very systematic approach has been taken to solve the class of problems in this chapter involving constant acceleration:

(1) Decide on the origin and positive direction of a coordinate system. A sketch is often helpful.

(2) Collect the given input data such as initial or final positions, velocities, and accelerations. Take care that the units are consistent. (For example, if position is in m, the velocity should be in m/s and the acceleration in m/s^2). Choose the signs of the input data in accordance with your choice in (1).

(3) Write down the equations of motion in terms of this input data.

$$x = x_0 + v_0 t + \frac{1}{2} at^2 \qquad v = v_0 + at$$

$$v^2 = v_0^2 + 2a(x - x_0)$$

Note which quantities are known and which are unknown.

(4) Solve the equations, which may be simultaneous in several unknowns, for the desired quantities.

(5) Check that your solution makes physical sense by reference to the coordinate system and sketch of (1). For example, if your coordinate system's positive axis points upward, a ball thrown upward falls back to earth with a negative velocity.

QUESTIONS AND ANSWERS

Question. A ball is thrown vertically upward. What is its velocity and acceleration at the very top of its trajectory?

Answer. At the top of the trajectory, the vertical velocity is zero while the acceleration is equal to g (9.8 m/s^2) pointing toward the earth.

Question. A grasshopper reaches a height of 1 meter on a vertical jump. A flea reaches a height of 25 cm on a vertical jump. What is different about the initial conditions they create at the starting point of their jumps? If the flea doubles its initial velocity, will it jump as high as the grasshopper?

Answer. The difference in their initial conditions is in the initial velocity, v_0. By doubling its initial velocity, the flea will rise as high as the grasshopper.

EXAMPLES AND SOLUTIONS

Example 1

Calculate the average velocity between $t_1 = 1$ s and t_2 for the sequence $t_2 = 2$ s, 3/2 s (1.5 s), 5/4 s (1.25 s), 9/8 s (1.125 s) and 17/16 s (1.0625 s) for a particle whose position as a function of time is given by $x(t) = 1/2\, gt^2$, where $g = 9.8$ m/s^2 (see Fig. 2-8a).

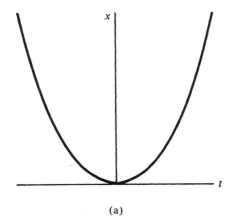

(a)

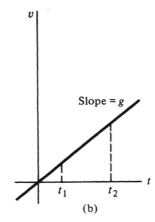

Slope = g

(b)

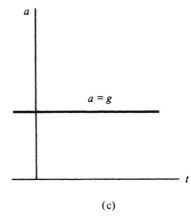

$a = g$

(c)

Figure 2-8

31

Solution:

$$v_{av} = \frac{x_2 - x_1}{t_2 - t_1} = \frac{x(t_2) - x(t_1)}{t_2 - t_1}$$

$$v_{av} = \frac{\frac{1}{2}gt_2^2 - \frac{1}{2}gt_1^2}{t_2 - t_1} = \frac{1}{2}g\frac{t_2^2 - t_1^2}{t_2 - t_1} = \frac{1}{2}g(t_2 + t_1)$$

Note we did not immediately substitute numbers for algebraic quantities. Evaluating this expression numerically at various times t_2,

for $t_2 = 2$ s, $\quad v_{av} = \frac{1}{2}g(2\text{ s} + 1\text{ s}) = \frac{3g}{2}\text{ s} = 14.7\text{ m/s}$

for $t_2 = 1.5$ s, $\quad v_{av} = \frac{1}{2}g(1.5\text{ s} + 1\text{ s}) = \frac{5g}{4}\text{ s} = 12.2\text{ m/s}$

for $t_2 = 1.25$ s, $\quad v_{av} = \frac{1}{2}g(1.25\text{ s} + 1\text{ s}) = \frac{9g}{8}\text{ s} = 11.0\text{ m/s}$

for $t_2 = 1.125$ s, $\quad v_{av} = \frac{1}{2}g(1.125\text{ s} + 1\text{ s}) = \frac{17g}{16}\text{ s} = 10.4\text{ m/s}$

for $t_2 = 1.0625$ s, $\quad v_{av} = \frac{1}{2}g(1.0625\text{ s} + 1\text{ s}) = \frac{33g}{32}\text{ s} = 10.2\text{ m/s}$

To find the limit of this sequence at $t_2 = t_1 = 1$ s, differentiate

$$x = \frac{1}{2}gt^2$$

and evaluate the result at $t = 1$ s. According to the rule

$$\frac{d}{dt}(ct^n) = nct^{n-1}$$

we find

$$v(t) = \frac{dx}{dt} = \frac{d}{dt}\left(\frac{1}{2}gt^2\right) = \frac{1}{2}g(2t) = gt$$

$$v(1\text{ s}) = g(1\text{ s}) = (9.8\text{ m/s}^2)(1\text{ s}) = 9.8\text{ m/s}$$

Example 2

Calculate, in the last example, the average acceleration over the same intervals and the instantaneous acceleration at t = 1 s.

Solution:

From the previous result, $v(t) = gt$,

$$a_{av} = \frac{v_2 - v_1}{t_2 - t_1} = \frac{v(t_2) - v(t_1)}{t_2 - t_1}$$

$$a_{av} = \frac{v(2) - v(1)}{2 - 1} = \frac{2g - g}{1} \qquad \text{in the first interval}$$

$$a_{av} = \frac{v(1.5) - v(1)}{1.5 - 1} = \frac{0.5\,g}{0.5} \qquad \text{in the second interval}$$

$$a_{av} = g \qquad \text{in all intervals.}$$

The average acceleration is the same in each interval. This is apparent from a sketch of v versus t, which is a straight line. See Fig. 2-8b. The chord joining v_1, t_1 to v_2, t_2 and the tangent to the curve v(t) are identical. In this special case the average acceleration is equal to the instantaneous acceleration, which is constant. To verify, calculate the derivative

$$a = \frac{dv}{dt} = \lim_{\Delta t \to 0} \left[\frac{v(t + \Delta t) - v(t)}{\Delta t} \right]$$

$$a = \lim_{\Delta t \to 0} \left[\frac{gt + g\Delta t - gt}{\Delta t} \right] = \lim_{\Delta t \to 0} \left[\frac{g\Delta t}{\Delta t} \right] = g$$

$$a = \lim_{\Delta t \to 0} \frac{gt + g\Delta t - gt}{\Delta t} = \lim_{\Delta t \to 0} \frac{g\Delta t}{\Delta t} = g$$

The acceleration is constant as shown in Fig. 2-8c.

Example 3

Find the position x(t) if $v = b + at$ and $x(t_2) = x_2$, $x(t_1) = x_1$.

Solution:

$$v = \frac{dx}{dt} = b + at$$

$$\int_{x_1}^{x_2} dx = \int_{t_1}^{t_2} (b + at)\, dt$$

The last line is shorthand for the limit of the sums

$$\Sigma\, \Delta x = \Sigma\, v\, \Delta t$$

which are areas under the curves, indicated in Fig. 2-9a for the left hand side, and Fig. 2-9b for the right hand side.

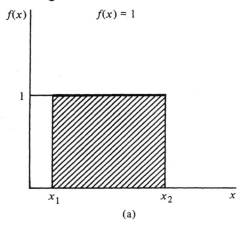
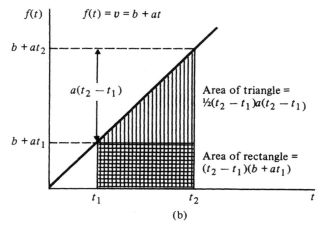

(a)   (b)

Figure 2-9

$$\text{Lim}_{\Delta x \to 0} \Sigma\, \Delta x = \int_{x_1}^{x_2} dx = x_2 - x_1 = \text{area of rectangle in Fig. } 2-9a.$$

$$\text{Lim}_{\Delta t \to 0} \Sigma\, v(\Delta t) = \int_{t_1}^{t_2} v\, dt$$

$$\text{Lim}_{\Delta t \to 0} \Sigma\, v(\Delta t) = (t_2 - t_1)(b + at_1) + \frac{1}{2}(t_2 - t_1)a(t_2 - t_1)$$

$$\text{Lim}_{\Delta t \to 0} \Sigma\, v(\Delta t) = v(t_1)(t_2 - t_1) + \frac{1}{2}a(t_2 - t_1)^2$$

= area under curve in Fig. 2-9b.

If $t_1 = 0$, $x_1 = x_0$, $v(t_1) = v_0$, $t_2 = t$ and $x_2 = x$, we have

$$x = x_0 + v_0 t + \frac{1}{2}at^2$$

Example 4

Calculate $\int_{x_1}^{x_2} x\, dx$ in Figure 2–10.

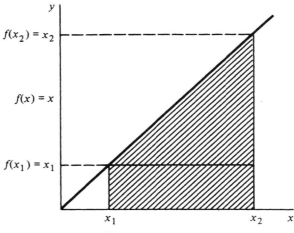

Figure 2-10

Solution:

This is the area under the curve in Fig. 2-10 between x_1 and x_2,

$$\int_{x_1}^{x_2} x\, dx = \left[\frac{x^2}{2}\right]_{x_1}^{x_2} = \frac{x_2^2}{2} - \frac{x_1^2}{2}$$

Example 5

If it takes 5 hours to drive from Boston to New York, a distance of 225 miles, what is the average velocity in mph and m/s?

Solution:

$$v_{av} = \frac{\Delta x}{\Delta t} = \frac{225\text{ mi}}{5\text{ hr}} = 45\text{ mph}$$

To convert from mph to m/s use:

$$45\text{ mph} = \frac{45\text{ mi}}{1\text{ hr}} = \left(\frac{45\text{ mi}}{1\text{ hr}}\right)\left(\frac{1.6\text{ km}}{\text{mi}}\right)\left(\frac{1\text{ hr}}{3600\text{ s}}\right)\left(\frac{10^3\text{ m}}{\text{km}}\right)$$

$$= 20\text{ m/s}$$

Example 6

Suppose x(t) is as given graphically in Fig. 2-11.
   (a) When is the velocity positive?
   (b) When is the acceleration positive?

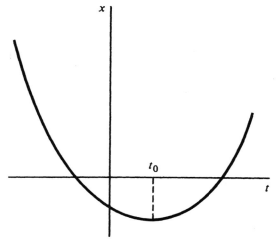

Figure 2-11

Solution:

The velocity is the slope of the tangent to the curve, which is positive for $t > t_0$. The acceleration is always positive. The velocity always increases as t increases. (The slope of the tangent curve always increases as x increases.)

Example 7

The position x(t) is given graphically in Fig. 2-12, with distance in meters and time in seconds.

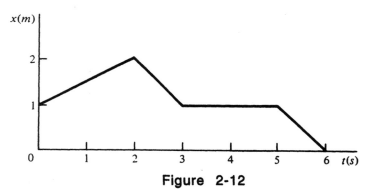

Figure 2-12

Find and plot v(t).

Solution:

Between t = 0 and t = 2 s

$$v = \frac{\Delta x}{\Delta t} = \frac{x(2) - x(0)}{2\,s - 0\,s} = \frac{2m - 1m}{2\,s} = 0.5\,m/s$$

Between t = 3 s and t = 2 s

$$v = \frac{\Delta x}{\Delta t} = \frac{x(3) - x(2)}{3\,s - 2\,s} = \frac{1\,m - 2m}{1\,s} = -1\,m/s$$

Between t = 5 s and t = 3 s

$$v = \frac{\Delta x}{\Delta t} = \frac{x(5) - x(3)}{5\,s - 3\,s} = \frac{1\,m - 1\,m}{2\,s} = 0$$

Between t = 6 s and t = 5 s

$$v = \frac{\Delta x}{\Delta t} = \frac{x(6) - x(5)}{6\,s - 5\,s} = \frac{0\,m - 1\,m}{1\,s} = -1\,m/s$$

A plot of v(t) is given in Fig. 2-13.

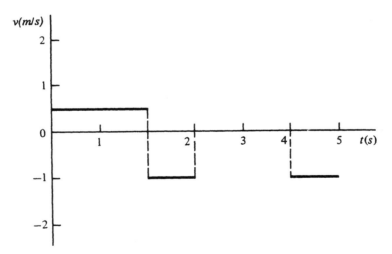

Figure 2-13

37

Example 8

A body undergoes an acceleration a(t) given in the graph of Fig. 2-14. Find and plot v(t) for t > 0. Take $v_0 = 0$.

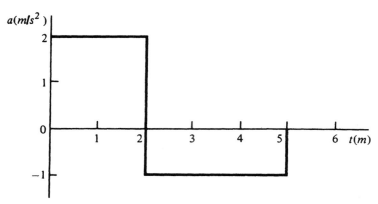

Figure 2-14

Solution:

In *each interval* of a single constant acceleration

$$v = v_0 + at$$

where v_0 is the initial velocity *in that interval* and t is the time measured *from the beginning of that interval.*

$\underline{0 < t < 2s}$ $a = 2m/s^2$ $v_0 = 0$

$$v = v_0 + at$$

$$v = 0 + 2 \text{ m/s}^2 \, t = (2 \text{ m/s}^2) \, t \quad \text{(in m/s if t in s)}$$

$$v(2 \text{ s}) = (2 \text{ m/s}^2)(2 \text{ s}) = 4 \text{m/s}$$

$\underline{2s < t < 5s}$ $a' = -1 \text{m/s}^2$ $v_0' = 4 \text{m/s}$

$v = v_0' + a't'$ $t' = t - 2 \text{ s} = $ time measured from beginning of second interval

$$= 4 - 1 \, (t - 2) = 6 - t$$

$$= 4 \text{ m/s} - (1 \text{ m/s}^2)(t - 2 \text{ s}) = 6 \text{ m/s} - (1 \text{ m/s}^2)t$$

$\underline{5 \text{ s} < t}$ $v_0'' = v(5) = 6 \text{ m/s} - (1 \text{ m/s}^2)(5 \text{ s}) = 1 \text{ m/s}$

$v = v_0'' + a''t''$ $a'' = 0$

$$v = 1 \text{ m/s}$$

v(t) is plotted in Fig. 2-15

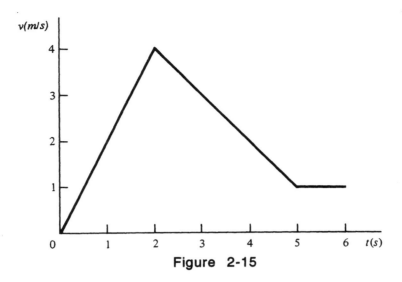

Figure 2-15

Example 9

A body undergoes the acceleration a = 2 m/s^2 for 0 <t < 2 and a = 0 for t > 2s. Find the position and velocity at t = 10 s.

Solution:

The initial position and velocity of the body are both zero, i.e., $x_0 = 0$, $v_0 = 0$.
For 0 < t < 2 s

$$v = v_0 + at = (2m/s^2)t$$

At the end of the acceleration interval, t = 2 s, and v = (2 m/s^2)(2s) = 4 m/s. In this interval the position is given by

$$x = x_0 + v_0t + \frac{1}{2} at^2 = \frac{1}{2}\left(2\,m/s^2\right)t^2$$

At t = 2s, x = (1/2)(2 m/s^2)(2 s)2 = 4 m .

Let us now reset the clock and the initial conditions to *start* the motion at t = 2s where

$x_0' = 4$ m, the initial position in this interval

$v_0' = 4$ m/s, the initial velocity for this interval

39

a' = 0 in this interval

t' = t - 2 s = time measured from beginning of this interval

In this interval we have

$$v = v_0' + a't' = 4 \text{ m/s}$$

$$x = x'_0 + v'_0 t' + \frac{1}{2} a't'^2 = 4 \text{ m} + (4 \text{ m/s}) t'$$

$$= 4 \text{ m} + 4 \text{ m/s}(t - 2 \text{ s})$$

Thus at t = 10 s,

$$x = 4 \text{ m} + 4 \text{ m/s } (10 \text{ s} - 2 \text{ s}) = 36 \text{ m}$$

$$v = 4 \text{ m/s}$$

Example 10

An automobile starts from rest and reaches 100 km/hr in 10 seconds.
   (a) Find the average acceleration. State your answer as a multiple of the acceleration due to gravity, g = 9.8 m/s^2.
   (b) Assuming constant acceleration, find the distance covered.

Solution:

(a) $a = \dfrac{\Delta v}{\Delta t} = \dfrac{(100 - 0) \text{ km/hr}}{10 \text{ s}} = \left(\dfrac{100 \text{ km/hr}}{10 \text{ s}} \right) \left(\dfrac{10^3 \text{ m/km}}{3600 \text{ s/hr}} \right)$

$$= 2.78 \text{ m/s}^2$$

$$\frac{a}{g} = \frac{2.79 \text{ m/s}^2}{9.8 \text{ m/s}^2} = 0.28; \text{ so that } a = 0.28 \text{ g}.$$

(b) For $a_{av} = a$ = constant and for initial position $x_0 = 0$ and initial velocity $v_0 = 0$,

$$x = x_0 + v_0 t + \frac{1}{2} at^2 = \frac{1}{2} at^2$$

$$x = \frac{1}{2} (2.78 \text{ m/s}^2)(10 \text{ s})^2 = 139 \text{ m}$$

Alternatively we may use the relation between displacement and the initial and final velocities:

$$v^2 = v_0^2 + 2ax = 2ax = \left[\frac{(100 \text{ km})(10^3 \text{ m/km})}{(\text{hr})(3600 \text{ s/hr})} \right]^2 = [27.8 \text{ m/s}]^2$$

$$x = \frac{v^2}{2a} = \frac{[27.8 \text{ m/s}]^2}{2(2.78 \text{ m/s}^2)} = 139 \text{ m}$$

Note we solve first for x in terms of v2 and a, and then substitute for the known v and a.

Example 11

A motorist traveling 60 km/hr brakes to a stop in 100 m.
   (a) Find the acceleration assuming it to be constant.
   (b) What would the stopping distance be if the acceleration were a = -g = -9.8 m/s2?
   (c) What is the stopping time in case (a)?

Solution:

Imagine a coordinate system in which the positive direction points toward the car's forward motion. The the initial velocity v_0 is positive and the final velocity is zero. The displacement will be positive also. The appropriate equation is the one relating acceleration, displacement, and velocity.

(a)   v2 = v02 + 2ax   v0 = 60 km/hr   v = 0

$$a = -\frac{v_0^2}{2x} = \frac{(-1)}{2(100 \text{ m})}\left(\frac{60 \times 10^3 \text{ m}}{3600 \text{ s}}\right) = -1.39 \text{ m/s}^2$$

The acceleration is negative because the velocity decreases. Note that the speed had to be converted to m/s to find the acceleration in m/s2.

(b)   v2 = v02 + 2ax; v0 = 60 km/hr; v = 0; a = -9.8 m/s2 .(Why is the acceleration negative?)

$$x = -\frac{v_0^2}{2a} = \frac{v_0^2}{2g} = \frac{(1)}{2(9.8 \text{ m/s}^2)}\left(\frac{60 \times 10^3 \text{ m}}{3600 \text{ s}}\right) = -1.39 \text{ m/s}^2$$

= 14.2 m

(c)   Now we must choose the equation which relates the initial and final velocities (v_0, v) to the unknown time t. We know that v = v0 + at where v = 0 and v0 = 60 km/hr. Also we calculated the acceleration and found: a = -1.39 m/s. Solving for time, t:

$$t = \frac{v - v_0}{a} = \frac{-v_0}{a} = \frac{(-1)}{2(-1.39 \text{ m/s}^2)}\left(\frac{60 \times 10^3 \text{ m}}{3600 \text{ s}}\right) = 12 \text{ s.}$$

Note we put in the actual numbers only at the last step.

Example 12

A car covers 150 m in 5 s, with constant acceleration. Its final velocity is 100 km/hr.
  (a) What is the acceleration?
    (b) What is the initial velocity?

Solution:

Choose the positive axis in the direction of the car's motion. Designate x = final position, x_0 = initial position with $x - x_0$ = 150 m, v = final velocity = 100 km/hr = 27.8m/s, v_0 = initial velocity; and t = 5 s. Then for constant acceleration,

$$x = x_0 + v_0 t + \frac{1}{2} at^2$$

$$v = v_0 + at$$

The two equations involve only the unknowns v_0 and a. We can solve for a by solving the second one for v_0,

$$v_0 = v - at$$

and substituting in the first

$$x - x_0 = (v - at)t + \frac{1}{2} at^2$$

$$x - x_0 = vt + a\left(\frac{1}{2}t^2 - t^2\right) = vt - \frac{1}{2} at^2$$

Solving for acceleration,

$$a = -\frac{2(x - x_0 - vt)}{t^2}$$

$$a = -\frac{2(150\,\text{m} - [27.8\,\text{m/s}][5\,\text{s}])}{(5\,\text{s})^2}$$

$$= -.88\ \text{m/s}^2$$

The initial velocity is

$$v_0 = v - at = 27.8 \text{ m/s} + 0.88 \text{ m/s}^2 (5 \text{ s})$$

$$= 32.2 \text{ m/s}.$$

Note we did not substitute numbers until the last step.

Example 13

A speeder traveling at a constant speed of 100 km/hr passes a waiting police car which immediately starts from rest with a constant acceleration of 2.5 m/s².
    (a) How long will it take the police car to catch the speeder?
    (b) How fast will the police car be traveling when it catches the speeder?
    (c) How far will the police car have traveled?

Solution:

We choose the positive direction along the car's forward motion, and the origin at the position of the waiting police car. Then the police car has $v_{0p} = 0$ and $x_{0p} = 0$, and the speeder has $a_S = 0$ because the speeder's velocity is constant.

$$\text{Police car: } x_p = x_{0p} + v_{0p}t + \frac{1}{2}at^2 = \frac{1}{2}at^2$$

$$v_p = v_{0p} + at = at$$

$$(a = 2.5 \text{ m/s}^2)$$

$$\text{Speeder: } x_S = x_{0S} + v_{0S}t + \frac{1}{2}a_St^2 = v_{0S}t$$

$$(x_{0S} = 0, a_S = 0)$$

$$v_S = v_{0S} + a_St = v_{0S}$$

$$v_{0S} = 100 \text{ km/hr} = 27.8 \text{ m/s}$$

The constants x_{0p} and x_{0S} were chosen so that the speeder and police car are at the same position at t = 0, ($x_{0p} = x_{0S} = 0$). We wish to know when this occurs again,

(a) $x_p = x_S$ implies that $\frac{1}{2}at^2 = v_{0S}t$

The equation for this time has roots t = 0 and

$$t = \frac{2v_{os}}{a} = \frac{2(27.8\,\text{m/s})}{2.5\,\text{m/s}^2}$$

$$= 22.2\,\text{s}.$$

(b) The velocity of the police car is then

$$v_p = at = (2.5\ \text{m/s}^2)(22.2\ \text{s}) = 55.5\ \text{m/s}$$

$$v_p = \frac{(55.5 \times 10^{-3}\,\text{km})}{(1/3600)\,\text{hr}} = 200\,\text{km/hr}$$

(c) The position is then

$$x_p = \frac{1}{2}\,at^2 = \frac{1}{2}(2.5\,\text{m/s}^2)(22.2\,\text{s})^2 = 616\,\text{m}$$

Alternatively, for the police car

$$v_p^2 = v_{op}^2 + 2ax_p \qquad v_{op} = 0$$

$$x_p = \frac{v_p^2}{2a} = \frac{(55.5\,\text{m/s})^2}{2(2.5\,\text{m/s}^2)} = 616\,\text{m}$$

Example 14

A body starts from an initial velocity of 1 m/s and reaches, under constant acceleration, a final velocity of 5 m/s. Its initial position is 20 m and its final position is 100 m.
   (a) What is its acceleration?
   (b) How much time does it take to reach its final velocity?

Solution:

(a) Since initial and final velocities and positions are known, the appropriate equation to use is

$$v^2 = v_0^2 + 2a(x - x_0)$$

where $v = 5$ m/s; $v_0 = 1$ m/s; $x = 100$ m; and $x_0 = 20$ m.

$$a = \frac{v^2 - v_0^2}{2(x - x_0)} = \frac{(5\,m/s)^2 - (1\,m/s)^2}{2(100\,m - 20\,m)} = 0.15\,m/s^2$$

(b) $v = v_0 + at$

$$t = \frac{v - v_0}{a} = \frac{(5\,m/s) - (1\,m/s)}{0.15\,m/s^2} = 26.7\ s.$$

Example 15

A car traveling 100 km/hr is 200 m away from a truck traveling 50 km/hr (in same direction). Assuming constant braking deceleration, what is the minimum deceleration the car must have if it is not to hit the truck?

Solution:

   We choose a coordinate system with its positive sense in the direction of the car's travel. First write the equations of motion for the car and the truck, incorporating initial conditions.

$$\text{Car: } x_C = x_{0C} + v_{0C}t + \frac{1}{2}at^2 = v_{0C}t + \frac{1}{2}at^2$$

(The coordinate system is chosen so the initial car position is zero.)

$$v_C = v_{0C} + at \qquad v_{0C} = 100\ km/hr = 27.8\ m/s$$

$$\text{Truck: } x_T = x_{0T} + v_{0T}t + \frac{1}{2}a_Tt^2 = x_{0T} + v_{0T}t$$

$$v_T = v_{0T} + a_Tt = v_{0T}$$

v_{0T} = 50 km/hr = 13.9 m/s

x_{0T} = 200 m   (This is 200 m ahead of the car.)

If the collision is to be barely avoided, then when they meet

$x_C = x_T$

the car and the truck will have zero relative velocity,

$v_C = v_T$.

Substituting for the car and truck final position and velocity, we have

$$v_{0C}t + \frac{1}{2}\, at^2 = x_{0T} + v_{0T}t$$

$$v_{0C} + at = v_{0T}$$

which are two equations in the unknowns a and t.

Solving the second for a and substituting in the first,

$$v_{0C}t + \frac{1}{2}\left(\frac{v_{0T} - v_{0C}}{t}\right)t^2 = x_{0T} + v_{0T}t$$

$$t = \frac{2\,x_{0T}}{v_{0T} - v_{0C}} = \frac{2\,(200\,\mathrm{m})}{(27.8\,\mathrm{m/s}) - (13.9\,\mathrm{m/s})}$$

$$= 28.8\ \mathrm{s}.$$

Thus

$$a = \frac{v_{0T} - v_{0C}}{t}$$

$$= -0.48\ \mathrm{m/s^2}.$$

(Why is this quantity negative?)

_____

Example 16

A ball is dropped from the roof of a building 100 m high. Find
    (a) the velocity when it hits the ground and
    (b) the time it takes to hit the ground.

Solution:

There are several ways of setting up this problem, as illustrated in Fig. 2-16.

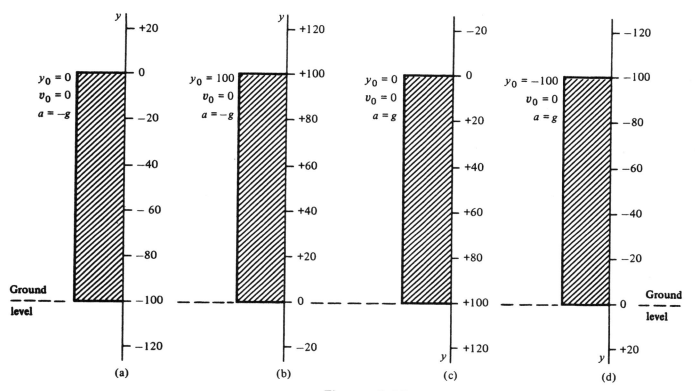

Figure 2-16

(a) $y = -\frac{1}{2} gt^2$ and $v = -gt$ (b) $y = 100 \text{ m} - \frac{1}{2} gt^2$ and $v = -gt$

(c) $y = \frac{1}{2} gt^2$ and $v = gt$ (d) $y = -100 \text{ m} + \frac{1}{2} gt^2$ and $v = gt$

In (a) we seek t when the ball is at $y = -100$ m:

$$-100 \text{ m} = -\frac{1}{2} gt^2, \text{ so that } t = 4.52 \text{ s.}$$

47

At this time

$$v = -gt = -9.8 \text{ m/s}^2(4.52 \text{ s}) = -44.3 \text{ m/s}.$$

The velocity is negative because the ball is headed along the negative y axis, in this coordinate system.

In (b) we seek t when the ball is at y = 0,

$$0 = 100 \text{ m} - \frac{1}{2} gt^2, \text{ or } t = \sqrt{\frac{2(100 \text{ m})}{9.8 \text{ m/s}^2}} = 4.52 \text{ s}.$$

$$v = -gt = -44.3 \text{ m/s}$$

In (c) we seek t when the ball is at +100 m,

$$100 \text{ m} = \frac{1}{2} gt^2, \text{ or again } t = 4.52 \text{ s}.$$

$$v = gt = 44.3 \text{ m/s}$$

The velocity is positive because the ball is headed in the positive y direction, in this coordinate system.

In (d) we seek t when y = 0,

$$0 = -100 \text{ m} + \frac{1}{2} gt^2, \quad t = 4.52 \text{ s}.$$

$$v = gt = 44.3 \text{ m/s}$$

Note the physical results are independent of the choice of origin and positive direction.

Example 17

A ball is thrown vertically upward from a 100 m high building with an initial velocity of 50 m/s.

    (a) How high does it rise?
    (b) How much time does it take to reach its maximum height?
    (c) How long does it take to return to the top of the building?
    (d) What is its velocity at this time?
    (e) How long does it take to reach the ground?
    (f) What is its velocity when it hits the ground?

Solution:

Using the coordinate system of Fig. 2-16b, we find

$$y = y_0 + v_0 t - \frac{1}{2} gt^2 = 100 \text{ m} + (50 \text{ m/s})t - \frac{1}{2}(9.8 \text{ m/s}^2)t^2$$

$$v = v_0 - gt = 50 \text{ m/s} - 9.8 \text{ m/s}^2 \, t$$

$$v^2 = v_0^2 - 2g(y - y_0) = (50 \text{ m/s})^2 - 2(9.8 \text{ m/s}^2)(y - y_0)$$

(a) $v^2 = 0$ at the topmost point, where

$$y - y_0 = \frac{v_0^2}{2g} = \frac{(50 \text{ m/s})^2}{2(9.8 \text{ m/s}^2)} = 128 \text{ m}$$

$$y = y_0 + 128 \text{ m} = 228 \text{ m}$$

Alternatively, at the topmost point, $v = 0$, and the time when this point is reached may be obtained from

$$0 = v = v_0 - gt = 50 \text{ m/s} - 9.8 \text{ m/s}^2 \, t$$

$$t = \left(\frac{50 \text{ m/s}}{9.8 \text{ m/s}^2}\right) = 5.1 \text{ s}$$

At this time,

$$y = y_0 + v_0 t - \frac{1}{2} gt^2$$

$$y = 100 \text{ m} + (50 \text{ m/s})(5.1 \text{ s}) - \frac{1}{2}(9.8 \text{ m/s}^2)(5.1 \text{ s})^2 = 228 \text{ m}$$

(b) see (a)

(c) At this time $y = 100$ m, or

$$100 \text{ m} = 100 \text{ m} + (50 \text{ m/s})t - \frac{1}{2}(9.8 \text{ m/s}^2)t^2$$

There are two roots; $t = 0$ and

$$t = \frac{2(50 \text{ m/s})}{9.8 \text{ m/s}^2} = 10.2 \text{ s} = 2(5.1 \text{ s})$$

The rise time is equal to the fall time.

(d) $v = 50$ m/s - $(9.8$ m/s$^2)t = 50$ m/s - $(9.8$ m/s$^2)(10.2)$

           $= -50$ m/s

(Same magnitude as initial velocity but opposite direction.)

(e) The ground is reached when $y = 0$

           $0 = y = 100$ m + $(50$ m/s$^2)t$ - $(4.9$ m/s$^2)t^2$

This is a quadratic equation of the form $at^2 + bt + c = 0$, with roots given in terms of a,b, and c:

$$t = \frac{-b \pm \sqrt{b^2 - 4ac}}{2a} = \frac{-50 \pm \sqrt{(50)^2 - 4(-4.9)(100)}}{2(-4.9)}$$

        $= -1.71$ s, 11.92 s

The first root is "extraneous" but has a simple physical interpretation. Consider a motion which starts the ball at a certain time from ground level such that it reaches the top of the building with a velocity of 50 m/s at time $t = 0$, matching the initial conditions of this problem. That earlier time is -1.71 s.

Example 18

A ball is dropped from a tall building. 1 s later a ball is thrown with a velocity of 30 m/s vertically down. Will the balls ever meet? If so, when and where? And what does this have to do with Superman?

Solution:

Using the coordinate system of Fig. 2-16(c) we have

$$y_1 = \frac{1}{2} at^2 = \frac{1}{2} gt^2 \qquad \textit{First ball}$$

$$y_2 = y_{02} + v_{02}t' + \frac{1}{2} g(t')^2; \text{ where } t' = t - 1 \text{ s} \qquad \textit{Second ball}$$

$$y_2 = y_{02} + v_{02}(t - 1 \text{ s}) + \frac{1}{2} g(t - 1 \text{ s})^2 = v_{02}(t - 1 \text{ s}) + \frac{1}{2} g(t - 1 \text{ s})^2$$

We have adjusted the time t' for the second ball so that when $t = 1$ s it is at the origin, in accordance with the initial conditions. ($y_2 = 0$ at $t = 1$ s.)

For the balls to meet, $y_1 = y_2$, so solve for the unknown time t when this happens.

50

$$\tfrac{1}{2} gt^2 = v_{02}(t - 1\,s) + \tfrac{1}{2} g(t - 1\,s)^2$$

The term gt²/2 cancels out leaving

$$t = \frac{v_{02}(1\,s) - \tfrac{1}{2} g(1\,s)^2}{v_{02} - g(1\,s)} = \frac{(30\,m/s)(1\,s) - \tfrac{1}{2}(9.8\,m/s^2)(1\,s)^2}{(30\,m/s) - (9.8\,m/s^2)(1\,s)} = 1.24\,s$$

At this time,

$$y_1 = y_2 = \tfrac{1}{2}(9.8\,m/s^2)(1.24\,s)^2 = 7.57\,m$$

More generally if Δt is the time interval between the first ball and the second ball,

$$y_1 = \tfrac{1}{2} gt^2$$

$$y_2 = v_{02}(t - \Delta t) + \tfrac{1}{2} g(t - \Delta t)^2$$

Setting $y_1 = y_2$, we obtain the following expression that can be solved for t .

$$0 = v_{02}(t - \Delta t) - gt(\Delta t) + \tfrac{1}{2} g(\Delta t)^2 = t(v_{02} - g\Delta t) - v_{02}\Delta t + \tfrac{1}{2} g(\Delta t)^2$$

Superman performs a similar calculation when he jumps from a 300 m high building to rescue a student who has been falling for 5 seconds. In that problem, the second ball is Superman, and $y_1 = 300$ m, $\Delta t = 5$ s, and v_{02} is the unknown. Thus from

$$y_1 = \tfrac{1}{2} gt^2, \text{ or } t = \sqrt{\frac{2y_1}{g}}$$

$$t = \sqrt{\frac{2(300\,m)}{(9.8\,m/s^2)}} = 7.82\,s$$

Solving the equation obtained by equating $y_1 = y_2$, we find that Superman's initial velocity to be

$$v_{02} = g(\Delta t)\left[\frac{t - (\Delta t/2)}{t - \Delta t}\right] = 92\,m/s^2$$

The limiting height, Y_L, in this problem, the height below which even an infinitely fast Superman misses the rescue, is the free fall distance in which student hits the ground as Superman jumps. This height is

$$Y_L = \frac{1}{2} g (\Delta t)^2 = \frac{1}{2} (9.8 \, \text{m/s}^2)(5 \, \text{s})^2 = 122.5 \, \text{m}$$

Example 19

The motion of a particle is given by

$$x = 2m + (6 \text{ m/s}^2)t^2 - (3 \text{ m/s}^4)t^4$$

Find the position, velocity, and acceleration at t = 2s.

Solution:

$$x(2) = 2 \text{ m} + (6 \text{ m/s}^2)(2 \text{ s})^2 - (3 \text{ m/s}^4)(2 \text{ s})^4$$

$$= -22 \text{ m}$$

$$v = (dx/dt) = 6 \text{ m/s}^2(2t) - (3 \text{ m/s}^4)(4t^3)$$

$$= (12 \text{ m/s}^2)t - (12 \text{ m/s}^4)t^3$$

$$v(2) = (12 \text{ m/s}^2)(2 \text{ s}) - (12 \text{ m/s}^4)(2 \text{ s})^3$$

$$= -72 \text{ m/s}$$

$$a = (dv/dt) = 12 \text{ m/s}^2 - (12 \text{ m/s}^4)3t^2$$

$$= 12 \text{ m/s}^2 - 36 \text{ m/s}^4 \, t^2$$

$$a(2) = 12 \text{ m/s}^2 - (36 \text{ m/s}^4)(2 \text{ s})^2$$

$$= -132 \text{ m/s}^2$$

Example 20

The acceleration of a particle is given by

$$a = (4\,\text{m/s}^4)t^2 = bt^2 = \frac{dv}{dt}$$

Find v and x as a function of t.

Solution:

$$\int_{v_0}^{v} dv = \int_{t_0}^{t} bt^2\, dt$$

The integral on the left is v - v₀ by the area rule as shown in Fig. 2-6a. To do the integral on the right we use the anti-derivative rule

$$\int_{t_0}^{t} bt^2\, dt = \left[\frac{bt^3}{3}\right]_{\text{at } t} - \left[\frac{bt^3}{3}\right]_{\text{at } t_0} = \frac{b}{3}\left(t^3 - t_0^3\right)$$

Thus

$$v - v_0 = \frac{b}{3}\left(t^3 - t_0^3\right)$$

or solving for the velocity, v, so that x can be obtained by integration,

$$v = \frac{dx}{dt} = v_0 + \frac{b}{3}\left(t^3 - t_0^3\right)$$

If we choose the constants v_0 and t_0 to be equal to zero, ($v_0 = 0$ and $t_0 = 0$), then

$$v = \frac{dx}{dt} = \frac{b}{3}t^3$$

so that for x = 0 at t = 0, one has:

$$\int_{0}^{x} dx = \int_{0}^{t} v\, dt = \int_{0}^{t} \frac{bt^3}{3}\, dt$$

$$x = \left(\frac{b}{3}\right)\left(\frac{t^4}{4}\right) = \frac{bt^4}{12}$$

QUIZ

1. An astronaut on the moon's surface throws a ball vertically upward at a velocity of 10 m/s and catches it 12 s later. What is the acceleration of gravity on the moon?

Answer: 1.7 m/s^2

2. A motorist traveling 60 km/hr brakes to a gentle stop in 200 m.
    a) Find his braking acceleration, assuming it to be constant.
    b) What would his stopping distance be if his acceleration were - g = - 9.8 m/s^2?
    c) What is his stopping time in case (a)?

Answer: - 0.7 m/s^2, 14.2 m, 24 s

3. A ball is thrown vertically upward from the top of a building 160 ft high. It hits the ground 5 s later.
    a) What is its initial velocity?
    b) How long does it take to reach its maximum height?
    c) What is its maximum height?

Answer: 48 ft/s, 1.5 s, 196 ft

4. An arrow shot vertically upward travels at 30 m/s when it first reaches a height of 40 m. How long after this does it take to reach the ground?

Answer: 7 s.

5. How long does it take a pen to hit the floor after falling off a desk 0.75 m high? What is its final velocity before hitting the floor?

Answer: 0.39 s, 3.83 m/s.

3
MOTION IN TWO OR THREE DIMENSIONS

OBJECTIVES

In the last chapter you described the motion of objects moving along in a straight line. Taking that straight line to be the x axis, you needed then to consider only one component (the x component) of the position, velocity, and acceleration. In this chapter you treat the more general case in which the motion is in a plane and the position must be described by two coordinates, the x and y components. Your objectives are to:

Describe the position of a particle by a *position vector* $\vec{r}$, its instantaneous velocity by a vector $\vec{v}$, and its instantaneous acceleration by a vector $\vec{a}$.

Resolve the vectors into *rectangular components*, and into components *parallel* (or tangent, t) and *perpendicular* (or normal, n) to the motion.

Analyze the motion of *projectiles* near the surface of the earth, objects with a constant downward acceleration.

Analyze objects in *circular motion* in horizontal and vertical planes.

Find the velocity of objects relative to different *frames of reference*.

COMMENTS on the OBJECTIVES

A review of the section of Chapter 1 concerning vectors may be necessary to achieve these objectives. A review of Chapter 2 may be necessary before attempting the projectile problems.

REVIEW and SUPPLEMENT

The motion of a particle in a plane is described by its position vector

$$\vec{r} = x\,\hat{\imath} + y\,\hat{\jmath}$$

where (x,y) are its rectangular coordinates, and $\hat{\imath}, \hat{\jmath}$ are unit vectors pointing in the positive x and y directions, as indicated in Fig. 3-1a. As the particle moves, the head of the position vector $\vec{r}$ traces out the trajectory curve illustrated in Fig. 3-1b.

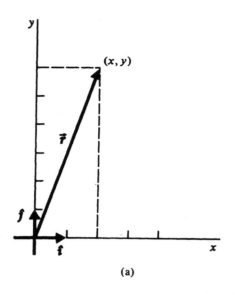 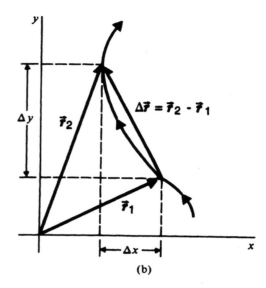

(a)   (b)

Figure 3-1

If the particle is at position $\vec{r}_1$ at t_1 and $\vec{r}_2$ at t_2, its displacement between times t_1 and t_2 is

$$\Delta\vec{r} = \vec{r}_2 - \vec{r}_1 = \Delta x\,\hat{\imath} + \Delta y\,\hat{\jmath}$$

and its average velocity is the vector

$$\vec{v}_{av} = \frac{\Delta\vec{r}}{\Delta t}$$

The instantaneous velocity $\vec{v}$ is the limit of the average velocity as Δt (and hence the length of $\Delta\vec{r}$) tends to zero,

$$\vec{v} = \underset{\Delta t \to 0}{\text{Lim}}\left(\frac{\Delta\vec{r}}{\Delta t}\right) = \underset{\Delta t \to 0}{\text{Lim}}\left(\frac{\Delta x}{\Delta t}\,\hat{\imath} + \frac{\Delta y}{\Delta t}\,\hat{\jmath}\right)$$

$$= \frac{dx}{dt}\,\hat{\imath} + \frac{dy}{dt}\,\hat{\jmath}$$

$$= v_x\,\hat{\imath} + v_y\,\hat{\jmath} = \frac{d\vec{r}}{dt}$$

The velocity $\vec{v}$ is always tangent to the trajectory curve as indicated in Fig. 3-2a.

56

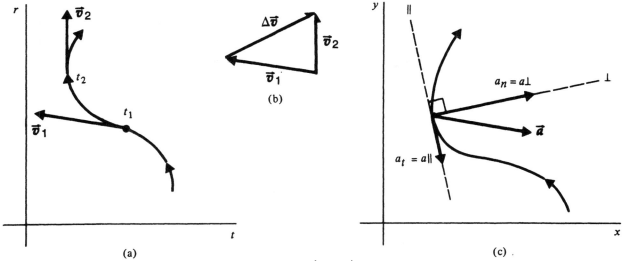

Figure 3-2

If the particle has velocity $\vec{v}_1$ at t_1 and $\vec{v}_2$ at t_2, the change in vector velocity is (Fig. 3-2b)

$$\Delta\vec{v} = \vec{v}_2 - \vec{v}_1 = (v_{2x} - v_{1x})\hat{\imath} + (v_{2y} - v_{1y})\hat{\jmath}$$

$$= \Delta v_x \hat{\imath} + \Delta v_y \hat{\jmath}$$

and the average acceleration is

$$\vec{a}_{av} = \frac{\Delta\vec{v}}{\Delta t} = \frac{\Delta v_x}{\Delta t}\hat{\imath} + \frac{\Delta v_y}{\Delta t}\hat{\jmath}$$

The instantaneous acceleration $\vec{a}$ is the limit of $\overline{a}$ as Δt becomes small

$$\vec{a} = \mathrm{Lim}_{\Delta t \to 0}\left(\frac{\Delta\vec{v}}{\Delta t}\right) = \frac{d\vec{v}}{dt} = \left(\frac{dv_x}{\Delta t}\hat{\imath} + \frac{dv_y}{\Delta t}\hat{\jmath}\right)$$

$$= a_x\hat{\imath} + a_y\hat{\jmath}.$$

Referring to Fig. 3-2, the velocity difference always points to the concave side of a curving trajectory; its limiting direction as $\vec{v}_2 \to \vec{v}_1$ is the direction of the acceleration $\vec{a}$. In general $\vec{a}$ is *not* tangent to the path.

In the above, you have resolved the acceleration and velocity vectors into rectangular components referred to a fixed coordinate system. Another useful resolution is in terms of a direction parallel or tangential (t) and a direction perpendicular or normal (n) to the trajectory. This coordinate system moves with the particle. The velocity vector is always parallel to the trajectory

$$v_t = v$$
$$v_n = 0$$

but the acceleration may have both parallel and perpendicular components as illustrated in Fig. 3-2c. In this case, since $|v_2| < |v_1|$ the particle *slowed* down in the interval and a_t is *negative*. (The positive sense here for a_t is the velocity direction.) The a_t component indicates that the particle is speeding up or slowing down *along* the trajectory. The a_n component indicates that it is changing direction. (The positive sense for a_n is toward the concave side of the curve.) $a_n = 0$ only when the motion is in a straight line.

Projectile Motion

A body given an initial velocity and then released, and acted upon only by the earth's gravity and perhaps air resistance, is called a projectile. When there is no air resistance the acceleration of the projectile is downward,

$$\vec{a} = - g\hat{j}$$

where the positive y (or $\hat{j}$) direction has been taken to be <u>up</u>. The equations of motion in component form are

$$a_x = 0 \qquad \text{and} \qquad a_y = - g$$

The x and y motions are independent since a_x is independent of y and a_y is independent of x. We have here two equations of motion already considered in Chapter 2, namely uniform motion ($a_x = 0$) and motion with constant acceleration ($a_y = - g$). The general solution, following the methods of Chapter 2, is

$$x = x(t) = x_0 + v_{0x}t \qquad v_x = v_{0x}$$

where x_0 is the x coordinate at $t = 0$ and v_{0x} is the x component of the velocity at $t = 0$. v_x is constant because there is no acceleration in the x direction. For the y motion we have

$$y = y(t) = y_0 + v_{0y}t - \frac{1}{2} gt^2$$

$$v_y = v_y(t) = v_{0y} - gt$$

where y_0 is the y coordinate at $t = 0$ and v_{0y} is the y component of the velocity at $t = 0$. As in Chapter 2, t may be eliminated between the last two equations, resulting in

$$v_y^2 = v_{0y}^2 - 2g(y - y_0)$$

The coordinate functions x(t) and y(t) when plotted for various values of time t trace out the trajectory; eliminating t between them yields the trajectory equation y(x),

$$y = y_0 + v_{0y}\left(\frac{x - x_0}{v_{0x}}\right) - \frac{1}{2} g\left(\frac{x - x_0}{v_{0x}}\right)^2$$

which is a concave down parabola, as shown in Fig. 3-3a for $x_0 = 0$, $y_0 > 0$, and $v_{0y} > 0$.

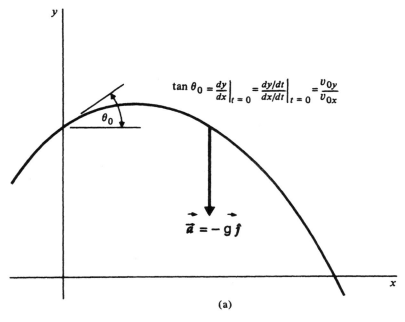

(a)

Figure 3-3a

When $y_0 = 0$ and $x_0 = 0$

$$y = x\left(\frac{v_{0y}}{v_{0x}}\right) - \frac{1}{2} g\left(\frac{x}{v_{0x}}\right)^2 = \left(\tan \theta_0\right)x - \frac{gx^2}{2v_{0x}^2 \cos^2 \theta_0}$$

where $v_{0x} = v_0 \cos \theta_0$, $v_{0y} = v_0 \sin \theta_0$ and θ_0 is the angle of initial projection as shown in Fig. 3-3a. Various motions possible are indicated in Fig. 3-3b. Since the trajectory equation does not change when v_{0y} is replaced by $- v_{0y}$ and v_{0x} is replaced by $- v_{0x}$, the same curve is followed, in the opposite direction, when the initial velocity vector is reversed. (See Fig. 3-3b.)

59

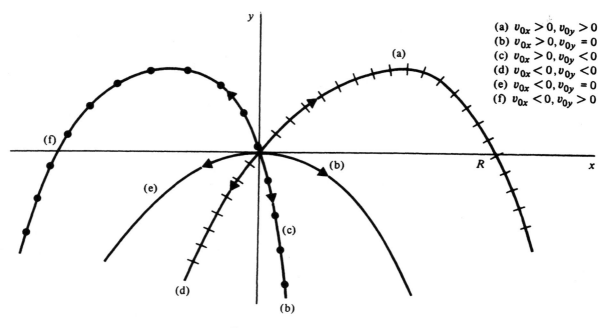

Figure 3-3b

To find the horizontal range R in Fig. 3-3b, set y = 0 in the last equation, resulting in the roots x = 0 and

$$x = R = (2v_0^2/g)\ \cos\ \theta\ \sin\ \theta = (v_0^2/g)\ \sin\ 2\theta$$

Circular Motion

When a body moves in a circle, the perpendicular or centripetal or *radial* component of its acceleration is given by

$$a_{rad} = \frac{v^2}{R}$$

directed inward towards the center. a_{rad} is called the centripetal or radial acceleration.

(This equation is also true for an arbitrary curve if R = radius of curvature.) The parallel component of acceleration is a_t = dv/dt where v is the *magnitude* of the velocity.

If the circular motion is "uniform", v = constant and a_t = 0. Then the only acceleration is a_n, acting to change the direction but not the magnitude of $\vec{v}$.

If a body in uniform circular motion of radius R completes one revolution in the *period* τ, its velocity is

$$v = \frac{2\pi R}{\tau}$$

60

Relative Velocity

Consider the velocity of a body as viewed from a frame E and a frame F. With WRT meaning "with respect to", we have,

$\vec{V}_{FE}$ = the velocity of F WRT E

$\vec{V}_{AE}$ = the velocity of the body WRT E

$\vec{V}_{AF}$ = the velocity of the body WRT F

Then
$$\vec{V}_{AE} = \vec{V}_{AF} + \vec{V}_{FE}$$

Note the subscript index of the intermediate frame F is repeated on the right and does not appear on the left. The first and last indices on the right appear on the left in the same order. This is the vector form of the similar one dimensional relation considered in the second chapter.

HINTS and PROBLEM-SOLVING STRATEGIES

Refer to the problem-solving strategies in the main text for projectile and relative velocity problems.

As in earlier chapters, a diagram and a coordinate system choice are always useful.

Write down all given information, in particular any knowledge you may have of the initial coordinates and velocities.

Write down the equations of motion and incorporate the initial coordinate and velocity information.

Solve for the remaining unknowns.

QUESTIONS AND ANSWERS

Question. A particle moves with constant speed. True or false: the acceleration of the particle is always perpendicular to the velocity, or it is zero.

Answer. This is true.

Question. A piece of baggage falls out of the luggage bay of an airplane that is flying at constant speed in a straight line. In what direction must the pilot look to follow the falling baggage and see where it lands. Does your answer depend on whether the plane is in level flight or not? Neglect air resistance.

Answer. The pilot should look immediately below the plane, if the plane is in level flight.

EXAMPLES and SOLUTIONS

Example 1

A particle's coordinates are given by

$$x = x(t) = (2 \text{ m/s})t$$

$$y = y(t) = 2m - (4.9 \text{ m/s}^2)t^2$$

Find the position, velocity, and acceleration at t = 0.5 s.

Solution:

The constants in the equation are chosen so that if t is given in s, x is in m. The position at time t = 0.5 s is

$$x(0.5 \text{ s}) = (2 \text{ m/s})(0.5 \text{ s}) = 1 \text{ m}$$

$$y(0.5 \text{ s}) = 2 \text{ m} - (4.9 \text{ m/s}^2)(0.5 \text{ s})^2 = 0.78 \text{ m}$$

$$\vec{r} = 1 \text{ m } \hat{\imath} + 0.78 \text{ m } \hat{\jmath}$$

$$r = (x^2 + y^2)^{1/2} = 1.3 \text{ m};$$

$$\theta = \text{arc tan } y/x = 38° \text{ (above the x axis)}$$

To find the velocities you must differentiate x and y with respect to t,

$$v_x = v_x(t) = \frac{dx}{dt} = 2 \text{ m/s}$$

$$v_y = v_y(t) = \frac{dy}{dt} = -4.9 \text{ m/s}^2 (2t) = -9.8 \text{ m/s}^2 (t)$$

$$\vec{v} = (2 \text{ m/s}) \hat{\imath} - (9.8 \text{ m/s}^2)t \; \hat{\jmath}$$

At t = 0.5 s,

$$v_x = 2 \text{ m/s}$$

$$v_y = -9.8 \text{ m/s}^2(0.5 \text{ s}) = -4.9 \text{ m/s}$$

$$v = (v_x^2 + v_y^2)^2 = 5.3 \text{ m/s}$$

$$\theta = \text{arc tan } \frac{v_y}{v_x} = -68° \quad (\text{below the x} - \text{axis})$$

To find the acceleration you must differentiate the velocities with respect to time

$$a_x = \frac{dv_x}{dt} = 0$$

$$a_y = \frac{dv_y}{dt} = -9.8 \, \text{m/s}^2$$

$\vec{a} = -g\hat{j}$ $|\vec{a}| = 9.8 \text{ m/s}^2 = g$ (down)

Example 2

A ball rolls off the edge of a tabletop 1 m above the floor with a horizontal velocity of 1 m/s. Find
 (a) the time it takes to hit the floor,
 (b) the horizontal distance it covers, and
 (c) the velocity when it hits the floor.

Solution:

This is a projectile problem, and following the text's problem-solving strategy, a convenient choice of coordinate system is illustrated in Fig. 3-4 as well as a sketch of the expected motion and the values of the initial position and velocity as culled from the statement of the problem. The parabolic trajectory is tangent to the x axis because the initial velocity is horizontal.

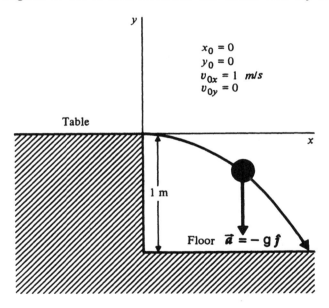

Figure 3-4

In projectile motion, there is no acceleration in the horizontal direction, $a_x = 0$, so the equations of motion for x and y are: ($x_0 = 0$ by choice of coordinate system.)

$$x = x_0 + v_{0x}t = v_{0x}t; \quad v = v_{0x} + a_x t = v_{0x}$$

and

$$y = y_0 + v_{0y}t - \frac{1}{2}gt^2 = -\frac{1}{2}gt^2$$

$$v_y = v_{0y} - gt = -gt$$

where ($a_y = -g$) has been used. We also take $v_{0y} = 0$ because the initial motion is in the $-x$ direction.

(a) The time t when the ball hits the floor may be found from the y equation of motion; this occurs for time t when $y = -1$ m.

$$y = -1 \text{ m} = -(1/2)(9.8 \text{ m/s}^2)t^2$$

$$t = \sqrt{\frac{2 \text{ m}}{9.8 \text{ m/s}^2}} = 0.45 \text{ s}$$

(b) In this time it covers a horizontal distance

$$x = v_{0x}t = (1 \text{ m/s})(0.45 \text{ s}) = 0.45 \text{ m}$$

(c) The velocity at this time is

$$v_x = v_{0x} = 1 \text{ m/s}$$

$$v_y = v_{0y} - gt = -(9.8 \text{ m/s}^2)(0.45 \text{ s}) = -4.4 \text{ m/s}$$

$$v = (v_x^2 + v_y^2)^{1/2} = 4.5 \text{ m/s}$$

$$\theta = \arctan \frac{v_y}{v_x} = -77° \quad (\text{below the } x-\text{axis})$$

Alternatively in (a) you could have solved for the trajectory by eliminating t in the pair of equations $x = v_{0x}t$; $y = -1/2(gt^2)$, resulting in

$$y = -\frac{1}{2}gt^2 = -\frac{1}{2}g\left(\frac{x}{v_{0x}}\right)^2 = -\frac{1}{2}g\frac{x^2}{v_{0x}^2}$$

$$x = \sqrt{\frac{-2yv_{0x}^2}{g}} = \sqrt{\frac{2(1 \text{ m})(1 \text{ m/s})^2}{9.8 \text{ m/s}^2}} = 0.45 \text{ m}$$

Example 3

An archer shoots an arrow into the air at an angle $\theta = 30°$ above the horizontal. It lands on a building 100 m away at a height of 20 m. What were the initial speed (in km/hr) and the time of flight?

Solution:

Refer to Fig. 3-5, where an appropriate system of coordinates is chosen, corresponding to the given initial input data.

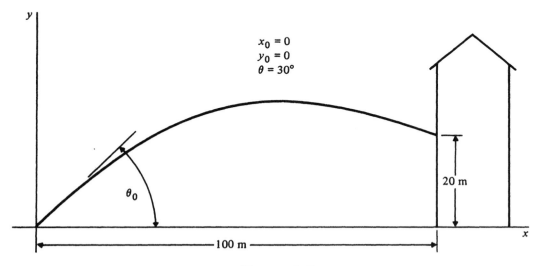

$x_0 = 0$
$y_0 = 0$
$\theta = 30°$

Figure 3-5

We see the pertinent information sketched on the figure. The equations of motion are:

$$x = x_0 + v_{0x}t = v_{0x}t = v_0 \cos \theta_0 t$$
$$y = y_0 + v_{0y}t - \frac{1}{2}gt^2 = \left(v_0 \sin \theta_0\right)t - \frac{1}{2}gt^2$$

These are two parametric equations in the parameter t. Eliminating t, we have the <u>trajectory equation</u> where y is a function of x:

$$y = x \tan \theta_0 - \frac{1}{2}g\left(\frac{x}{v_0 \cos \theta_0}\right)^2$$

which can be solved for the initial velocity v_0 in terms of x and y.

$$v_0 = \sqrt{\frac{gx^2}{2 \cos^2 \theta_0 \left(x \tan \theta_0 - y\right)}}$$

65

This equation is satisfied by all x,y on the trajectory, in particular by the terminal point x = 100 m, y = 20 m.

$$v_0 = \sqrt{\frac{(9.8 \text{ m/s}^2)(100 \text{ m})^2}{2\,(0.87)^2\,[(100 \text{ m})(0.58) - (20 \text{ m})]}}$$

$$v_0 = 41 \text{ m/s} = (41 \text{ m/s})(10^{-3} \text{ km/m})(3600 \text{ s/hr})$$

$$= 150 \text{ km/hr}$$

The time of flight is

$$t = \frac{x}{v_0 \cos \theta_0} = \frac{(100 \text{ m})}{(41 \text{ m/s})(0.87)} = 2.8 \text{ s}$$

Note that here again it is convenient to substitute numbers for symbols only at the last step, after the algebra is completed.

___

Example 4

A soccer ball is kicked 25 m. Its maximum height is 6 m. Find the initial velocity and the time of flight.

Solution:

Referring to Fig. 3-6, we have

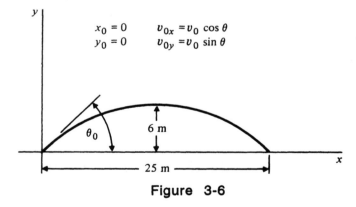

Figure 3-6

The given initial conditions are: $x_0 = 0$; $v_{0x} = v_0 \cos \theta$ (an unknown); $y_0 = 0$; and $v_{0y} = v_0 \sin \theta$ (also an unknown). We solve for the initial velocity and the time in terms of the known maximum height and the projectile range.

$$x = v_{0x}t = v_0 (\cos \theta_0)t$$

$$v_x = v_{0x}$$

$$y = v_{0y}t - \frac{1}{2}gt^2 = \left(v_0 \sin\theta_0\right)t - \frac{1}{2}gt^2$$

$$v_y = v_{0y} - gt = v_0 \sin\theta - gt$$

At the maximum height $y = 6$ m and $v_y = 0$; substituting these in the last two equations we have

$$6\,\text{m} = \left(v_0 \sin\theta_0\right)t - \frac{1}{2}gt^2$$

$$0 = v_{0y} - gt = v_0 (\sin\theta_0) - gt$$

These equations may be solved simultaneously for t and $v_0 \sin\theta_0$:

$$v_0 \sin\theta_0 = 10.8 \text{ m/s}$$

$$t = 1.1 \text{ s}$$

The other information is that at maximum range $x = 25$ m. Since the orbit, a parabola, is symmetric about its top most point, the time of flight is twice the time to reach the midpoint, or 2.2 s. Thus from $x = v_0(\cos\theta_0)t$ we have

$$v_0 \cos\theta_0 = (x/t) = (25 \text{ m}/2.2 \text{ s}) = 11.3 \text{ m/s}$$

Combining this with

$$v_0 \sin\theta_0 = 10.8 \text{ m/s}$$

yields

$$\theta_0 = \arctan(10.8/11.3) = 44°$$

and then

$$v_0 = (10.8 \text{ m/s})/(\sin\theta_0) = 15.6 \text{ m/s}$$

Alternatively the time of flight can be calculated by setting $y = 0$ in the equation of motion for y,

$$y = 0 = \left(v_0 \sin\theta_0\right)t - \frac{1}{2}gt^2 = (10.8 \text{ m/s})t - \left(4.9 \text{ m/s}^2\right)t^2$$

This quadratic equation for t has two roots, the starting time $t = 0$ and time of flight,

$$t = 2.2 \text{ s}$$

Example 5

A boy throws a ball from a cliff at a departure angle of $\theta = 30°$ and an initial velocity 10 m/s. It lands 100 m from the base of the cliff.
  (a) How high is the cliff?
  (b) What is the time of flight?
  (c) Find its velocity upon impact. See Fig. 3-7.

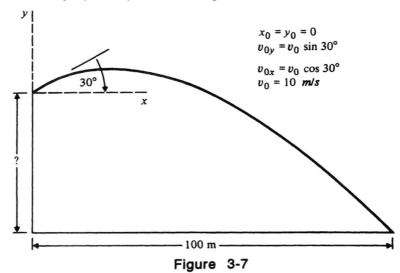

Figure 3-7

Solution:

The initial data is

$$x_0 = y_0 = 0 \qquad v_{0x} = v_0 \cos 30°$$

$$v_{0y} = v_0 \sin 30° \qquad v_0 = 10 \text{ m/s}$$

By methods which should now be familiar, the trajectory is found to be

$$y = x \tan \theta - \frac{1}{2} g \left[\frac{x^2}{v_0^2 \cos^2 \theta} \right]$$

At impact x = 100 m and

$$y = (100 \text{ m})(0.58) - \left[\frac{(9.8 \text{ m/s}^2)(100 \text{ m})^2}{2(10 \text{ m/s})^2 (0.75)} \right]$$

$$= 58 \text{ m} - 653 \text{ m} = -595 \text{ m}$$

68

(a) The cliff is 595 m high. (b) The time of flight is given, from $x = (v_0 \cos \theta)t$, by

$$t = (x/v_0 \cos \theta) = [(100 \text{ m})/(10 \text{ m/s})(0.87)] = 11.5 \text{ s}$$

(c) The velocity at impact is

$$v_x = v_0 \cos \theta = (10 \text{ m/s})(.87) = 8.7 \text{ m/s}$$

$$v_y = v_0 \sin \theta = - gt = (10 \text{ m/s})(0.5) - (9.8 \text{ m/s}^2)(11.5 \text{ s})$$

$$= - 107 \text{ m/s}$$

$$v = (v_x^2 + v_y^2)^{1/2} = 108 \text{ m/s}$$

$$\theta = \text{arc tan} (- 107/8.7) = - 85° \quad (85° \text{ below x axis})$$

Example 6

Lois playfully pushes Clark Kent horizontally off a 500 m building, giving him an initial velocity of 15 m/s. Clark must save himself by changing into his Superman suit before hitting the ground. How much time does Clark have?

Solution:

$$y = - \frac{1}{2} gt^2 = - 500 \text{ m}$$

Solving for the time t yields:

$$t = \sqrt{\frac{2(500 \text{ m})}{9.8 \text{ m/s}^2}} = 10.1 \text{ s}$$

Note: Lois' playful horizontal push is irrelevant, as long as she does not impart to Clark a vertical component of velocity.

Example 7

A particle's coordinates are given by

$$x = A \cos \omega t$$

$$y = A \sin \omega t$$

69

(a) Find the trajectory equation.
(b) Find v_x, v_y, and v.
(c) Find a_x, a_y, and a.

Solution:

(a) $r^2 = x^2 + y^2 = A^2(\cos^2 \omega t + \sin^2 \omega t) = A^2$

Hence

$$\vec{r} = x\hat{\imath} + y\hat{\jmath} = A(\cos \omega t \,\hat{\imath} + \sin \omega t \,\hat{\jmath})$$

traces out a circle of radius $r^2 = x^2 + y^2 = A^2$, describing uniform circular motion, with

(b) $v_x = \dfrac{dx}{dt} = -A\omega \sin \omega t$

$v_y = \dfrac{dy}{dt} = A\omega \cos \omega t$

The velocity vector

$$\vec{v} = -A\omega \sin \omega t \,\hat{\imath} + A\omega \cos \omega t \,\hat{\jmath}$$

has constant magnitude

$$v^2 = \omega^2 A^2 \sin^2 \omega t + \omega^2 A^2 \cos^2 \omega t$$

$$= A^2\omega^2 = r^2\omega^2; \quad v = r\omega.$$

(c) $a_x = \dfrac{dv_x}{dt} = \dfrac{d}{dt}\left(-A\omega \sin \omega t\right) = -A\omega^2 \cos \omega t$

$$= -\omega^2 x$$

$a_y = \dfrac{dv_y}{dt} = \dfrac{d}{dt}\left(A\omega \cos \omega t\right) = -A\omega^2 \sin \omega t$

$$= -\omega^2 y$$

$$\vec{a} = (-\omega^2)x\,\hat{\imath} + (-\omega^2)y\hat{\jmath} = (-\omega^2)\vec{r}$$

$$|\vec{a}| = a = \omega^2 r = v^2/r$$

Note the velocity and position vector have zero inner (scalar) product, that is,

$$\vec{r} \cdot \vec{v} = r_x v_x + r_y v_y = x v_x + y v_y$$

$$\vec{r} \cdot \vec{v} = A \cos \omega t(- A\omega \sin \omega t) + A \sin \omega t(A\omega \cos \omega t)$$

$$= 0$$

indicating that $\vec{r}$ and $\vec{v}$ are always perpendicular. See Fig. 3-8.

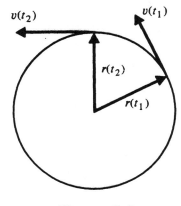

Figure 3-8

Example 8

You are a ship of the line steaming due north, in formation, 1000 m astern of the flagship, at 10 km/hr. You are ordered to come 1000 m abeam of her. Your maximum speed is 20 km/hr. What is your course to come to station in best time?

Solution:

Referring to Fig. 3-9, where you are A, and F is the flagship, we see that

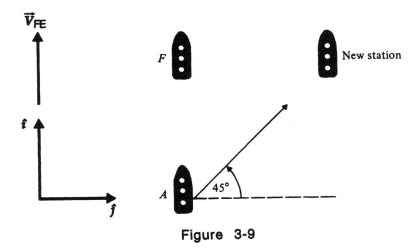

Figure 3-9

71

$$\vec{v}_{AF} = v_{AF} \sin 45°\hat{\imath} + v_{AF} \cos 45°\hat{\jmath}$$

where $\vec{v}_{AF}$ is your velocity relative to the flagship F, with v_{AF} the unknown magnitude.

The velocity of the flagship relative to the earth is

$$\vec{v}_{FE} = 10 \text{ km/hr } \hat{\imath} = v_{FE}\hat{\imath}$$

If $\vec{v}_{AE}$ is your velocity relative to the earth, then

$$\vec{v}_{AE} = \vec{v}_{AF} + \vec{v}_{FE}$$

$$\vec{v}_{AE} = \frac{v_{AF}}{\sqrt{2}} \hat{\imath} + \frac{v_{AF}}{\sqrt{2}} \hat{\jmath} + v_{FE} \hat{\imath}$$

$$\vec{v}_{AE} = \left(\frac{v_{AF}}{\sqrt{2}} + v_{FE}\right) \hat{\imath} + \frac{v_{AF}}{\sqrt{2}} \hat{\jmath}$$

To use best speed, set the magnitude of this vector equal to your maximum speed

$$v_{AE}^2 = \left(\frac{v_{AF}}{\sqrt{2}} + v_{FE}\right)^2 + \left(\frac{v_{AF}}{\sqrt{2}}\right)^2 = \left[20 \text{ km/hr}\right]^2$$

Since $v_{FE} = 10$ km/hr is known, this is a quadratic equation for v_{AF} with the solution

$$v_{AF} = 11 \text{ km/hr.}$$

Thus

$$\vec{v}_{AE} = \left[\left(\frac{11}{\sqrt{2}} + 10\right) \hat{\imath} + \frac{11}{\sqrt{2}} \hat{\jmath}\right] \text{ km/hr}$$

$$= 18 \text{ km/hr } \hat{\imath} + 8 \text{ km/hr } \hat{\jmath}$$

Your true heading is thus

$$\theta = \arctan (8/18) = 24° \text{ east of north.}$$

Example 9

A ship wishes to reach a point 10 km due north. It steams at 20 km/hr. The current is 5 km/hr due east. What should its heading be?

Solution:

Defining the directions $\hat{\imath}$ = north and $\hat{\jmath}$ = east and the velocities

$\vec{v}_{AE}$ = ship relative to earth

$\vec{v}_{AW}$ = ship relative to water

$\vec{v}_{WE}$ = water relative to earth

we have

$$\vec{v}_{AE} = \vec{v}_{AW} + \vec{v}_{WE}$$

We seek a solution in which the velocity $\vec{v}_{AE}$ is due north

$$\vec{v}_{AE} = v_{AE}\hat{\imath}$$

and in which v_{AW} = 20 km/hr,

$$\vec{v}_{AW} = v_{AW} \cos\theta\, \hat{\imath} + v_{AW} \sin\theta\, \hat{\jmath}$$

where θ is the heading angle. The current is

$$\vec{v}_{WE} = 5 \text{ km/hr}\, \hat{\jmath}$$

Thus we have

$$v_{AE}\, \hat{\imath} = v_{AW} \cos\theta\, \hat{\imath} + v_{AW} \sin\theta\, \hat{\jmath} + v_{WE}\, \hat{\jmath}$$

$$v_{AE}\, \hat{\imath} = 20 \text{ km/hr} \cos\theta\, \hat{\imath} + [5 \text{ km/hr} + (20 \text{ km/hr})\sin\theta]\, \hat{\jmath}$$

Equating the x and y components separately on each side,

$$v_{AE} = 20 \text{ km/hr} \cos\theta$$

$$5 \text{ km/hr} + (20 \text{ km/hr}) \sin\theta = 0$$

$$\sin\theta = - (5/20) = - (1/4) \quad \theta = - \ 14.5°$$

Since θ is negative, the heading is 14.5° away from $\hat{\imath}$ in the negative $\hat{\jmath}$ direction, or 14.5° west of north. See Fig. 3-10.

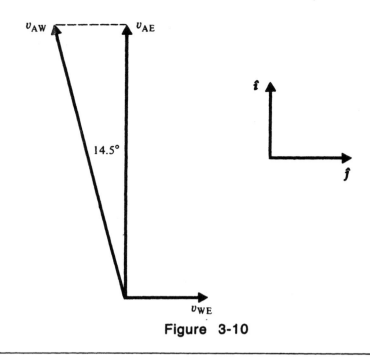

Figure 3-10

QUIZ

1. A batter hits a ball which lands on the top of a 15.3 m high fence 122 m away. The angle of departure is 45°. Find (a) Find the initial velocity; (b) Find the time of flight; and (c) Find the maximum height of the ball.

Answer: (a) 36.9 m/s, (b) 4.7 s, (c) 34.8 m

2. What is the centripetal acceleration of a person standing on the equator of the earth?

Answer: 0.03 m/s^2

3. A marble rolls on a laboratory table 1 m above the floor and falls off the table, landing a horizontal distance of 0.5 m from where it fell off. What was its initial velocity as it fell off the table?

Answer: 1.1 m/s

4. An artillery shell is fired at an elevation angle of 30° and lands 1000 m away on level ground. What was its maximum height?

Answer: 140 m

5. A race car driver takes a 100 m radius curve at 150 km/hr. Show that his acceleration in terms of g = 9.8 m/s^2 is a = 1.8 g.

4

NEWTON'S LAWS OF MOTION

OBJECTIVES

In Chapters 2 and 3 you studied the acceleration of a body moving along a line or in a plane, without specifying the cause of the acceleration. In Chapter 4, *force*, a vector quantity is identified as the means by which bodies influence each others motion and the cause of acceleration. Your objectives are to:

Identify the *forces* acting between bodies and calculate the resultant or total force, the vector sum of all forces acting on a single body.

Observe that a body continues at rest or in uniform motion along a straight line if no net force acts on it (*Newton's first law*).

Define *equilibrium* as a state in which the total or net force acting on a body is zero.

Recognize an *inertial frame of reference* as one in which Newton's first law is valid.

Define *mass* as the proportionality constant relating force and acceleration.

Determine the *acceleration* $\vec{a}$ of a body in terms of the total force $\Sigma \vec{F}$ acting on it: $\Sigma \vec{F} = m\vec{a}$. (*Newton's second law.*)

Calculate forces, masses, and accelerations in the SI, cgs, and engineering system of *units*.

Relate the downward force or *weight* w of objects near the surface of the earth to the acceleration of gravity g.

Observe that the mutual forces between two bodies are equal in magnitude and opposite in direction. (*Newton's third law.*)

Apply Newton's laws to *simple systems*; more complicated ones will be encountered in Chapters 5 and 6.

Draw <u>Free-Body diagrams</u> to help in calculating the total or net force acting on a body.

REVIEW

According to Newton's first law, a body moves in a straight line with constant velocity, that is, with zero acceleration, if the net force acting on it is zero. If the net force is not zero, the body accelerates. By Newton's second law, the acceleration is proportional to the force, according to the vector relation,

$$\vec{F} = m\vec{a}$$

where the constant m is called the *mass* of the body. The units of mass are those of force divided by acceleration. In the SI the unit of force is the newton (N) and the unit of mass is

$$N / (m/s^2) = N{\cdot}s^2/m = kg \text{ (kilogram)}$$

$$= 10^3 \text{ g (gram)}$$

In the British system the unit of force is the pound (lb) and the unit of mass is:

$$\frac{lb}{ft/s^2} = \frac{lb{\cdot}s^2}{ft} = slug$$

Newton's third law states that when body A exerts a force $\vec{F}$ on body B, body B must exert a force on A equal in magnitude but opposite in direction: $-\vec{F}$. Newton's third law is useful when analyzing the forces acting between different parts of a mechanical system. Detailed examples are given in the next chapter of this Study Guide.

Forces, if unbalanced, tend to change the state of motion of a body. *A particle is said to be in equilibrium* if it is at rest or moves at constant speed in a straight line. The condition for equilibrium of a particle is that the total force acting on a body be zero,

$$\vec{R} = \vec{F}_1 + \vec{F}_2 + \vec{F}_3 + ... = \Sigma \vec{F} = 0$$

This condition is a statement of *Newton's First Law of motion*, "Every body continues in its state of rest, or of uniform motion, in a straight line, unless it is compelled to change that state by forces impressed on it."

This law is true in "inertial frames of reference". The classroom, fixed with respect to the earth, is approximately an inertial system; a car moving smoothly at constant speed in a straight line with respect to the earth is another inertial system. Any system in uniform motion in a straight line with respect to an inertial system is another inertial system.

An accelerating car or a car moving at constant speed around a curve is not an inertial system; a marble on a smooth horizontal table will *not* remain at rest in these non-inertial systems even when there are no horizontal forces acting on it.

As for the first Law, Newton's second law is valid only in an *inertial frame of reference*: if Newton's laws are valid to observer A, then they are also valid in the frame of reference of observer B who moves with constant velocity with respect to A.

The force that makes projectiles accelerate downward is the force of their weight,

$$w = mg$$

All bodies on the surface of the earth are subject to the downward force of their weight.

HINTS AND PROBLEM-SOLVING STRATEGIES

Remember all the previous hints and problem-solving strategies still apply--there are no statutes of limitations for physics.

1. Choose a convenient coordinate system, decide on the positive direction, and stick to your choice.

2. Try to carry all results forward algebraically, substituting numbers only at the end.

To these, let's now add another--DRAW A FREE-BODY DIAGRAM FOR THE BODY IN QUESTION, INDICATING ALL FORCES ACTING ON IT, AND NO OTHER FORCES.

Typical forces that act on a body could be its weight (earth pulls body down), the tension of a cable pulling on it, or the force of contact at a floor or wall.

QUESTIONS AND ANSWERS

Question. Why does the discussion of motion stop when acceleration is introduced? Why not continue and introduce the third and fourth derivatives of position with respect to time?

Answer. For physical descriptions, it is required to obtain the position as a function of time. Newton's second law gives the acceleration of a particle when the forces are specified making it possible to find the velocity as a function of time and then the position as a function of time. Thus when the acceleration is known and the values of the initial velocity (v_0) and initial position are known (x_0), the position as a function of time is known.

Question. A sack of potatoes is dragged across a rough floor by a rope making an angle with the horizontal. What forces act on the sack? What are the reactions to these forces?

The forces on the sack are: its weight, the tension in the rope, the contact force (normal) between the sack and the rough floor and friction between the sack and the floor. The reactions to these forces are: the gravitational force on the earth due to the sack of potatoes; the force on the rope due to pulling the sack, the contact force on the floor due to the sack (equal and opposite to the normal); and a friction force equal and opposite to the friction force on the sack.

EXAMPLES AND SOLUTIONS

Example 1

A wagon is pushed up a steep driveway with an incline angle of 20° by a horizontal force of 25 N. What is the component of the pushing force in the direction of motion?

Solution

As shown in Fig. 4-1, we choose a coordinate system with the x-axis along the driveway. Then the component of $\vec{F}$ along the driveway is

$$F_x = F \cos \theta = 25 \text{ N} \cos 20° = 23 \text{ N}$$

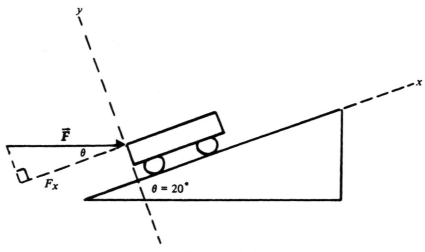

Figure 4-1

Example 2

A cat weighs 20 N on the surface of the earth, where the acceleration of gravity is 9.8 m/s². What is its mass?

Solution:

$$w = mg \quad \text{so } m = \frac{w}{g} = \frac{(20 \text{ N})}{(9.8 \text{ m/s}^2)} = 2.04 \text{ kg.}$$

Example 3

A rock of mass 4 kg is suspended by a wire from a tree branch. When a horizontal force of 29.4 N is applied to the rock, it assumes an equilibrium position when the wire makes an angle θ *with the vertical*. Find the angle θ and the force the wire exerts on the rock.

Solution:

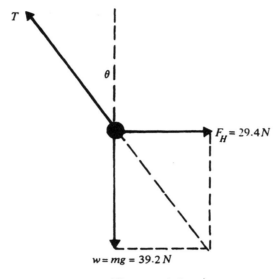

Figure 4-2

     Your first step is to draw a free-body diagram showing <u>all</u> forces acting <u>on</u> the rock (and no others). Referring to Fig. 4-2, we see that the rock is in equilibrium under the action of <u>three</u> forces: its weight, acting down as shown,

$$w = mg = 4 \text{ kg } (9.8 \text{ m/s}^2) = 39.2 \text{ N},$$

the horizontal applied force F_H, and the wire tension T. In equilibrium, <u>the sum of all forces, and force components, is zero</u>,

$$\Sigma F_x = 0 \qquad \Sigma F_y = 0$$

For the horizontal or x components we find

$$0 = \Sigma F_x = F_H - T \sin \theta = 29.4 \text{ N} - T \sin \theta$$

For the vertical or y components we find

$$0 = \Sigma F_y = T \cos \theta - 39.2 \text{ N}$$

or

$$T \sin \theta = 29.4 \text{ N}$$

$$T \cos \theta = 39.2 \text{ N}$$

Dividing these last two equations we find

$$\frac{\sin \theta}{\cos \theta} = \tan \theta = \frac{29.4}{39.2} = 0.75$$

$$\theta = \tan^{-1}(0.75) = 37°$$

Thus the tension is

$$T = \frac{29.4 \text{ N}}{\sin 37°} = 49 \text{ N}$$

Example 4

(a) An iceskater pushes off the wall of the rink. What is the reaction to the force of the skater against the wall?

(b) A book rests on a table. What is the reaction to the upward force of the table on the book?

(c) In Example 3, what is the reaction to the tension force on the rock?

Solution:

   (a) The wall pushes against the skater, setting her in motion.

   (b) The book exerts a downward force on the table.

   (c) The wire is stretched between the rock and the tree branch. The branch pulls on the rock. The reaction of this force is the rock pulling the branch.

Note that in all cases, that action and reaction forces act on <u>different</u> bodies.

Example 5

A 3.1 g bullet traveling at 300 m/s penetrates a block of wood to a depth of 15 cm. Assuming a constant retarding force, find the magnitude of the force and the time required for the bullet to come to rest.

Solution:

When the force is constant, the acceleration is constant, by Newton's second law: F = ma. Thus we may use our earlier (Chapter 2) results for <u>constant acceleration</u>:

$$v^2 = v_0^2 + 2ax$$

where v is the final velocity (v = 0), v_0 is the initial velocity (v_0 = 300 m/s), and x is the distance traveled between initial and final times, x = 5 cm. Thus the acceleration is

$$a = \frac{\left(v^2 - v_0^2\right)}{2x} = \frac{-\left(300 \text{ m/s}\right)^2}{2\left(0.15 \text{ m}\right)}$$

$$= -3 \times 10^5 \text{ m/s}^2$$

Newton's second law indicates that the force is then

$$F = ma = (3.1 \times 10^{-3} \text{ kg})(-3 \times 10^5 \text{ m/s}^2)$$

$$= -930 \text{ N}$$

The stopping time is given by the constant acceleration relation

$$v = v_0 + at$$

with the final velocity v = 0:

$$t = \frac{v - v_0}{a} = \frac{\left(-300 \text{ m/s}\right)}{\left(-3 \times 10^5 \text{ m/s}^2\right)} = 10^{-3} \text{ s}$$

Example 6

The coordinates of a body are given by:

$$x = A \cos \omega t, \quad y = \text{constant},$$

where A = 0.2 m and ω = 1 rad/s = 1 /s. Find the total force on the body when t = 0.5 s.

Solution:

The motion is in two dimensions, so x and y components of force must be considered. Since F = ma, we write for the components,

$$F_x = ma_x = m\frac{dv_x}{dt} = m\frac{d^2x}{dt^2}$$

$$F_y = ma_y = m\frac{dv_y}{dt} = m\frac{d^2y}{dt^2} = 0.$$

Thus the force is in the x-direction with

$$F_x = m\frac{d^2}{dt^2}\left(A\cos\omega t\right) = m\frac{d}{dt}\left[\frac{d}{dt}\left(A\cos\omega t\right)\right]$$

$$F_x = m\frac{d}{dt}\left[-A\omega\sin\omega t\right] = -m\omega^2 A\cos\omega t$$

$$= -(2\text{ kg})(1\text{ /s})^2\ (0.2\text{ m})\ \cos[(1\text{ /s})\cdot(0.5\text{ s})]$$

$$= 0.35\text{ N}.$$

Note in the expression $\cos\omega t$, ωt is in radians. The expression

$$\frac{d}{dt}\left[\sin\omega t\right] = \omega\cos\omega t$$

is true only when ωt is in radian measure.

Example 7

A 1000 kg elevator rises with an upward acceleration equal to g. What is the tension in the supporting cable? (See Fig. 4-3)

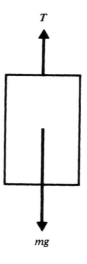

Figure 4-3

Solution:

Note the first step is to draw a free-body diagram, as shown, indicating all forces acting on the elevator. If the up direction is taken as positive, we have

$$T - mg = ma = mg$$

$$T = 2\ mg = 2(1000\ kg)(9.8\ m/s^2)$$

$$= 19,600\ N$$

Example 8

An elevator weighing 4000 lbs falls with a *downward* acceleration of magnitude (1/2)g. What is the tension in the supporting cable?

Solution:

Again, the positive direction is taken to be up, so the acceleration is <u>negative</u> in sign. Referring to Fig. 4-3, we set the *total* force equal to the mass times the acceleration

$$T - mg = ma = m\left(\frac{-g}{2}\right)$$

Thus the tension is

$$T = mg - m\left(\frac{g}{2}\right) = m\left(\frac{g}{2}\right) = \left(\frac{w}{2}\right) = \left(\frac{4000\ lbs}{2}\right) = 2000\ lbs.$$

Example 9

A block, starting from rest, is acted upon by gravity and an upward force of 5 N. Its acceleration is 3 m/s², upward.
   (a) What is its mass? (See Fig. 4-3, with T the upward force.)
   (b) If the upward force acts for 5 s, how high does the block rise before it starts its descent?

Solution:

The free-body diagram is again supplied by Fig. 4-3, with the up direction is taken as positive.

(a) $T - mg = ma$;   so $m(a + g) = T$,

$$m = \frac{T}{a + g} = \frac{(5\ N)}{(3\ m/s^2) + (9.8\ m/s^2)} = 0.39\ kg$$

(b)   While the upward force acts, the acceleration is in an upward direction,

$$x = x_0 + v_0 t + \frac{1}{2} at^2$$

where $x_0 = 0$, $v_0 = 0$, and $a = 3$ m/s^2.

At t = 5 s,

$$x = \frac{1}{2} at^2 = \frac{1}{2}\left(3 \,\text{m/s}^2\right)(5 \,\text{s})^2 = 37.5 \,\text{m},$$

and

$$v = v_0 + at = (3 \text{ m/s}^2)(5 \text{ s}) = 15 \text{ m/s}.$$

At this point the upward force ceases to act and only a downward weight acts, producing a downward acceleration of magnitude g. The above results form the *initial* conditions for the subsequent motion: *After* t = 5 s, a = − g, and the coordinate and velocity are given by

$$x = x_0 + v_0 t - \frac{1}{2} gt^2 \quad \text{with } x_0 = 37.5 \text{ m},$$

$$v = v_0 - gt \text{ where } v_0 = 15 \text{ m/s}.$$

At the maximum height, v = 0 and

$$t = \frac{v_0}{g} = \frac{15 \,\text{m/s}}{9.8 \,\text{m/s}^2} = 1.53 \text{ s}.$$

$$x = 37.5 \text{ m} + (15 \text{ m/s})(1.53 \text{ s}) - (1/2)(9.8 \text{ m/s}^2)(1.53 \text{ s})^2$$

$$x = 49 \text{ m}$$

Alternatively, use

$$v^2 = v_0^2 + 2a(x - x_0) = v_0^2 - 2g(x - x_0) = 0$$

with v = 0 at the maximum height. Solving for x − x₀ we have

$$x - x_0 = \frac{v_0^2}{2g} = \frac{\left(15 \,\text{m/s}\right)^2}{2\left(9.8 \,\text{m/s}^2\right)} = 11.5 \text{ m}.$$

$$x = 37.5 \text{ m} + 11.5 \text{ m} = 49 \text{ m}.$$

Example 10

A block of mass 10 kg is suspended by a string. What is the string tension when:
  (a) The block is at rest;
  (b) The block is accelerating upward at 5 m/s²;
  (c) The block is accelerating downward at 5 m/s²;
  (d) The block is in free fall.

Solution: (See Fig. 4-3.) Again we take the up direction as positive.

(a) $T - mg = ma = 0$; so $T = mg = (10$ kg$)(9.8$ m/s²$) = 98$ N.

(b) $T - mg = ma$; so $T = m(a + g) = (10$ kg$)(5 + 9.8)$m/s² $= 148$ N

(c) $T - mg = ma$; so $T = m(a + g) = (10$ kg$)(- 5 + 9.8)$m/s² $= 48$ N

(d) $T - mg = ma = - mg$; so $T = 0$. ("Free fall" means that only the gravitational force acts and $a = - g$.)

Example 11

A 100 kg man stands on a bathroom scale in an elevator. What is the acceleration of the elevator when the scale reading is (a) 150 kg; (b) 100 kg; and (c) 50 kg.

Solution:

Referring to Fig. 4-4, where a free-body diagram <u>for the man</u> is given to the right, F_s is the force of the scale <u>on the man</u> and mg is the force of gravity <u>on the man</u>. The acceleration of the man is equal to the acceleration of the elevator a,

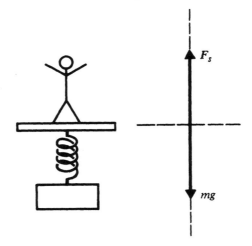

Figure 4-4

$$\Sigma F = F_s - mg = ma$$
$$a = \frac{F_s - mg}{m}.$$

The scale measures force but is calibrated so that it reads 1 kg when the force is equal to the weight of a 1 kg mass, that is, $F_s = m_s g$ where m_s is the *scale reading* as given above. Thus

$$a = \frac{m_s g - mg}{m} = \left(\frac{m_s}{m} - 1\right)g$$

Note again, that we do the algebra first, then substitute the numbers.

(a) $\quad a = \left(\dfrac{150}{100} - 1\right)g = 4.9\,\text{m/s}^2$

(b) $\quad a = \left(\dfrac{100}{100} - 1\right)g = 0$

(c) $\quad a = \left(\dfrac{50}{100} - 1\right)g = -4.9\,\text{m/s}^2$

Example 12

A ball slides without friction off the elevated inclined plane shown in Fig. 4-5a, starting from rest at the top of the plane. Find the distance $\bar{x}$ in Fig. 4-5a.

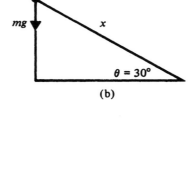

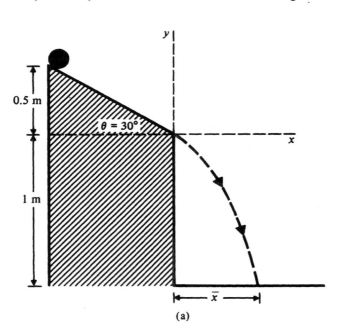

(a)

Figure 4-5a,b

Solution:

This problem is really two problems: first, to find the velocity at the bottom of the inclined plane, and second, to use this velocity as an initial velocity for the projectile part of the motion. Refer to Fig. 4-5b for the first part of the motion (while the ball is on the plane). During this time, the ball is acted upon by two forces, its weight and the normal force of the plane. By Newton's second law, the total force along the plane gives the mass times the acceleration along the plane:

$$F_x = mg \sin \theta = ma_x \; ; \; a_x = g \sin \theta$$

$$F_y = N - mg \cos \theta = 0$$

(A convenient, tilted, coordinate system has been chosen so the x-axis points <u>down</u> the inclined plane.) Thus while on the plane we have constant acceleration in the tilted x-direction,

$$x = x_0 + v_{0x}t + \frac{1}{2} a_x t^2 = \frac{1}{2} \left(g \sin \theta \right) t^2 \; ;$$

$$t = \sqrt{\frac{2x}{g \sin\theta}}$$

$$v_x = v_{0x} + a_x t = g \sin \theta \, t$$

where the initial conditions are $v_{0x} = 0$, $x = 0$ for the *tilted coordinate system* with origin at the top of the plane. The time to reach the bottom of the plane at $x = 0.5$ m/sin θ is

$$t = \sqrt{\frac{2x}{g \sin \theta}} = \sqrt{\frac{2(0.5 \, \text{m})}{(9.8 \, \text{m/s}^2)\left(\sin \theta \right)^2}} = \frac{0.32 \, \text{s}}{\sin \theta}$$

and hence the velocity is

$$v_x = (g \sin \theta) \, t = (9.8 \, \text{m/s}^2)(0.32 \; \text{s})$$

$$= 3.1 \, \text{m/s}.$$

Alternatively we can use

$$v_x{}^2 = v_{0x}{}^2 + 2a_x(x - x_0).$$

With $x = (.5 \, \text{m})/(\sin \theta)$, $v_{0x} = 0$ and $a_x = g \sin \theta$ we have

$$v_x{}^2 = 2(g \sin \theta)(0.5 \; \text{m})/(\sin \theta)$$
$$v_x = 3.1 \, \text{m/s}$$

independent of the incline angle θ.

You are now ready to do the second part of the problem, resetting your clock, coordinate system, and initial conditions, as shown in the x,y system of Fig. 4-5a. The initial conditions are: $x_0 = 0$ $y_0 = 0$

$v_{0x} = v_0 \cos \theta$ $v_{0y} = -v_0 \sin \theta$ where $v_0 = 3.1$ m/s.

and thus

$$x = x_0 + v_{0x}t = v_0 \cos \theta \, t$$

$$y = y_0 + v_{0y}t + \frac{1}{2} a_y t^2 = -\left(v_0 \sin \theta\right)t - \frac{1}{2}gt^2 \;;$$

The time the ball hits the floor is when y = -1 m. Solving the last equation for t,

$$\frac{1}{2}gt^2 + \left(v_0 \sin \theta\right)t + y = 0 = at^2 + bt + c$$

$$t = \frac{-b \pm \sqrt{b^2 - 4ac}}{2a}$$

$$t = \frac{-\left(v_0 \sin \theta\right) \pm \sqrt{\left(v_0 \sin \theta\right)^2 - 4gy}}{2g}$$

$$= -0.63 \text{ s, or } + 0.31 \text{ s}$$

(The correct time is the positive root, but the negative root also has a physical interpretation. If the inclined plane were not present, then a possible motion is one in which the ball follows the full trajectory as indicated in Fig. 4-5c, starting at the floor at t = - 0.63 s at negative x and arriving at the origin at t = 0.)

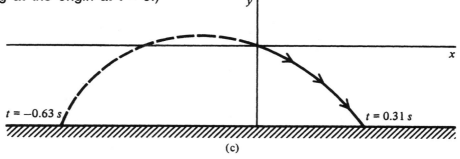

(c)

Figure 4-5c

The distance $\bar{x}$ is now given as

$$\bar{x} = x(0.31 \text{ s}) = v_0 \cos 30° \, (.31 \text{ s}) = 3.1 \text{ m/s}(.87)(.31 \text{ s}) = .84 \text{ m}$$

The distance x(–) at t = – 0.63 s is

$$x(–) = (3.1 \text{ m/s})(.87)(– 0.63 \text{ s}) = – 1.7 \text{ m}$$

QUIZ

1. A 1000 kg automobile is pulling a trailer of mass 500 kg.
   (a) When the automobile accelerates at 2.0 m/s^2, what is the tension in the trailer hitch?
   (b) What is the horizontal force of the automobile's wheels against the track?

Answer: (a) 1000 N   (b) 3000 N

2. A bathroom scale in a rising elevator registers 100 kg when a man of mass 85 kg stands on it. What is the acceleration of the elevator?

Answer: 1.73 m/s^2, upward

3. A 20 kg sled on smooth ice is pushed from rest for 3 s by a constant horizontal force of 10 N.
   (a) What is its final velocity?
   (b) How far does it move during this time?

Answer: (a) 1.5 m/s   (b) 2.3 m

4. An offensive tackle of mass 100 kg is subject to a down field force of 300 N and a cross field force of 400 N.
   (a) What force must he exert against the ground to insure he remains at rest?
   (b) What is his acceleration if he slips and exerts no force on the ground?

Answer: (a) 500 N, at a 53° angle from the upfield direction.
        (b) 5 m/s^2, at an angle of 53° from downfield.

5. What is the net force on a race car driving 150 km/hr around a curve of 100 m radius if its mass is 2500 kg? Compare the force with the weight of the car.

Answer: 26,000 N or 1.8 times the weight of the car.

89

5
APPLYING NEWTON'S LAWS

OBJECTIVES

In this chapter you will apply Newton's laws of motion to objects in *equilibrium* (at rest or in uniform motion) and *not in equilibrium* (in accelerated motion.) Your objectives are to:

Use Newton's first law to *insure equilibrium* of a body.

Use Newton's third law to relate forces acting between two bodies.

Analyze simple systems by identifying the forces acting on each *component*. Examples are blocks on inclined planes, and systems of weights and pulleys.

Draw *free-body diagrams* illustrating all forces acting on a body.

Recognize common mechanical forces such as *tension*, *contact*, and *weight*.

Use Newton's second law to find the acceleration of, or the force on, a body.

Analyze the force, velocity, and acceleration of objects in *circular motion*.

Apply this analysis to motion in a *vertical circle*.

REVIEW and SUPPLEMENT

As noted in Chapter 4, forces, such as $\vec{F}$, $\vec{w}$, $\vec{n}$, or $\vec{f}$ in Fig. 5-1, are vector quantities, with magnitude and direction, exerted on a body at a point, along some *line of action*. Fig. 5-1 illustrates four common examples of forces you will repeatedly encounter in the next two chapters. $\vec{F}$ is a pull exerted by, say, a man *on the box* as he attempts to drag it to the right. $\vec{w}$ is the weight of the box, a downward force due to the gravitational attraction of the earth *on the box*. $\vec{n}$ is the normal force of the earth pushing upward *on the box*. It is always normal (perpendicular) to the two surfaces in contact. $\vec{f}$ is the frictional force or drag exerted *on the box* by the earth, opposing the motion which the force $\vec{F}$ tends to create if the system is initially at rest. The frictional force is parallel to the surface, perpendicular to the normal force. $\vec{n}$ and *f* are components of the total contact force.

It is crucial in this and later chapters to distinguish between (1) forces exerted *on a body by its surroundings* and (2) forces exerted *on its surroundings by the body*. In the example of Fig. 5-1 the indicated forces are those acting *on the box*. Fig. 5-1a is a *"free body diagram"*.

As noted in the last chapter, the unit of force in the SI is a newton (N) and in the British system a pound. The conversion factor is 1 pound = 4.448 N.

Bathroom scales measure force, for example, the weight of your body, in units of pounds. Similarly spring balances may be calibrated in newtons to measure the forces such as the force $\vec{F}$ in Fig. 5-1.

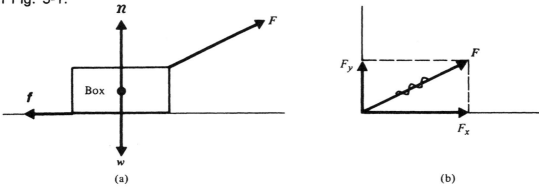

(a)   (b)

Figure 5-1

It is often useful to resolve forces into their rectangular coordinates as in Fig. 5-1b. Experimentally $\vec{F}_x$ and $\vec{F}_y$ acting at the same point of application as $\vec{F}$ have the same effect as $\vec{F}$, consistent with the equality $\vec{F}_x + \vec{F}_y = \vec{F}$. Note in Fig. 5-1b we have followed the text's convention of drawing a snake over $\vec{F}$ to indicate that it is not a force additional to its components: work either with $\vec{F}$ or with $\vec{F}_x$ and $\vec{F}_y$, but not with all three.

Similarly if two forces $\vec{F}_1$ and $\vec{F}_2$ act on an object at the same point, their vector sum $\vec{R} = \vec{F}_1 + \vec{F}_2$ has the same effect as $\vec{F}_1$ and $\vec{F}_2$.

The two techniques of vector summation and resolution of forces into components will be central to most of the problems of the next few chapters. The rectangular system of coordinates need not be the familiar vertical and horizontal axes; in the inclined plane example, you will indeed find it more convenient to resolve the forces into components along the plane and perpendicular to the plane, as shown in Fig. 5-2 and Fig. 5-3

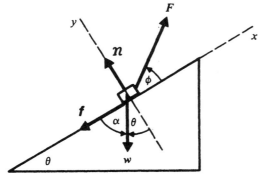

Figure 5-2

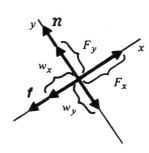

Figure 5-3

in which a force $\vec{F}$ drags a box up the plane against the friction force $\vec{f}$. Note the incline angle θ of the plane is equal to the angle between $\vec{w}$ and the negative y axis because they are complements of the same angle α. We have the components

$F_x = F \cos \phi$ $w_x = - w \sin \theta$

$F_y = F \sin \phi$ $w_y = - w \cos \theta$

$n_x = 0$ $f_x = - f$

$n_y = n$ $f_y = 0$

We have stressed above that forces may be divided into those exerted *on* a body A *by* its surroundings and those exerted *by* the surroundings *on* the body. Its surroundings are simply other bodies (B, C, D,...). By Newton's third law, if B exerts a force on A ($\vec{F}_{AB}$) then A exerts a force on B, which is equal in magnitude and opposite in direction, $\vec{F}_{AB} = - \vec{F}_{BA}$. The equal and opposite forces occur *on* different bodies (A on B, B on A) and are called action, reaction pairs. *They are equal and opposite whether or not the bodies are in equilibrium.*

The application of Newton's Third Law will be critical in your analysis of the simple mechanical systems of the next few chapters. Return to our original example of the man pulling the box in Fig. 5-1, where we showed all forces acting on the box. To each "action" $\vec{F}, \vec{n}, \vec{w}, \vec{f}$ corresponds an equal and opposite "reaction", a force of the box on the components of its surroundings. Some action-reaction pairs of forces are given in the table below and illustrated in Fig. 5-4.

"Action"	"Reaction"
$\vec{F}$ (force of man on box)	$\vec{F}' = - \vec{F}$ (force of box on man)
$\vec{n}$ (force of contact of earth's surface against box)	$\vec{n}' = - \vec{n}$ (force of contact of box against earth's surface)
$\vec{w}$ (gravitational force of earth on box)	$\vec{w}' = - \vec{w}$ (gravitational force of box on earth)
$\vec{f}$ (friction force of earth against box tending to prevent box from moving to right)	$\vec{f}' = - \vec{f}$ (force of box on earth tending to drag earth to right)

ACTION   REACTION

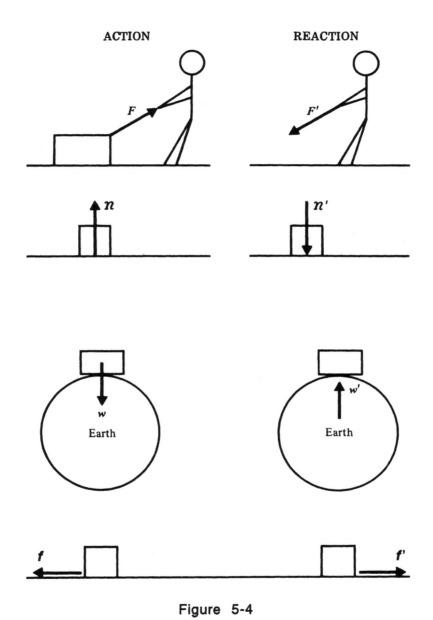

Figure 5-4

Like inclined planes, pulleys and ropes are components of systems useful in the next few chapters to illustrate basic principles. The tension T at a point in a rope is the magnitude of the force of one part of a rope on the other part as shown in Fig. 5-5, where T_{AB} and T_{BA} are action-reaction pairs.

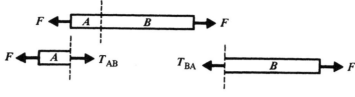

Figure 5-5

A rope, and any two of its segments A and B, are in equilibrium only if there are equal and opposite forces at their ends. The segment A is in equilibrium; hence the force of B on A at the division, the tension T, is equal in magnitude to F. (Note that the tension is NOT 2F.) The tension in this example is uniform, that is, the same at any point because our argument did not depend on where we choose to divide the rope into segments. This is true in an ideal massless rope whether or not it is in equilibrium. You will usually, consider rope of this kind; it transmits force without change of magnitude, even around corners as with pulleys (provided the pulley is massless or not accelerating) as indicated in Fig. 5-6.

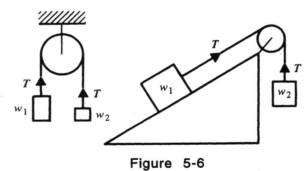

Figure 5-6

For an accelerating rope with mass (not in equilibrium) the tension is not uniform. For a stationary massive rope hanging in a gravitational field the tension is also non-uniform.

The friction force f between two surfaces is directed along the surfaces in a direction to oppose the relative motion which is occurring (kinetic friction) or the motion which would occur if the friction were not there (static friction.)

The static friction force is equal to or less than some maximum value. This maximum value is proportional (for simple surfaces) to the normal force n between the surfaces:

$$f_s < \mu_s \, n$$

The proportionality constant μ_s, the coefficient of static friction, is characteristic of the surfaces. When the external force, tending to move the surfaces over each other, exceeds $\mu_s n$, the surfaces slip and the *kinetic* friction force is now given by

$$f_k = \mu_k \, n$$

We will discover in the examples that $\mu_k < \mu_s$ for a simple pair of surfaces. Note there is no reason why $\mu < 1$.

When a body moves in a circle, the radial component of its acceleration is given by

$$a_{rad} = \frac{v^2}{R}$$

directed toward the center.

(This equation is also true for an arbitrary curve if R = radius of curvature.) The parallel (or tangential) component of acceleration is $a_{tan} = dv/dt$ where v is the magnitude of the velocity. The equations of motion are

$$F_{tan} = m\frac{dv}{dt}$$

$$F_{rad} = ma_{rad} = m\frac{v^2}{R}$$

F is called the centripetal force and v^2/R the centripetal acceleration. They both point toward the center of the motion.

If the circular motion is "uniform", v = constant and $a_{tan} = 0$, $F_{tan} = 0$. The only acceleration is a_{rad}, acting to change the direction but not the magnitude of $\vec{v}$.

Motion in a Vertical Circle

Referring to Fig. 5-7, an object on the end of a string of tension T is undergoing motion in a vertical circle of radius R.

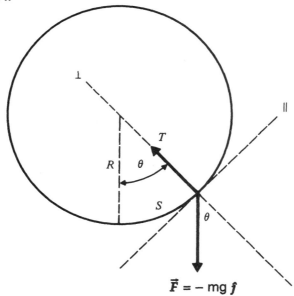

$$\vec{F} = -\, mg \, \hat{j}$$

Figure 5-7

The equations of motion are

$$F_{tan} = ma_{tan} = m\frac{dv}{dt} = -mg\sin\theta$$

$$F_{rad} = T - mg\cos\theta = m\frac{v^2}{R}$$

The tension T is obtained from the last equation.

$$T = mg\cos\theta + m\frac{v^2}{R}$$

T depends on the velocity v and the angle θ. Intuitively we expect the body to speed up at the bottom and slow down at the top, so that v is *not* a constant.

HINTS and PROBLEM-SOLVING STRATEGIES

An elaborate problem-solving strategy for equilibrium problems (Newton's first law) and Newton's second law problems is contained in the main text and should be reviewed now.

A very systematic approach has been used to solve all the problems met in this chapter. Briefly, you should always:

(1) Draw a sketch of the problem situation. Isolate the various bodies in the problem and make a FREE-BODY DIAGRAM for each body showing all the forces acting *on* the body.

(2) Choose a suitable coordinate system (i.e. y axis is vertical and x axis is horizontal except in situations like the inclined plane where a different choice is simpler) and resolve the various force vectors into their x and y components.

(3) Apply the Newton's law condition for translational equilibrium and write
$\Sigma F_x = 0$; $\Sigma F_y = 0$; and $\Sigma F_z = 0$, if the body is at rest or moving uniformly.

(4) Apply Newton's second law in the case of general motion,
$\Sigma F_x = ma_x$, $\Sigma F_y = ma_y$, and $\Sigma F_z = ma_z$.

(5) Solve the set of equations obtained in step (3) or (4) for the unknown quantities.

QUESTIONS AND ANSWERS

Question. Can the coefficient of kinetic friction between a body and a surface be greater than the corresponding coefficient of static friction?

Answer. No. Consider the situation where the object is just on the verge of moving and the static frictional force is at its maximum value. If the coefficient of kinetic friction was greater than the coefficient of static friction, the object would not move when the applied net force exceeded the maximum static frictional force.

Question. What is the <u>direction</u> of the force acting on a pendulum bob at the end of its swing? At the bottom of the swing?

Answer. At the end of the swing, the velocity is zero so the net force is tangent to the path. At the bottom, the bob is in motion on a circle so the net force points toward the center of the circle.

EXAMPLES AND SOLUTIONS

Example 1

Consider a block of weight w at rest or in uniform motion in a straight line on a smooth frictionless surface. Find all forces acting on the block.

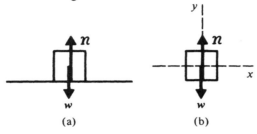

(a)   (b)

Solution: **Figure 5-8**

The block is in <u>equilibrium</u>, and hence $\Sigma \vec{F} = 0$. The sketch on the left in Fig. 5-8 shows the table and the block. The sketch on the right is a "free-body diagram" showing all forces acting on the block.

In order to avoid the temptation of adding an extraneous force acting on other bodies, such as the <u>downward</u> contact force of the block on the earth, it is helpful to "isolate" the body as in Fig. 5-8b, where we have also drawn a system of coordinate axes useful in resolving components. We have the equilibrium condition $\Sigma \vec{F} = 0 = \vec{w} + \vec{n}$. In terms of x and y components,

$$\Sigma F_x = 0$$

$$\Sigma F_y = 0 = n - w$$

and we learn that the normal force is equal and opposite to the weight. *This is not always the case*, as we will see below, in Example 3. Note how the problem-solving strategy of the text has been followed.

Example 2

Consider now the same block pulled with constant velocity along a surface with friction, as shown in Fig. 5-9, by a force $\vec{F}$. Find the force F in terms of μ_k and w.

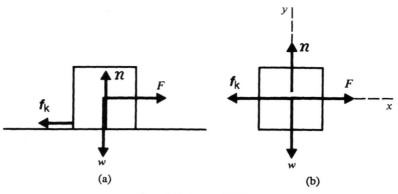

(a) (b)

Solution: Figure 5-9

In Fig. 5-9b, the "free-body diagram", we have moved all vectors so that their tails are at the origin (see rules for addition, Chapter 1). The point of application of the force and its line of action *are relevant* but not in connection with the condition of equilibrium for particles, $\Sigma \vec{F} = 0$.

Since the box is in equilibrium, (it moves with constant velocity)

$$\Sigma F_x = 0 = F - f_k = F - \mu_k n$$

$$\Sigma F_y = 0 = n - w.$$

Solving the first equation for F, and substituting $n = w$ from the second, we have

$$F = \mu_k n = \mu_k w$$

Note we have considered here the general case of arbitrary weight w and coefficient of friction μ_k, in terms of which we found the force F. The analysis answers every possible variation on this problem, such as

(i) What is the coefficient of friction μ_k if the weight is $w = 30$ N and the external force F necessary to make the box move uniformly is 10 N?

$$\mu_k = \frac{F}{w} = \frac{(10\,N)}{(30\,N)} = \frac{1}{3}$$

(ii) What is force F if the coefficient of friction is 0.25 and weight is 30 N?

$$F = \mu_k w = 0.25\,(30\,N) = 7.5\,N$$

(iii) How heavy is the body if the force $F = 15$ N and $\mu_k = 0.20$?

$$w = \frac{F}{\mu_k} = \frac{(15\,N)}{(0.2)} = 75\,N$$

(Notice that by working out the general case in terms of algebraic quantities, we easily answer all numerical questions by substituting numbers only at the end.)

Example 3

Consider now the case when the block is moving uniformly against friction but the force F is not horizontal, as in Fig. 5-10. Find F given the weight w and the coefficient of friction μ_k.

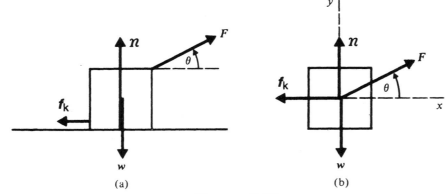

Figure 5-10

Solution:

Note the sketch and free-body diagram indicating forces and parameters. Applying the conditions of equilibrium, insuring that the total force is zero, we find

$$\Sigma F_x = 0 = F \cos \theta - f_k = F \cos \theta - \mu_k n \ ,$$

$$\Sigma F_y = 0 = n + F \sin \theta - w.$$

Notice that $n \neq w$. (This was true in Example 1, but is <u>not</u> a general rule.) Solving the second equation for n and substituting the result in the first equation, we have

$$0 = F \cos \theta - \mu_k(w - F \sin \theta)$$

$$= F(\cos \theta + \mu_k \sin \theta) - \mu_k w$$

or

$$F = \mu_k \left[\frac{w}{\cos \theta + \mu_k \sin \theta} \right]$$

You can check this result by putting $\theta = 0$. In this limit it should reduce to the previous result, $F = \mu_k w$, as it does. Note also $F = 0$ if $\mu_k = 0$.

Note that this is a "symbols only" problem--no numbers to substitute. The result is valid for any angle θ and any coefficient of friction you choose.

Example 4

Consider now a body of weight w moving uniformly up a plane inclined at 30° with coefficient of friction $\mu_k = 0.30$, under the influence of an external force F pulling parallel to the plane. Find the forces F and n .

Solution:

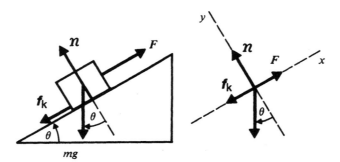

Figure 5-11

Again, we must first draw a sketch and a free body diagram indicating all forces acting on the body, as illustrated in Fig. 5-11. Here it is convenient to orient the x axis parallel to the incline. Imposing the equilibrium condition, we find

$$0 = \Sigma F_x = - f_k + F - w \sin \theta = - \mu_k n + F - w \sin \theta$$

$$0 = \Sigma F_y = n - w \cos \theta$$

It is worthwhile to think about each sign in the above equations. Physics is as unforgiving as a bank concerning sign errors. The x component of the weight is *negative* because the projection of $\vec{w}$ on the x axis points down the plane in the negative x direction. Correspondingly the y component of the weight is negative.

Note also that the distinction between sin θ and cos θ is of essence. Return to Chapter 1 for pertinent comments if you are not sure why $w_x = - w \sin \theta$ and $w_y = - w \cos \theta$ with θ defined in Fig. 5-11.

There is no point yet in substituting the specific values for θ, μ_k, or w.

We now can solve the two simultaneous equations for F and n . The second yields

$$n = w \cos \theta$$

substituting this result in the first yields

$$0 = -\mu_k \, w \cos\theta + F - w \sin\theta$$

or

$$F = w(\sin\theta + \mu_k \cos\theta)$$

yielding F and n for all w, θ, and μ_k.

Now if you were asked, what are the values of F and n if $w = 10$ N, the results would simplify as follows:

$$n = (10 \text{ N}) \cos 30° = 8.7 \text{ N}$$

$$F = (10 \text{ N})(\sin 30° + 0.30 \cos 30°)$$

$$= 7.6 \text{ N}$$

Example 5

A block of weight w is at rest on a rough surface of coefficient of static friction μ_s under the influence of a horizontal force F. Find the range of values possible for F before the block moves.

Solution:

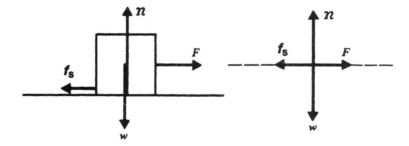

Figure 5-12

Since the values of w and μ_s have not been supplied, one must assume that the desired range of F is to be given algebraically in terms of w and μ_s. In Fig. 5-12 we have a sketch and free-body diagram. The equilibrium conditions are

$$\Sigma F_x = 0 = F - f_s \quad \text{but} \quad f_s \le \mu_s \, n$$

$$\Sigma F_y = 0 = n - w$$

Note it is incorrect to conclude that $f_s = \mu_s n$. The static friction f_s depends on F. If F = 0 the block just sits there and $f_s = 0$. The correct conclusion is f_s = F when the block does not move, that is when

$$f_s \leq \mu_s n = \mu_s w$$

or since f_s = F when the block is in equilibrium

$$F \leq \mu_s w$$

Static equilibrium is maintained as long as F does not exceed $\mu_s w$. Beyond that point the block moves and the friction force is $f_k = \mu_k n = \mu_k w$. This force of kinetic friction must be smaller than the maximum static friction, or the block would never move, that is

$$\mu_k w < \mu_s w \text{ or } \mu_k < \mu_s.$$

Example 6

Suppose a block of weight w is at rest on an adjustable inclined plane of angle θ and static coefficient of friction μ_s. Find the range of possible values of θ for which the block remains at rest.

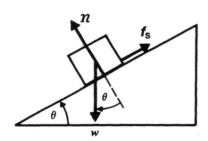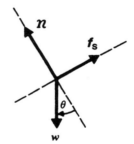

Figure 5-13

Solution:

The obligatory sketch and free-body diagram are indicated in Fig. 5-13. Both the angles θ are complimentary to ϕ. The conditions of equilibrium are

$$0 = \Sigma F_x = f_s - w \sin \theta$$

$$0 = \Sigma F_y = n - w \cos \theta$$

The condition $f_s < \mu_s n$ yields,

102

$$w \sin \theta < \mu_s \, w \cos \theta$$

or

$$\frac{\sin \theta}{\cos \theta} = \tan \theta \leq \mu_s$$

Alternately we may observe that the critical case occurs when $f_s = \mu_s \, n$ or
$w \sin \theta_c = \mu_s \, w \cos \theta_c$; $\mu_s = \tan \theta_c$.

At the angle of impending motion, $\tan \theta_c = \mu_s$, and $f_s = \mu_s$ ($w \sin \theta_c$) is at its maximum.

Example 7

In the pulley and weight system shown (Fig. 5-14a) the pulley is frictionless and the rope is massless. What force P applied to the free end of the rope is required to support the weight w?

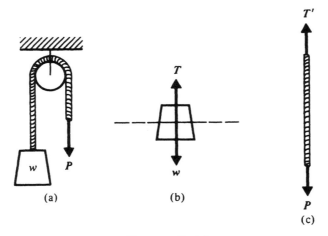

Figure 5-14

Solution:

In Fig. 5-14b we have a free-body diagram of the weight. In Fig. 5-14c we have a free-body diagram of a segment of the rope. Since the rope is massless (see supplementary section) T' = T. Since the segment of rope is in equilibrium P = T'. Hence P = T' = T = w. <u>Massless ropes have uniform tension throughout.</u>

103

Example 8

Consider the system shown in Fig. 5-15 where the weight $w_1 = 5$ N moves at constant speed on a rough table under the influence of the weight $w_2 = 2$ N. What is the coefficient of kinetic friction?

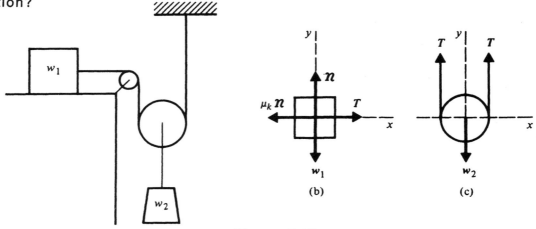

Figure 5-15

Solution:

In Fig. 5-15b and 5-15c we have the free-body diagrams for the forces acting on the two components of the system, w_1 and w_2 respectively. For w_1 the equilibrium conditions are

$$0 = \Sigma F_x = T - \mu_k n$$

$$0 = \Sigma F_y = n - w_1$$

from which we learn $T = \mu_k w_1$. For w_2 we have

$$0 = \Sigma F_x \quad \text{and} \quad 0 = \Sigma F_y = 2T - w_2$$

or $T = w_2/2$ from which we conclude

$$\mu_k = \frac{T}{w_1} = \frac{w_2}{2w_1} = \frac{(2\,\text{N})}{2(5\,\text{N})} = 0.20$$

Suppose you had substituted immediately the input data $w_1 = 5$ N and $w_2 = 2$ N. Then we have

$$0 = T - \mu_k n \quad \text{and} \quad 0 = n - 5\,\text{N} \;;\; \text{which yields } n = 5\,\text{N}$$

for w_1 and

$$0 = 2T - w_2 = 2T - 2\,\text{N} \;;\; T = 1\,\text{N}$$

for w_2. Then the first relation yields

$$\mu_k = \frac{T}{n} = \frac{(1\,N)}{(5\,N)} = 0.20$$

It is usually a labor saving device to carry through an algebraic solution, plugging in the data only at the end.

This problem is one of many coupled systems you will deal with in the next few chapters. The motion of and forces on w_1 are related to those of w_2 by the coupling through the rope and pulley. A similar system is considered in the next problem.

Example 9

Consider the rough incline and pulley system of Fig. 5-16. Find the weight w_2 necessary to keep w_1 moving up the plane at constant velocity.

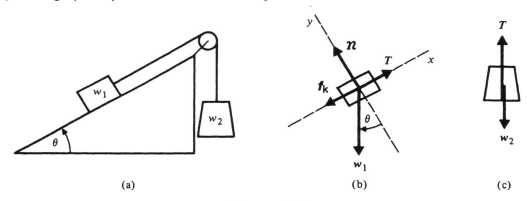

(a) (b) (c)

Figure 5-16

Solution:

"Rough" is the signal that you must consider friction for w_1. "...at constant velocity" is the signal that this is an equilibrium problem. Since w_1 moves up the plane, the force of kinetic friction points down the plane as indicated in w_1's free-body diagram (Fig. 5-16b). Note, the tilted coordinate system is convenient for w_1 but a vertical one is convenient for w_2. The equilibrium condition for w_1 is

$$0 = \Sigma F_x = T - f_k - w_1 \sin \theta = T - \mu_k\, n - w_1 \sin \theta$$

$$0 = \Sigma F_y = n - w_1 \cos \theta \quad \text{Note: } n \neq w_1$$

and for w_2 (here we use a different coordinate system from the one used for w_1)

105

$$0 = \Sigma \, F_x$$

$$0 = \Sigma \, F_y = T - w_2$$

using the last result in the first two,

$$0 = w_2 - \mu_k \, n \ - w_1 \sin \theta$$

$$0 = n \ - w_1 \cos \theta$$

Eliminating n we have

$$w_2 = \mu_k w_1 \cos \theta + w_1 \sin \theta$$

which, with $w_1 = w$ should be compared to Example 4. w_2 corresponds to F.

Example   10

Consider the weight w hung from the ceiling by two equal length strings as in Fig. 5-17. Find the tension in each string.

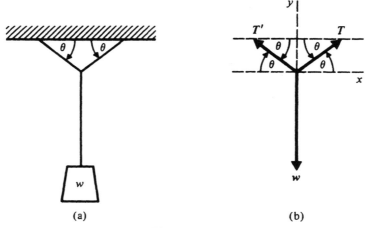

(a)   (b)

Figure   5-17

Solution:

A little geometric thought is necessary to get the angles straight. The free-body diagram for the knot where the 3 ropes join is shown in Fig. 5-17b. The tension in the vertical string is the weight w. The knot's equilibrium conditions are then

$$0 = \Sigma \, F_x = T \cos \theta - T' \cos \theta$$

$$0 = \Sigma \, F_y = T \sin \theta + T' \sin \theta - w$$

106

The first tells us T = T', which could have been guessed from the symmetry of the geometry; the system doesn't know left from right. The second yields

$$T = \frac{w}{2 \sin \theta}$$

As θ approaches 0, T approaches ∞. Infinite tension is required to keep a loaded cable straight.

Example 11

Consider the trigonometrically more challenging case in Fig. 5-18 where the two strings at the top are not of equal length. Find the tensions T_1 and T_2 of each string.

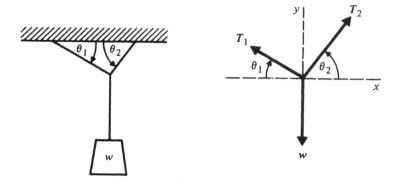

Figure 5-18

Solution:

The equilibrium equations are

$$0 = \Sigma F_x = T_2 \cos \theta_2 - T_1 \cos \theta_1$$

$$0 = \Sigma F_y = T_2 \sin \theta_2 + T_1 \sin \theta_1 - w$$

The first equation tells us the ratio of tensions

$$\frac{T_2}{T_1} = \frac{\cos \theta_1}{\cos \theta_2} .$$

The second then yields

$$T_1 = \frac{w}{\sin \theta_1 + \left(\cos \theta_1 / \cos \theta_2\right) \sin \theta_2}$$

$$T_1 = \frac{w \cos \theta_2}{\sin \left(\theta_1 + \theta_2\right)}$$

If $\theta_1 = \theta_2 = \theta$,

$$T_1 = T_2 = \frac{w}{\sin \theta}$$

as in the last problem.

If $\theta_1 + \theta_2 = 90°$ then

$$T_1 = w \sin \theta_1 = w \cos \theta_2$$

$$T_2 = w \cos \theta_1 = w \sin \theta_2$$

Note again how a good general treatment leads to an algebraic solution covering many alternatives.

Example 12

Consider the weightless horizontal boom and cable configuration in Fig. 5-19. Suppose that the only force of the wall on the boom is a perpendicular force n , as if the boom and wall were unattached and perfectly frictionless. The system is in equilibrium. Find the tension T and normal force n if $\theta = 60°$ and $w = 100$ N.

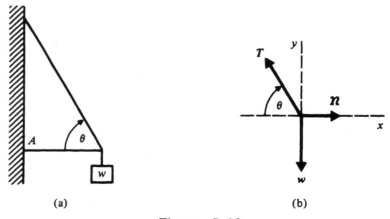

(a)   (b)

Solution: **Figure 5-19**

Fig. 5-19b can be interpreted as a free-body diagram of the boom, showing the normal force of the wall on the boom, the tension T of the cable on boom, and the weight w.

Alternatively Fig. 5-19b can be interpreted as the free-body diagram for the point on the boom where the slanted and vertical cables meet, with the understanding that the normal force of the wall on the boom is transmitted by the compressive forces within the boom) to the cable joint. The equilibrium equations are

$$0 = \Sigma F_x = n - T \cos \theta$$

$$0 = \Sigma F_y = T \sin \theta - w$$

from which we conclude

$$n = T \cos \theta = \frac{w \cos \theta}{\sin \theta} = \frac{w}{\tan \theta} = \frac{(100 \text{ N})}{(\tan 60°)} = 58 \text{ N}$$

$$T = \frac{w}{\sin \theta} = \frac{(100 \text{ N})}{(\sin 60°)} = 115 \text{ N}$$

Example 13

A body of mass 10 kg, initially at rest on a smooth horizontal surface, experiences a horizontal force of 20 N.
   (a) Find its acceleration.
   (b) How far does it move in 5 s?
   (c) What is its velocity at t = 10 s?

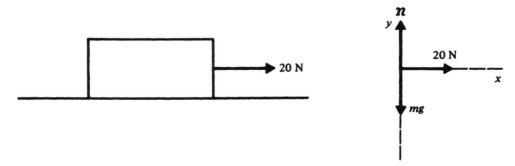

Figure 5-20

Solution:

   Is this an equilibrium problem? No--the box has an unbalanced horizontal force, so it accelerates. However, referring to Fig. 5-20 we do see that the vertical forces cancel because the body can't accelerate in the vertical direction, indicating that

$$F_y = n - mg = ma_y = 0.$$

(a) The equation of motion for the horizontal motion (a_x = a) is

$$F_x = ma_x = ma$$

$$a = \frac{F_x}{m} = \frac{(20\ N)}{(10\ kg)} = 2\ m/s^2.$$

Now that the acceleration is known (and constant), you may apply the rules for constant acceleration,

(b) $\qquad$ x =5 at2 = 5 (2 m/s2)(5 s)2 = 25 m

(c) $\qquad$ v = at = (2 m/s2)(10 s) = 20 m/s

Example 14

A body is dragged by a constant force F *equal to its weight* over a rough surface with coefficient of friction μ. (a) Find its acceleration in terms of μ. (b) What is its acceleration if μ = 0.25?

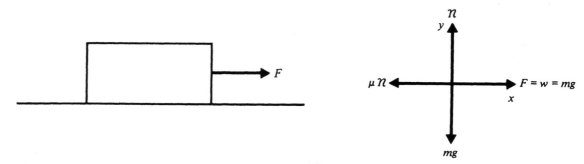

Figure 5-21

Solution:

(a) Referring to the free body diagram and coordinate system of Fig. 5-21, we note the force F is equal to the body's weight and thus F = w = mg. Thus

$$\Sigma\ F_x = ma = F - \mu\ n\ ; \quad a = (F - \mu\ n\ )/m$$

$$\Sigma\ F_y = 0 = n - mg; \quad n = mg.$$

Yielding

$$a = \frac{F - \mu\ n}{m} = \frac{mg - \mu mg}{m}$$

$$a = (1 - \mu)g = (3/4)g$$

(b) $a = 7.35 \text{ m/s}^2.$

If $\mu > 1$ it appears from this answer that a is negative and the box accelerates in a direction opposite to the applied force. This conclusion is false. If $\mu > 1$ the force necessary to drag the box, i.e. break the static friction, is greater than the body's weight, and the motion described above is not possible with the given force. Note that nothing in principle excludes $\mu > 1$.

Example 15

A 5 kg body slides down a smooth inclined plane of angle $\theta = 30°$. Find the normal force on the body and its acceleration. (See Fig. 5-22)

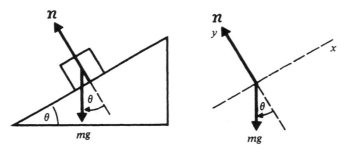

Figure 5-22

Solution:

For convenience the axes have been chosen along the perpendicular to the incline, as shown in Fig. 5-22. It is extremely tedious to work this problem with any other system of coordinates. The equations of motion are:

$$\Sigma F_x = ma = - mg \sin \theta$$

$$\Sigma F_y = 0 = n - mg \cos \theta.$$

The first equation may be solved for a and the second one for n :

$$a = - g \sin \theta = - (9.8 \text{ m/s}^2)(0.5) = - 4.9 \text{ m/s}^2$$

$$n = mg \cos \theta = (5 \text{ kg})(9.8 \text{ m/s}^2)(0.87) = 42.4 \text{ N}$$

The acceleration is negative because the coordinate system has a positive x axis that points up the plane. If the positive x axis had been chosen to point down the plane, the acceleration would be positive. The sign of acceleration depends on your choice of coordinate system.

Example 16

A body slides down a rough inclined plane of angle θ = 30° with acceleration a = 4 m/s². Find the coefficient of kinetic friction μ.

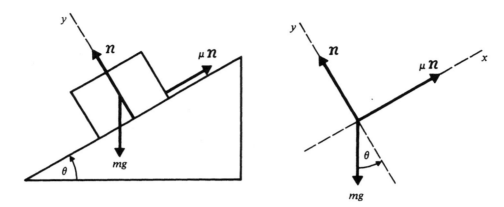

Figure 5-23

Solution:

Again, your best strategy is to solve for the coefficient of friction μ in terms of general θ and a, and substitute numbers only at the end. Referring to the free body diagram and coordinate system of Fig. 5-23, we have

$$\Sigma \, F_x = \mu \, n \, - mg \sin \theta = ma$$

$$\Sigma \, F_y = n \, - mg \cos \theta = 0; \quad n = mg \cos \theta.$$

The last equation may be solved for n and the result substituted in the first:

$$\mu \, mg \cos \theta - mg \sin \theta = ma$$

$$\mu = \frac{1}{g \cos \theta} \left(g \sin \theta + a \right).$$

With the x axis pointing up the plane, the acceleration down is *negative* so

$$\mu = \frac{1}{(9.8 \, \text{m/s}^2)(0.87)} \left[(9.8 \, \text{m/s}^2)(0.5) - (4 \, \text{m/s}^2) \right]$$

$$\mu = 0.11.$$

Had we chosen the x axis in Fig. 5-23 to point down the plane, we would have obtained

$$\Sigma\, F_x = mg \sin\theta - \mu\, n = ma$$

$$\Sigma\, F_y = n - mg \cos\theta; \quad n = mg \cos\theta$$

and thus

$$mg \sin\theta - \mu\, mg \cos\theta = ma$$

or

$$\mu = \frac{g \sin\theta - a}{g \cos\theta}.$$

With respect to this choice, $a = 4$ m/s^2 and the final result for μ is unchanged.

Example 17

A packing case is kicked, giving it an initial velocity of 2 m/s. It slides across a rough floor and comes to a stop 1 m from its initial position. Find the coefficient of friction. (See Fig. 5-24.)

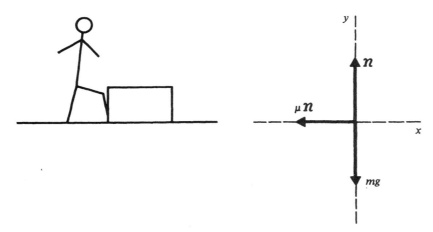

Figure 5-24

Again, we have a sketch to remind us of the physical situation, and a free-body diagram of forces <u>on</u> the box (and no other forces!).

Solution:

This is a two step problem. First you find the acceleration a from Newton's Second Law.

$$\sum F_x = -\mu n = ma; \quad a = -\frac{\mu n}{m}$$

$$\Sigma F_y = n - mg = 0; \quad n = mg$$

The acceleration is thus $a = -\mu g$. Then you apply results of the previous chapter for constant acceleration

$$v^2 = v_0^2 + 2ax;$$

With input data, $a = -\mu g$, $x = 1$ m, $v = 0$, $v_0 = 2$ m/s, we have

$$a = -\frac{v_0^2}{2x} = -\mu g$$

$$\mu = \frac{v_0^2}{2xg} = \frac{(2\,\text{m/s})^2}{2(1\,\text{m})(9.8\,\text{m/s}^2)} = 0.2$$

Example 18

A body of mass 10 kg moves with constant velocity 5 m/s on a horizontal surface under the influence of a horizontal force of 20 N.

    (a) Find the coefficient of kinetic friction.

    (b) If the horizontal force is removed (see Fig. 5-25), how long will it take to stop?

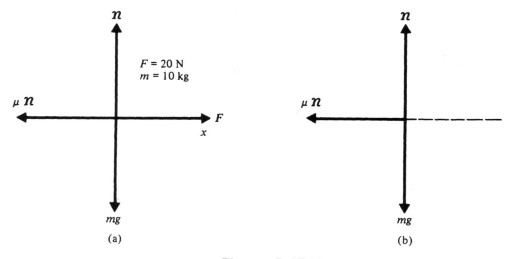

Figure 5-25

Solution:

(a) Because a = 0, we have v = constant, and the body is in equilibrium. Hence (see Fig. 5-25a)

$$\Sigma F_x = F - \mu n = 0$$

$$\Sigma F_y = n - mg = 0$$

and we have

$$\mu n = -\mu mg = F.$$
$$\mu = \frac{F}{mg} = \frac{(20\,N)}{(10\,kg)(9.8\,m/s^2)} = 0.2$$

(b) Now the horizontal force is removed and there is only frictional drag:

$$\Sigma F_x = -\mu n = ma; \quad a = -\frac{\mu n}{m}$$

The acceleration is negative, slowing the box until v = 0. Since v = (v₀ + at) and v₀ = 5 m/s,

$$t = \frac{-v_0}{a} = \frac{-v_0}{-\mu g} = \frac{(5\,m/s)}{(0.2)(9.8\,m/s^2)} = 2.5\,s$$

Example 19

A chain of mass 10 kg and length 10 m lies on a smooth horizontal surface and is pulled by a force at one end which gives it an acceleration of 1 m/s². Find the tension in the chain at any point along its length.

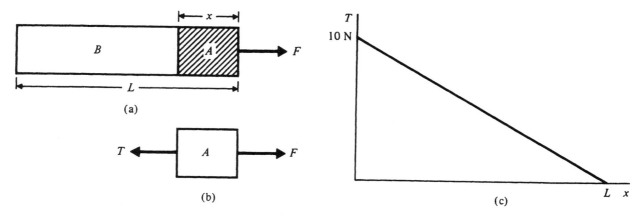

Figure 5-26

115

Solution:

In Fig. 5-26a, the chain is somewhat unartistically represented by the rectangle mathematically dividing the chain into two segments A and B. Consider the equation of motion of segment A (see Fig. 5-26b) acted upon by the pulling force F and the tension T at x in the chain.

If m is the mass of the entire chain, $F = ma = (10 \text{ kg})(1 \text{ m/s}^2) = 10 \text{ N}$.

With the notation, $m_A = (x/L)m$, $L = 10$ m, and $m = 10$ kg, the equation of motion for chain segment A is:

$$F - T = m_A a \quad \text{or } T = F - m_A a = F - m_A a = F - \left(\frac{x}{L}\right)ma$$

$$T = F - \left(\frac{x}{L}\right)ma = F - \left(\frac{x}{L}\right)F = F\left(1 - \frac{x}{L}\right) = (10 \text{ N})\left(1 - \frac{x}{L}\right)$$

The tension is greatest at the pulling end (x = 0) and drops linearly to zero at the free end, as shown in Fig. 5-26c.

Example 20

Find the acceleration of the frictionless system in Fig. 5-27.

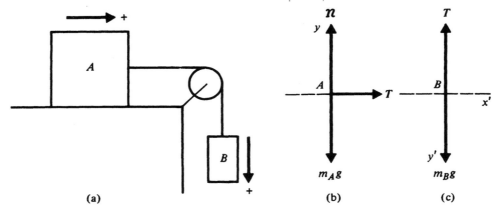

Figure 5-27

Solution:

This is a coupled system and you must apply Newton's Laws to each body. If A moves a distance $\Delta\ell$ in a time t, a length $\Delta\ell$ of the rope passes over the pulley and B moves the same distance in the same time. Thus $v_A = \Delta\ell/\Delta t = v_B$. The signs of the velocities are the same provided an appropriate sign convention is chosen. For box A velocity to the right is positive as indicated in Fig. 5-27A and by the coordinate system choice of Fig. 5-27b. For box B, the downward direction is taken to be positive as indicated in Fig. 5-27c. Since the velocities v_A and v_B are always the same, the accelerations a_A and a_B are also the same, $a_A = a_B$. Since the rope is massless, the tension is uniform. The equations of motion for A are:

$$\Sigma F_x = T = m_A a$$

$$\Sigma F_y = n - m_A g = 0; \quad n = m_A g$$

and for B is

$$\Sigma F_{y'} = m_B g - T = m_B a.$$

The last two equations may be solved simultaneously for a and T. (Solve one for T and substitute it into the other; this yields a. Then use this result to find T from either equation, giving,)

$$a = \frac{m_B}{m_A + m_B} g$$

$$T = \frac{m_A m_B}{m_A + m_B} g.$$

Note a is always less than g and T is always less that the weight $m_B g$. (A common error is to assume $T = m_B g$. That is only true if B is in equilibrium. *Here B is not in equilibrium.*)

___

Example 21

Find the acceleration of the frictionless system in Fig. 5-28.

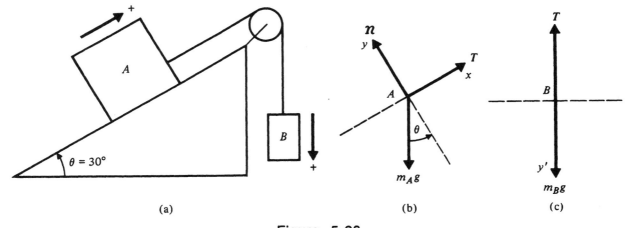

(a)   (b)   (c)

Figure 5-28

117

Solution:

Referring to Fig. 5-28b and c, the equations of motion for A are:

$$\Sigma F_x = T - m_A g \sin \theta = m_A a$$

$$\Sigma F_y = n - m_A g \cos \theta = 0$$

and for B is:

$$\Sigma F_{y'} = m_B g - T = m_B a.$$

The first and last equation may be solved simultaneously for a and T:

$$a = \left(\frac{m_B - m_A \sin \theta}{m_A + m_B} \right) g$$

$$T = \left(\frac{m_A m_B}{m_A + m_B} g \right) \left(1 + \sin \theta \right)$$

Again note $T \neq m_B g$.

If $\theta = 0$, the plane degenerates into the table of the previous example, verifying the results in Example 18.

If $m_A = 5$ kg, $m_B = 2$ kg, and $\theta = 30°$, then

$$a = \left[\frac{(2 \text{ kg}) - (5 \text{ kg})(0.50)}{(5 \text{ kg} + 2 \text{ kg})} \right] (9.8 \text{ m/s}^2) = -0.7 \text{ m/s}^2$$

The negative sign indicates that for this mass and this angle the system accelerates with A falling down the plane and B rising.

Example 22

Find the acceleration of the Atwood's machine in Fig. 5-29 if m_A = 2 kg and m_B = 5 kg.

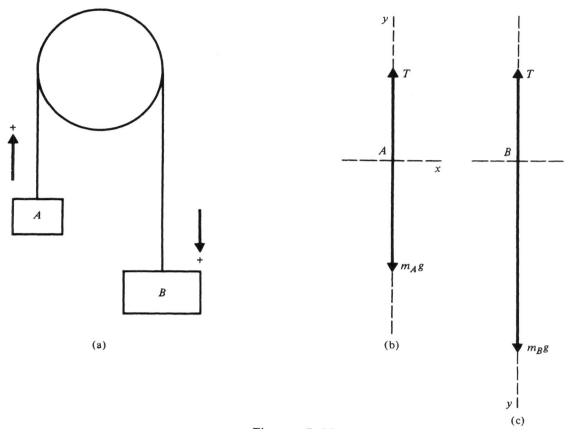

Figure 5-29

Solution:

A and B have the same accelerations if the positive directions are those indicated in Fig. 5-29b and c. The equations of motion for each mass are thus

$$T - m_A g = m_A a$$

$$m_B g - T = m_B a$$

Adding these equations eliminates T and yields,

$$a = \left(\frac{m_B - m_A}{m_A + m_B} \right) g \ .$$

Note this is the limit of Example 19 with $\theta = 90°$.

$$a = \left[\frac{(5\,\text{kg}) - (2\,\text{kg})}{7\,\text{kg}}\right](9.8\,\text{m/s}^2) = 4.2\,\text{m/s}^2$$

Example 23

Find the tensions and acceleration in the frictionless system of Fig. 5-30.

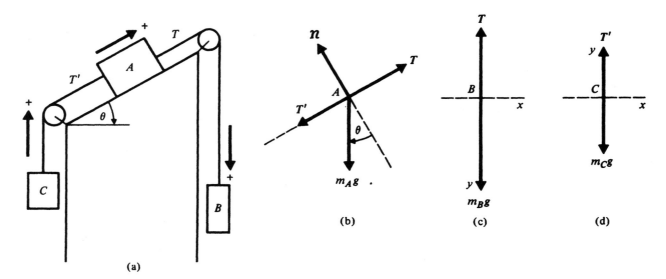

Figure 5-30

Solution:

The equations of motion are

A: $T - T' - m_A g \sin \theta = m_A a$ (Fig. 5-30b)

B: $m_B g - T = m_B a$ (Fig. 5-30c)

C: $T' - m_C g = m_C a$ (Fig. 5-30d)

Adding all three equations eliminates T and T' and yields

$$a = \left(\frac{m_B - m_A \sin \theta - m_C}{m_A + m_B + m_C}\right)g \; .$$

120

The remaining equations may be solved for T and T',

$$T' = m_C(a + g) = \left(\frac{2m_C m_B + m_C m_A(1 - \sin\theta)}{m_A + m_B + m_C}\right)g .$$

$$T = m_B(g - a)$$

$$T = \left(\frac{2m_C m_B + m_B m_A(1 + \sin\theta)}{m_A + m_B + m_C}\right)g .$$

In the limit m_C approaches 0 we retrieve the results of Example 21.

Example 24

In the last example, if $\theta = 0°$ you should retrieve the results for a horizontal table. Take this limit and check your results.

Solution:

Setting $\theta = 0°$ in the above equations gives:

$$a = \left(\frac{m_B - m_C}{m_A + m_B + m_C}\right)g$$

$$T = \left(\frac{2m_C m_B + m_B m_A}{m_A + m_B + m_C}\right)g$$

$$T' = \left(\frac{2m_C m_B + m_C m_A}{m_A + m_B + m_C}\right)g$$

Example 25

A wedge car as shown in Fig. 5-31 is accelerating to the left. The coefficient of static friction between the block and the inclined plane is μ. Find the maximum acceleration the wedge car can have before the block slides up the plane.

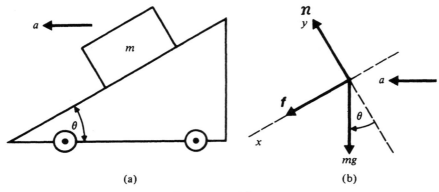

(a)   (b)

Solution: Figure 5-31

Just before slipping up the plane the block feels a friction force pointing down the plane as shown in Fig. 5-31b. When not slipping, the block has the same acceleration as the car, resolved in the inclined system as,

$$a_x = +a \cos \theta$$

$$a_y = +a \sin \theta$$

The equations of motion are

$$\Sigma F_x = F + mg \sin \theta = ma_x = ma \cos \theta$$

$$F = ma \cos \theta - mg \sin \theta,$$

and

$$\Sigma F_y = n - mg \cos \theta = ma \sin \theta$$

$$n = ma \sin \theta + mg \cos \theta.$$

The no slip condition is $F \leq \mu n$ or

$$ma \cos \theta - mg \sin \theta \leq \mu(ma \sin \theta + mg \cos \theta)$$

$$a(\cos \theta - \mu \sin \theta) \leq g(\sin \theta + \mu \cos \theta)$$

Solving for the acceleration that produces maximum static friction, for friction down the wedge.

$$a \leq g \frac{\left(\sin \theta + \mu \cos \theta\right)}{\left(\cos \theta - \mu \sin \theta\right)}$$

Thus the acceleration is finite unless the denominator becomes equal to zero, so:

$$\mu_s < \frac{\cos \theta}{\sin \theta}$$

As θ approaches 0° (flat car)

$$a \leq \mu\, g.$$

Example 26

The deflection of the accelerometer in Fig. 5-32 is 30°. Find the magnitude and direction of the cart's acceleration.

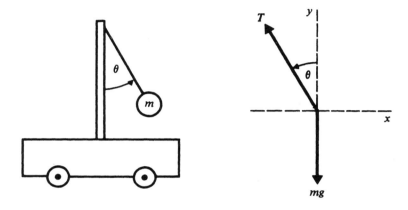

Figure 5-32

Solution:

The cart is accelerating to the left, so we expect its acceleration to be negative with respect to the coordinate system chosen. The equations of motion are <u>of the mass</u> m are:

$$\Sigma F_x = -\ T \sin \theta = ma_x = ma$$

$$\Sigma F_y = T \cos \theta - mg = 0.\ (a_y = 0 \text{ because } \theta \text{ is constant.})$$

Dividing the first equation by the second to eliminate T gives:

$$a = -\ g \tan \theta = -\ (9.8 \text{ m/s}^2)(0.58) = -\ 5.7 \text{ m/s}^2.$$

The acceleration is in the negative x direction, as expected.

Example 26

A 4000 kg locomotive "coasts" at a constant speed of 100 km/hr about a circular section of track of radius 800 m.
    (a) Find the force of the locomotive against the tracks.
    (b) How does this force compare with the weight of the locomotive?

Solution:

(a) The force of the locomotive against the tracks is equal but opposite to the centripetal force of the tracks against the locomotive, of magnitude

$$F = ma = m\frac{v^2}{R}$$

where v is the speed of the locomotive and R the radius of the track:

$$F = (4000 \text{ kg})\frac{\left(100 \text{ x } 10^3 \text{m}/3600 \text{ s}\right)^2}{(800 \text{ m})}$$

$$= 3900 \text{ N}$$

(b) The weight of the locomotive is

$$w = mg = 4000 \text{ kg}(9.8 \text{ m/s}^2) = 39,000 \text{ N},$$

ten times the centripetal force.

Example 27

The Unbanked Curve: A car is driven in a circle of radius R = 500 m on a horizontal track with coefficient of static friction μ = 0.5 between the tires and the track. Find the maximum speed the car can have before skidding off the track.

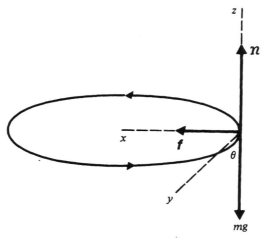

Figure 5-33

Solution:

The frictional force of the tires on the car points perpendicular to the tires toward the center of the circle. Referring to Fig. 5-33, the equations of motion for the car are

$$F_x = F_{rad} = ma_{rad} = m\frac{v^2}{R} = f_s \quad \text{(perpendicular to car's motion)}$$

$$F_y = F_{tan} = ma_{tan} = 0 \quad \text{(tangential to car's motion)}$$

$$F_z = ma_z = 0 = n - mg; \; n = mg \quad \text{(no motion in vertical direction)}$$

where f_s is the static force of friction of the track on the tires of the car. The static friction condition $f_s < \mu\, n$ is here

$$m\frac{v^2}{R} \leq \mu mg$$

or

$$v \leq (\mu Rg)^{1/2}$$

If v exceeds this limit the lateral static friction force of the road against the tires is not strong enough to supply the centripetal acceleration, and the car skids out of the circle.

Example 28

The Banked Curve: To avoid the previous mishap, curves are banked so that a component of the normal force helps in supplying the centripetal force. Find the maximum velocity before skidding for a track of radius R banked at an angle θ.

Solution:

Referring to Fig. 5-34, we bank the track, viewing the car end-on.

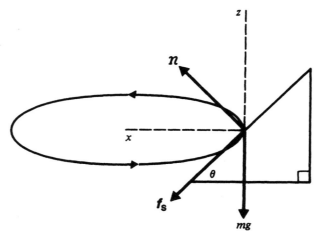

Figure 5-34

The equations of motion in the xz plane are

$$F_x = F_{rad} = m\frac{v^2}{R} = n \sin\theta + f_s \cos\theta$$

$$F_z = n \cos\theta - f_s \sin\theta - mg = 0$$

Note the x component is the centripetal component in this case. We have assumed the velocity is large enough that f_s points down the plane. These equations may be solved for n and f_s,

$$n = m\frac{v^2}{R}\sin\theta + mg\cos\theta$$

$$f_s = m\frac{v^2}{R}\cos\theta - mg\sin\theta$$

The condition that $f_s \leq \mu_s n$ then implies

$$v^2 \leq Rg \left[\frac{\mu \cos \theta + \sin \theta}{\cos \theta - \mu \sin \theta} \right]$$

Even when there is no friction at all ($\mu = 0$) a judicious banking angle will provide the right centripetal acceleration for a given velocity,

$$m \frac{v^2}{R} = n \sin \theta$$

$$n \cos \theta - mg = 0$$

implying (in agreement with the above result when $\mu = 0$)

$$v^2 = Rg \tan \theta$$

Example 29

A plastic bottle of negligible mass contains one liter of water. The bottle is tied to a string of length 0.8 m and swung in a vertical circle, reaching a speed of 2 m/s at the bottom of its swing. Find the tension in the string at the bottom of the swing.

Solution:

The total force on the bottle at the bottom of the swing is the upward tension minus the downward weight,

$$F = T - mg \quad \text{direction: up}$$

At the bottom of the swing the acceleration of the pail is centripetal (directed upward, toward the center of the circle) and of magnitude v^2/R. By Newton's second law we have

$$T - mg = m \frac{v^2}{R}$$

or

$$T = mg + m \frac{v^2}{R}$$

$$T = mg + m \frac{v^2}{R} = m \left(g + \frac{v^2}{R} \right) = (1 \text{ kg}) \left(9.8 \text{ m/s}^2 + \frac{(2 \text{ m/s})^2}{(0.8 \text{ m})} \right)$$

$$= 14.8 \text{ N}$$

Example   30

The apparent weight w' of an object is the upward force exerted by its supporting surface. A passenger in a Ferris wheel rotates at 3 m/s in a vertical circle of radius 5 m. Find the ratio of his apparent weight to his true weight at the top and the bottom of the circle.

Solution:

Referring to Fig. 5-35,

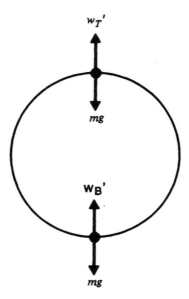

Figure 5-35

we write the equations of motion as

$$\text{Top:} \quad -w_T' + mg = F_n = m\frac{v^2}{R}$$

$$\text{Bottom:} \quad w_B' - mg = F_n = m\frac{v^2}{R}$$

yielding

$$w_T' = mg - m\frac{v^2}{R}$$

$$\frac{w_T'}{w} = 1 - \frac{v^2}{Rg} = 1 - \frac{(3 \text{ m/s})^2}{(5 \text{ m})(9.8 \text{ m/s}^2)} = 0.82$$

$$w_B' = mg + m\frac{v^2}{R}$$

$$\frac{w_B'}{w} = 1 + \frac{v^2}{Rg} = 1 + \frac{(3 \text{ m/s})^2}{(5 \text{ m})(9.8 \text{ m/s}^2)} = 1.18$$

Example 31

A pilot pulls a plane out of a dive by flying in an arc of a vertical circle of radius 1 km at a speed of 500 km/hr. What is his apparent weight at the bottom of the circle?

Solution:

Referring to Fig. 5-36,

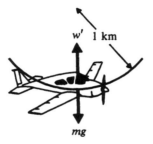

Figure 5-36

we write the radial equation of motion as

$$F_{rad} = w' - mg = m\frac{v^2}{R}; \qquad w' = mg + m\frac{v^2}{R}$$

$$\frac{w'}{w} = \frac{w'}{mg} = 1 + \frac{v^2}{Rg}$$

$$\frac{w'}{w} = 1 + \frac{(500 \times 10^3 \text{ m}/3600 \text{ s})^2}{(10^3 \text{ m})(9.8 \text{ m/s}^2)} = 3.0$$

The pilot is said to be experiencing "forces" of 2g, that is, his apparent weight is as if the acceleration of gravity had been increased by 2g from g to 3g.

129

Example 32

A 100 kg man stands on the outer edge of a carousel of 4 m radius. The coefficient of friction of his shoes against the floor is $\mu = 0.3$. How small a period of rotation may the carousel have before the man slips off?

Solution:

Referring to Fig. 5-37,

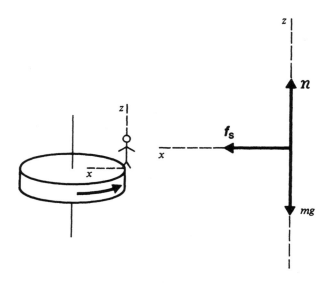

Figure 5-37

we write the equations of motion as

$$F_x = f_s = m\frac{v^2}{R}$$

$$F_z = n - mg = 0$$

The static friction condition is

$$f_s < \mu_s n$$

$$m\frac{v^2}{R} \le \mu_s mg \quad \text{or} \quad v^2 \le \mu_s gR = v_c^2$$

For any velocity smaller than a critical velocity v_c the man does not slip.

$$v < (\mu_s g R)^{1/2} = v_c = [(0.3)(9.8 \text{ m/s}^2)(\ 4 \text{ m})]^{1/2} = 3.4 \text{ m/s}$$

The period corresponding to this critical velocity, $T_c = 2\pi R/v_c$ and thus

$$T_c = \frac{2\pi (4 \text{ m})}{(3.4 \text{ m/s})} = 7.4 \text{ s.}$$

If the period of rotation is made smaller than 7.4 seconds, the man will slip.

Example 33

In the previous problem the carousel speeds up and the man grabs a post. His maximum gripping strength for any reasonable length of time is twice his weight. What is the maximum velocity before he falls off? What is the period and frequency of rotation at this point?

Solution:

The gripping force F < 2 mg so the radial equation of motion yields

$$m\frac{v^2}{R} = F \le 2 \text{ mg}$$

$$v^2 < 2Rg; \quad v < (2 \cdot 4 \text{ m} \cdot 9.8 \text{ m/s}^2)^{1/2} = 8.8 \text{ m/s}$$

The period at the critical point is

$$T = \frac{2\pi R}{v} = \frac{2\pi (4 \text{ m})}{(8.8 \text{ m/s})} = 2.9 \text{ s.}$$

and the frequency of rotation is

$$f = \frac{1}{T} = 0.35 \text{ /s} = 0.35 \text{ Hz.}$$

Example 34

A mass attached to a vertical post by two strings, rotates in a circle at constant speed v. (See Fig. 5-38.) At high enough speeds both strings are taut but below a critical velocity the lower string slackens. The mass is 1 kg and the radius is 1.5 m. Find the critical velocity and the tension in the upper string at the critical velocity.

Solution:

Referring to Fig. 5-38,

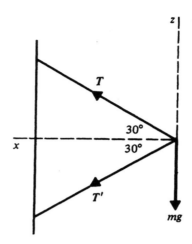

Figure 5-38

the equations of motion are

$$F_x = F_{rad} = m\frac{v^2}{R} = (T + T')\cos 30°$$

$$F_z = (T - T')\sin 30° - mg = 0.$$

When v^2 is known these are two equations which can be solved for T and T':

$$T = \frac{mv^2}{2R\cos 30°} + \frac{mg}{2\sin 30°}$$

$$T' = \frac{mv^2}{2R\cos 30°} - \frac{mg}{2\sin 30°}$$

If the strings were rods, capable of delivering both compressive (T, T' < 0) and tensile forces (T, T' > 0) then any velocity is possible. A string however must have T > 0, T' > 0 or it slackens. Then we require

T' > 0 or v^2 > gR/(tan θ)

$$v > \sqrt{\frac{(9.8 \text{ m/s}^2)(1.5 \text{ m})}{(0.58)}} = 5 \text{ m/s}$$

132

At the critical velocity T' = 0, $v^2 = gR/(\tan \theta)$, and

$$T = \frac{m}{2R \cos 30°} \frac{gR}{\tan 30°} + \frac{mg}{2 \sin 30°} = \frac{mg}{2 \sin 30°} + \frac{mg}{2 \sin 30°}$$

$$T = \frac{mg}{\sin 30°} = \frac{(1 \text{ kg})(9.8 \text{ m/s}^2)}{(0.5)} = 20 \text{ N.}$$

Example 35

A rowboat of mass 200 kg, including its rower, is rowed at a constant velocity of 4 km/hr. When the rower stops rowing he finds that he coasts to half his original velocity in 3 m. The drag of the water, he estimates, is proportional to the square of his velocity.

$$F_d = -bv^2$$

    (a) Find the coefficient b.
    (b) How far will the boat coast before reducing its speed to 1/10 the original value? 1/100 the original value?
    (c) Find the force necessary to propel him at the constant speed of 4 km/hr.

Solution:

(a) The equation of motion when the rower is coasting is

$$F_x = -bv^2 = ma = m\frac{dv}{dt} = m\frac{dv}{dx}\frac{dx}{dt}$$

$$F_x = m\frac{dv}{dx}v = mv\left(\frac{dv}{dx}\right)$$

$$m\left(\frac{dv}{dx}\right) = -bv$$

We then separate the variables and integrate the equation of motion,

$$\int_{v_0}^{v} \frac{dv}{v} = -\int_0^x \frac{b\,dx}{m}$$

between the initial velocity v_0 when x = 0 to a later velocity when the displacement from the initial position is x. The result is

$$\ln\left(\frac{v}{v_0}\right) = -\frac{b}{m}x \quad \text{or} \quad \ln\left(\frac{v_0}{v}\right) = \frac{b}{m}x$$

133

We now solve for the parameter b;

$$b = -\frac{m}{x} \ln \left(\frac{v}{v_0}\right) = \frac{m}{x} \ln \left(\frac{v_0}{v}\right) = \frac{(200 \text{ kg})}{(3 \text{ m})} \ln (2)$$

b = 46.2 kg/m

(b) Thus

At $v = 0.1 \, v_0$, $x = (4.3 \text{ m}) \ln 10 = 9.9$ m.

At $v = 0.01 \, v_0$, $x = (4.3 \text{ m}) \ln 100 = 19.8$ m.

(c) At constant velocity, the total force is zero, $F_R + F_d = 0$ where F_R is the force supplied by the rower.

$$v = 4 \text{ km/hr} = \frac{4 \times 10^3 \text{ m}}{(3600 \text{ s})}.$$

$$F_R = -F_d = bv^2 = (46.2 \text{ kg/m}) \left[\frac{4 \times 10^3 \text{ m}}{(3600 \text{ s})}\right]^2$$

$$= 57 \text{ N}$$

QUIZ

1. A horizontal rope is stretched between two buildings 20 m apart. When a man weighing 800 N hangs from the center of the rope, the center is 4 m lower that the ends. What is the tension in the rope?

Answer: 1081 N

2. Two ropes attached to the ceiling support a 100 N weight. One rope makes an angle of 30° with the ceiling and the other makes an angle of 45° with the ceiling. Find the tension in each rope.

Answer: 73 N, 90 N

3. 20 lb and 10 lb weights attached by a cord hang from a frictionless pulley, the 20 lb weight 10 ft above the ground and the 10 lb weight on the ground. The system is released from rest.
   (a) Find the acceleration.
   (b) Find the tension in the rope.
   (c) Find the time it takes the 20 lb weight to hit the floor.
   (d) Find the velocity when it hits the floor.

Answer: 10.7 ft/s, 13.3 lb, 1.37 s, 14.6 s

4. A box is kicked along a rough floor, giving it an initial velocity of 10 m/s. It comes to rest 20 m from the place where it was kicked. What is the coefficient of friction?

Answer: 0.25

5. A turntable rotates at 45 rev/min. A coin placed on the turntable will revolve without slipping provided it is less than 2.25 inches from the center. Find the coefficient of friction.

Answer: 0.13

6. A 60 kg person stands on a bathroom scale on a rotating Ferris wheel of radius 5 m. How many revolutions per minute is the Ferris wheel making if the person's apparent "weight" is 65 kg at the bottom of the rotation?

Answer: 3.8 rpm

135

6
WORK AND KINETIC ENERGY

OBJECTIVES

In this chapter Newton's laws are used to develop the ideas of work, kinetic energy, and power. This will provide you with a powerful tool for analyzing the motion of mechanical systems. Your objectives are to:

Calculate the *work* done by constant and varying forces.

Define and calculate the change in *kinetic energy* of a body, given the work done by the total force acting on the body.

Define and calculate the *power* associated with motion under the influence of force.

REVIEW and SUPPLEMENT
Work and Kinetic Energy

The work done by a constant force $\vec{F}$ acting on a body which moves in a straight line through a displacement **s** is

$$W = \vec{F} \cdot \vec{s} = Fs \cos \theta$$

where θ is the direction between F and s as shown in Fig. 6-1a. The SI unit of work is the N·m = joule; the British unit is the ft-lb = 1.356J. Work is a *scalar* quantity, and can be positive, negative, or zero.

The work is positive if the force acts in the direction of displacement, $\theta < 90°$, and negative if $90° < \theta < 180°$. The work is zero if $\vec{F}$ and **s** are perpendicular. If $\vec{F}$ is the total force on the body, then W is the total work done *on the body* by $\vec{F}$. If $\vec{F} = \vec{F}_A + \vec{F}_B$, $W = \vec{F}_A \cdot \vec{s} + \vec{F}_B \cdot \vec{s} = W_A + W_B$; the total work is the sum of the partial work due to $\vec{F}_A$ and $\vec{F}_B$.

If the force varies or the displacement changes direction, then the displacement curve must be divided into small increments Δs_i as shown in Fig. 6-1b. In each increment $\vec{F}$ does not vary significantly if the increments are small enough, and the work within that displacement increment is well approximated by $\vec{F}_i \cdot \Delta \vec{s}$. The total work is the sum of all such increments

$$W = \Sigma_i \, \vec{F}_i \cdot \vec{s}_i$$

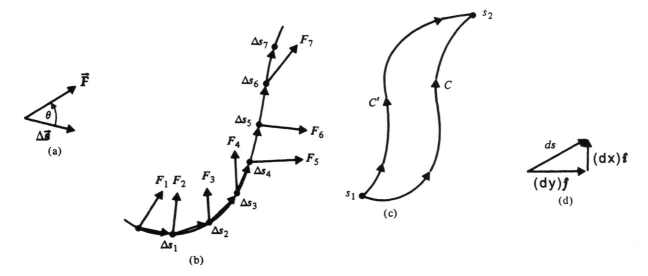

Figure 6-1

The limit of this sum as $|\Delta \vec{s}_i|$ approaches 0 defines the work in terms of a *line integral*,

$$W = \lim_{\Delta s_i \to 0} \Sigma_i \vec{F}_i \cdot \Delta \vec{s}_i = \int_C \vec{F} \cdot d\vec{s}$$

$$W = \int_C \left(F \cos \theta \right) ds$$

The work in general depends on the path C connecting the end points of the motion; its value may be different for the path C' in Fig. 6-1c.

The *kinetic energy* K of a body with mass m moving with velocity v is defined as

$$K = \frac{1}{2} mv^2$$

The total work W done *on a body* is equal to the increase in kinetic energy of the body,

$$W = \Delta K = \frac{1}{2} mv_2^2 - \frac{1}{2} mv_1^2 = K_2 - K_1$$

where v_1 is the initial and v_2 the final velocity.

A force acting in the direction of displacement does positive work and hence increases the kinetic energy: an example is a ball falling in a gravitational field. Here the force of gravity and the displacement are in the same direction. The work is positive. The kinetic energy increases.

137

Power

The average power P_{av} of a force is defined as

$$P_{avg} = \frac{\Delta W}{\Delta t}$$

where ΔW is work done by the force in an interval Δt.

Instantaneous power is

$$P = \text{Lim}_{\Delta t \to 0} \left(\frac{\Delta W}{\Delta t} \right) = \frac{dW}{dt}$$

In SI, the unit of power is the joule/s = watt. (1 J/s = 1 W). In the British system the unit is the ft/lb·s. A horsepower (hp) is

1 hp = 550 ft/lb·s = 746 W

Another useful form is given by

$$P_{avg} = \frac{\Delta W}{\Delta t} = F\left(\cos \theta\right)\frac{\Delta s}{\Delta t} = F\left(\cos \theta\right)v$$

$$= F_t v = \vec{F} \cdot \mathbf{v}$$

As Δt approaches 0, P approaches P_{avg} with

$$P = F (\cos \theta) v = F_t v = \vec{F} \cdot \mathbf{v}.$$

PROBLEM-SOLVING STRATEGIES

Note the work, a scalar, may be positive or negative. A force acting perpendicular to the path of a particle (such as the tension of a pendulum bob) does no work. Such forces do not contribute to the work W in the work energy relation.

Now is a good time to review the problem-solving strategy for calculations with work and kinetic energy in the main text. All the previous strategies about coordinate systems, free-body diagrams, and careful sign conventions continue to reward you.

QUESTIONS AND ANSWERS

Question. Two sleds, one of mass 25 kg and the other of mass 50 kg, are pulled from rest to final speeds of 2 m/s and 1 m/s respectively over a smooth snow field. From the information given can you determine (a) which sled first reached final speed? (b) how much distance it took each sled to reach final speed? (c) how much work was required to bring each sled to final speed?

Answer. (a) No. (b) No. (c) For the 25 kg sled, 50 J are required while for the 50 kg sled, 25 J are required.

Question. How much work is done by the force of gravity when a ball of mass 0.5 kg rolls off a shelf and falls to the floor 1.5 m below? What is the increase in its kinetic energy? How do your answers depend on the speed of the ball as it left the edge of the shelf?

Answer. The work done by gravity is equal to the product of the force and displacement (since they are parallel) and equal to mgh or 7.35 J. The increase in kinetic energy is equal to the work done so D(KE) = 7.35 J. Neither of these answers depend on the speed of the ball as it leaves the shelf.

EXAMPLES AND SOLUTIONS

Example1

A 10 kg box is pulled through a distance of 2 m by a horizontal force F of magnitude 20 N acting in the same direction as the displacement. Find the work done by F.

Solution:

$$W = Fs \cos \theta; \qquad F = 20 \text{ N}, s = 2 \text{ m}, \cos \theta = 1$$

$$W = (20 \text{ N}) \cdot (2 \text{ m}) \cdot (1) = 40 \text{ J}$$

Note the work done by F doesn't depend on such things as the mass, velocity or acceleration of the box, or whether or not there are other forces acting on the box.

Example 2

The same box starts from rest on a smooth surface, acted upon by the same total force F = 20 N.
  (a) Find the final kinetic energy after s = 2 m, using the methods of Ch. 2 and compare it to the work done by the total force F.
  (b) Find the final velocity v.

Solution:

Because the force is constant, the acceleration is also constant, and we have

(a)

$$v^2 = v_0^2 + 2as \qquad\qquad a = F/m \qquad v_0 = 0$$

$$v^2 = 2\,as = 2\frac{Fs}{m}$$

$$K = \frac{1}{2}mv^2 = \frac{1}{2}m\left(2\frac{Fs}{m}\right) = Fs = W = 40\ J$$

This illustrates the work-energy theorem for this specific case.

(b)

$$v = \sqrt{\left(2\frac{Fs}{m}\right)} = \sqrt{\left(2\frac{(20\ N)(2\ m)}{(10\ kg)}\right)} = 2.83\ m/s$$

Note that although a numerical answer was requested, numbers were not substituted until the last step.

Example 3

The same box undergoes the same displacement on a rough floor at constant velocity. Find
   (a) The work done by F;
   (b) The coefficient of friction μ_k;
   (c) The work done by the friction force;
   (d) The work done by gravity;
   (e) The work done by the normal force; and
   (f) The work done by the total force acting on the body

Solution:

Refer to Fig. 6-2, where all the forces acting <u>on the box</u> are shown:

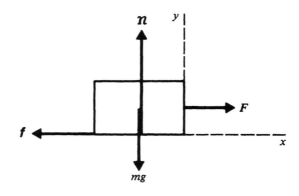

Figure 6-2

140

The body is in equilibrium because v = constant. Thus

$$\Sigma F_x = 0 = F - f = F - \mu n; \quad F = \mu n$$

$$\Sigma F_y = 0 = n - mg; \quad n = mg$$

and

$$\mu_k = \frac{F}{n} = \frac{F}{mg} = \frac{(20\ N)}{(10\ kg)(9.8\ m/s^2)} = 0.22$$

(a) W_F = Fs cos θ = (20 N)·(2 m)·1 = 40 J = the work of the pulling force.

(b) μ_k = 0.22

(c) W = fs cos θ = (20 N)·(2 m)·(– 1) = -40 J = the work of the frictional force.

The work done by the friction force is <u>negative</u> because f is in a direction opposite to the displacement, cos θ = – 1.

(d) The work done by gravity, W_g, is zero because the displacement is perpendicular to the force, cos θ = 0.

(e) Similarly W_n = 0.   (Work of the normal force.)

(f) The total force $\vec{F}_t$ is

$$\vec{F}_t = \vec{F} + \vec{f} + \vec{n} + \vec{w} \text{ (weight)} = 0$$

and hence the total work W is

$$W = W_F + W_f + W_n + W_g = 40\ J - 40\ J + 0 + 0$$

$$W = 0$$

Since the total work is zero, the increase in kinetic energy is also zero.

Example 4

The same box is pushed 2 m up a *smooth* inclined plane of angle θ = 30° by a 100 N horizontal force.
        (a) Find the work done on the box and
        (b) the increase in its kinetic energy.

Solution:

Refer to Fig. 6-3, where a free-body diagram is illustrated and a coordinate system is chosen following our problem-solving strategies.

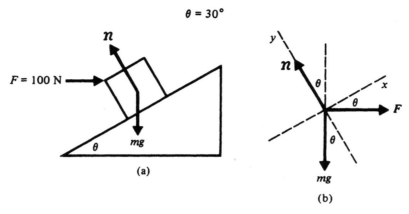

Figure 6-3

We have from Newton's Second Law,

$$\Sigma F_x = F\cos \theta - mg \sin \theta = ma_x$$

$$\Sigma F_y = n - mg \cos \theta - F \sin \theta$$

(a)　　$W_F = Fs \cos \theta = 100 \text{ N}\cdot 2 \text{ m}\cdot\cos 30°$

$$= 173 \text{ J} = \text{work of force } F$$

$W_n = 0$　(Normal force does no work because $\hat{n}\cdot d\vec{s} = 0$)

$$W_g = F_g \cos \theta\, s = -mg \sin \theta\, s$$

$$= -(10 \text{ kg})\cdot(9.8 \text{ m/s}^2)\cdot\sin 30°\cdot 2 \text{ m}$$

$$= -98 \text{ J} = \text{negative work of gravitational force.}$$

$$W = W_F + W_g = [Fs \cos \theta - mgs \sin \theta] = 75 \text{ J}$$

(b)

$$a_x = \frac{1}{m}\left[F \cos \theta - mg \sin \theta\right] = \frac{F \cos \theta}{m} - g \sin \theta$$

$$a_x = \frac{(100 \text{ N}\cdot\cos 30°)}{(10 \text{ kg})} - (9.8 \text{ m/s}^2)\sin 30° = 3.76 \text{ m/s}^2$$

$$v_x^2 = v_{ox}^2 + 2a_x x ; \qquad v_{ox} = 0$$

$$v_x^2 = 2a_x x = 2(3.76 \text{ m/s}^2)2 \text{ m} = 15.0 \text{ m}^2/\text{s}^2$$

$$K = \frac{1}{2}mv_x^2 = ma_x s = \left[Fs \cos \theta - mgs \sin \theta\right] = \frac{1}{2}(10 \text{ kg})(15 \text{ m}^2/\text{s}^2) = 75 \text{ J}$$

Verified here is the work-energy relation

$$W = \Delta K$$

Note W is the *total* work done by all the forces acting on the body. The work-energy relation has been verified for this case both numerically and algebraically (in the square [..] brackets).

Example 5

A body sliding on a rough surface is given an initial velocity of 3 m/s and comes to a stop in 1 m. Using the work energy relation, find the coefficient of friction.

Solution:

$$K_1 = \frac{1}{2}m(3 \text{ m/s})^2 \qquad K_2 = 0$$

$$W = \Delta K = K_2 - K_1; \qquad W = -fx = -\mu n \, x$$

$$W = -\mu nx = -\mu mgx = 0 - \frac{1}{2}mv^2$$

$$\mu = \frac{v^2}{2 \, gx} = \frac{(3 \text{ m/s})^2}{2 \,(9.8 \text{ m/s}^2)(1 \text{ m})} = 0.5$$

Note K_2 is the *final* kinetic energy, K_1 the initial so that ΔK is the *change* in kinetic energy, negative in this case; correspondingly the work is negative in this case because the force F is opposite to the displacement.

Example 6

A body is released from rest at the top of a rough inclined plane of height y and angle θ. The coefficient of friction is μ. Use the work-energy relation to find the velocity of the body at the bottom of the incline in terms of m, y, μ and θ.

Solution:

Referring to the free-body diagram of Fig. 6-4, we have

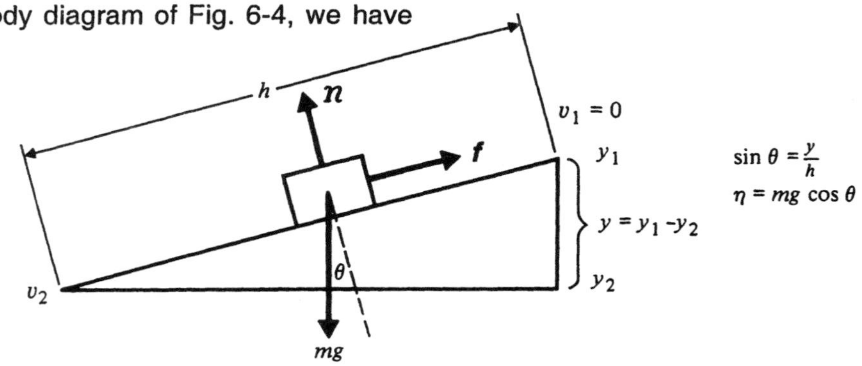

$$\sin \theta = \frac{y}{h}$$
$$n = mg \cos \theta$$

Figure 6-4

$$K_1 = 0 \qquad K_2 = \frac{1}{2}mv^2$$

$$W = W_g + W_f + W_n = K_2 - K_1 = \tfrac{1}{2}mv^2; \quad K_1 = 0$$

where W_g = work of gravity, W_f = work of friction, and W_n = work of the normal force. Now we calculate the work of all forces:

$$W_g = \left(mg \sin \theta \right)h = \left(mg \sin \theta \right)\left(\frac{y}{\sin \theta} \right) = mgy$$

This is positive because the component of the force of gravity along the plane is in the direction of motion.

$$W_f = -fh = -\mu Nh = -\mu \left(mg \cos \theta \right)h$$

$$W_f = -\mu\, mgy \left(\frac{\cos \theta}{\sin \theta} \right) = -\left(\frac{\mu\, mgy}{\tan \theta} \right)$$

This is negative because the frictional force opposes the motion.

$$W = mgy \left(1 - \frac{\mu}{\tan \theta} \right) = \frac{1}{2}mv_2^2$$

(For slippage $\tan \theta > \mu_s$, where μ_s is the static coefficient; since $\mu_s > \mu$, the sliding coefficient of friction, $\tan \theta > \mu$ and $W > 0$ in the last expression.) Then we have

$$v_2 = \sqrt{2\, gy \left(1 - \frac{\mu}{\tan \theta} \right)}$$

Example 7

What average power is required, in watts and hp, if an escalator is to lift twenty 100 kg people, 3 m high, in one minute? Does it matter whether they are lifted straight up or at an angle?

Solution:

$$P = Fv = \frac{\Delta W}{\Delta t} \quad \text{but we have } \Delta W = mgh; \text{ so that}$$

$$P = \frac{(20 \times 100 \text{ kg})(9.8 \text{ m/s}^2)(3 \text{ m})}{(60 \text{ s})} = 980 \text{ W}$$

$$P = 0.98 \text{ kW} = \frac{(0.98 \text{ kW})(1 \text{ hp})}{(0.746 \text{ kw})} = 1.31 \text{ hp}$$

The power is independent of the angle because gravitational work depends only on the vertical elevation.

Example 8

An automobile has a 150 hp engine. Its top speed is 100 mph. Assuming half the power is delivered by the engine to the tires on the road, find the drag of the air resistance and all other dissipative forces.

Solution:

At top (and constant) speed, the force F delivered to the tires is equal to the drag force of friction and the air resistance. F is related to the power delivered to the wheels by:

$$P = Fv; \quad \text{so } F = \frac{P}{v} \quad \text{and}$$

$$100 \text{ mph} = \frac{(100 \text{ mi/hr})(5280 \text{ ft/mi})}{(3600 \text{ s/hr})} = 147 \text{ ft/s} \quad (\text{convert mph to ft/s})$$

Thus

$$F = \frac{(75 \text{ hp})(550 \text{ ft} \cdot \text{lbs/s} \cdot \text{hp})}{(147 \text{ hp})}$$

$$= 281 \text{ lbs}$$

(Only half the power was delivered to the wheels, hence the 75 hp in the numerator.)

145

QUIZ

1. Find the kinetic energy of a 4 kg bowling ball traveling at 4.43 m/s.

Answer: 39 J.

2. What is the kinetic energy of a 4 kg bowling ball just before it hits the floor when dropped from a height of 1 m? What is the work done by the gravitational force on the ball?

Answer: 39 J; 39 J.

3. A 50 kg ice skater is coasting at 2 m/s. What constant force is necessary to stop the skater in a distance of 0.5 m?

Answer: 200 N

4. A spring attached to a 0.5 kg mass is compressed 5 cm and released from rest. When the mass is 2 cm from equilibrium it moves with a velocity of 0.3 m/s. Find the spring constant.

Answer: 21 N/m

7

POTENTIAL ENERGY and ENERGY CONSERVATION

OBJECTIVES

In this chapter Newton's laws are used to develop the ideas of potential energy and the conservation of mechanical energy. This will provide you with another powerful tool for analyzing the motion of mechanical systems. Your objectives are to:

Identify and distinguish between *conservative* and *dissipative* forces.

Define the *potential energy* of a body acted upon by conservative forces. Calculate potential energy for the force of gravity and the elastic force of a spring.

Formulate the *principle of conservation of mechanical energy* and apply it to a variety of conservative systems.

Find the force on an object from its potential energy function.

Discuss the *qualitative motion* of a system, given its potential energy function.

REVIEW and SUPPLEMENT

Conservative and Dissipative Forces; Conservation of Mechanical Energy

If the work done by a force on a body moving between two points does not depend on the path taken, the force is *conservative*; otherwise it is *dissipative*. When only conservative forces act, positive and negative parts of the work about a closed path cancel.

A familiar dissipative force is the force of sliding friction. To displace a box a distance s to the right one can take a straight path with work of friction $W_F = -f_s s$, where the work is negative because the force is opposite to the displacement (See Fig. 7-1a); or one can displace the box a distance 2s to the right and then a distance s to the left arriving at the same endpoint. The work for the second displacement is

$$W_F' = -2f_s s - f_s s = -3f_s s.$$

(See Fig. 7-1b) The work depends on the path.

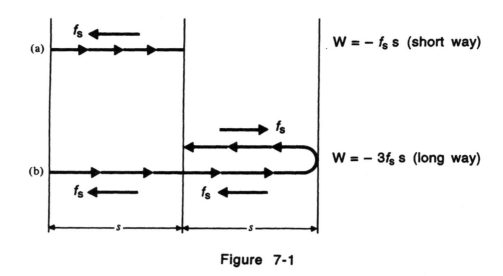

$W = - f_s s$ (short way)

$W = - 3 f_s s$ (long way)

Figure 7-1

Constant Gravitational Field and the Conservation of Energy

The gravitational force is a conservative force. To show this refer to Fig. 7-2 and consider the curved path connecting two points (x_1, y_1) and (x_2, y_2) in a constant gravitational field.

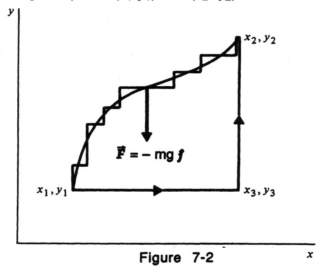

Figure 7-2

The path has been approximated by small horizontal and vertical staircase increments. The work of the weight is zero along the horizontal increments; no work is done in the horizontal displacements because

$$\vec{F} \cdot \vec{s} = 0$$

here. Along a vertical segment the work is $\Delta W_i = -F \Delta y_i = -mg \Delta y_i$, negative because the

148

displacement and $\vec{F}$ are *anti*parallel. The total gravitational work is the sum of all these vertical increments

$$W_g = \Sigma_i \, (\text{-} \, mg\Delta y_i) = \text{-} \, mg \, \Sigma \, \Delta y_i = \text{-} \, mg(y_2 \, \text{-} \, y_1)$$

Any other path connecting the two end points, e.g. the 1 to 3 to 2 path of Fig. 7-3 has the same total ascent Δy and hence the same gravitational work. If the gravitational potential energy is defined as

$$U(y) = mgy$$

we find that the gravitational work is

$$W_g = mgy_1 \, \text{-} \, mgy_2 = U_1 \, \text{-} \, U_2$$

$$= \text{-} \, \Delta U$$

The sum K + U is the *total mechanical energy* E for this system, and changes according to the rule

$$W' = E_2 \, \text{-} \, E_1 = \Delta E$$

where W' is the work of all forces other than gravity. If W' is zero, the total mechanical energy is constant and is said to be conserved,

$$W' = 0 = E_1 \, \text{-} \, E_2; \quad E_1 = E_2;$$
$$\frac{1}{2} \, mv_1^2 + mgy_1 = \frac{1}{2} \, mv_2^2 + mgy_2$$

Elastic Forces and the Conservation of Energy

Any force which is conservative has a potential energy function. Its work can be grouped with the kinetic energy term, in an analogous way to the gravitational force considered in the last section. For example, consider the *elastic* force of a spring on a mass as shown in Fig. 7-3.

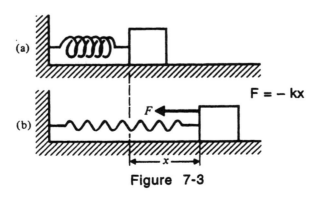

Figure 7-3

149

If the mass is displaced from x_1 to x_2, the work done by the elastic force is

$$W_e = \int_{x_1}^{x_2} \vec{F} \cdot d\vec{s} = \int_{x_1}^{x_2} F_x \, dx = -k \int_{x_1}^{x_2} x \, dx$$

$$W_e = -\left[\frac{kx^2}{2}\right]_{x_1}^{x_2} = \frac{kx_1^2}{2} - \frac{kx_2^2}{2} = U_1 - U_2$$

Here the elastic potential energy has been defined as

$$U = \frac{kx^2}{2}$$

For this system the mechanical energy is conserved

$$E = K + U = \frac{mv^2}{2} + \frac{kx^2}{2}$$

provided the work of all the other forces, W', is zero.

Finding the Force From the Potential Energy Function

Consider the spring again in Fig. 7-3. The work done by the elastic force over a small displacement Δx is:

$$dW_e \approx F_x \, \Delta x = - (U_2 - U_1) = - \Delta U$$

where ΔU is the *increase* in potential energy. Thus

$$F_x \approx - \frac{\Delta U}{\Delta x}$$

and in the limit $\Delta x \to 0$, we have

$$F_x = - \frac{d}{dx} U(x)$$

In general, we can obtain the component of a force on an object by taking the derivative of the potential energy function with respect to that coordinate.

Potential Energy and the Qualitative Description of Motion When Mechanical Energy is Conserved

In Fig. 7-4 the potential function U(x) is sketched for a mass m on a smooth horizontal surface attached to a wall by a spring of force constant k.

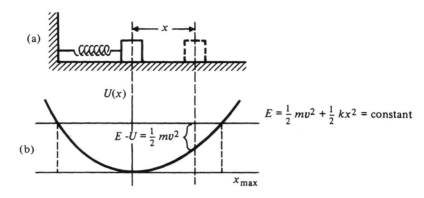

Figure 7-4

All other forces acting on the mass (e.g., gravity and the normal forces) do no work because they are perpendicular to the displacement, so W' = 0 and the mechanical energy E is conserved. This constant energy is indicated by the straight line E = constant in Fig. 7-4b. The difference of E and U is the kinetic energy

$$K = \frac{mv^2}{2} \, ,$$

also indicated in the figure. Since

$$K = \frac{mv^2}{2} = E - U \geq 0$$

motion is only allowed between the extreme coordinates $\pm x_{max}$ where K = 0 and

$$E = U\left(x_{max}\right) = \frac{k\left(x_{max}\right)^2}{2}$$

The maximum velocity and the maximum kinetic energy occur when U = 0 at x = 0, where

$$E = K\left(v_{max}\right) = \frac{m\left(v_{max}\right)^2}{2}$$

The total energy E is entirely potential energy at the endpoints and entirely kinetic energy at the equilibrium position x = 0. At other points the energy is split between the potential and kinetic parts, summing to the constant E.

If the spring is initially released from rest at x_{max}, the mass recoils toward $x = 0$, gaining maximum velocity at $x = 0$. As it overshoots $x = 0$, the spring pulls it back, decreasing its speed until it is zero again at $x = -x_{max}$, whereupon it reverses its direction, speeding up toward the origin, then slowing down as it approaches the initial point. The potential energy curve provides a quick qualitative description of the motion, its limits, and the velocity at each point x, if E is known.

The value of the constant E is determined by how the mass is set into motion, that is how much energy it is initially given. It may be evaluated at any point where x and v are both known.

PROBLEM-SOLVING STRATEGIES

Now is time to review the Problem-Solving strategy in the text for this chapter. When applying energy conservation, make a list of the initial kinetic and potential energies (K_1, U_1) and the final kinetic and potential energies (K_2, U_2). Note the knowns and the unknowns. Solve for the unknowns.

Note again that the work, a scalar, may be positive or negative. A force acting perpendicular to the path of a particle (such as the tension of a pendulum bob) does no work. Such forces do not contribute to the work W or W' in the work energy relation.

QUESTIONS AND ANSWERS

Question. Tarzan crosses a jungle gorge by starting from rest and swinging from one vine to another. Can he, by this method, ever reach the other side if it is higher than the side on which he starts? Does a running jump help matters?

Answer. Yes, with running jump, he can swing to a point higher than his starting point.

Question. A ball of "silly putty" is dropped from a height (h) and sticks to the floor. Discuss the changes in the kinetic, potential, and internal energy of the ball during the fall.

Answer. The initial potential energy at height (h) is converted into an equal amount of kinetic energy just before impact. Upon impact, all the kinetic energy is dissipated into heat (internal energy) as the putty ball has no velocity after impact.

EXAMPLES AND SOLUTIONS

Example 1

A body in a plane is acted upon a force

$$\vec{F} = -bx^2 \hat{i} \qquad\qquad b = 20 \text{ N/m}^2 \qquad \text{so that } F_x = -bx^2 \qquad \text{and} \qquad F_y = 0$$

What work is done by the force F when a body is displaced along the paths $\vec{s}_1$, $\vec{s}_2$, $\vec{s}_3$ of Fig. 7-5?

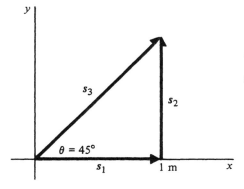

$$\cos \theta = \frac{dx}{ds_3}$$

Figure 7-5

Solution:

Along $\vec{s}_1$, $\vec{F}$ and $d\vec{s}$ are parallel, $\cos \theta = 1$, and

$$W_1 = \int \mathbf{F} \cdot d\mathbf{s}_1 = \int (F \cos \theta) \, ds_1 = \int_0^{1\,m} F_x \, dx$$

$$W_1 = -b \int_0^{1\,m} x^2 \, dx = -(20 \text{ N/m}^2) \left[\frac{x^3}{3} \right]_0^{1\,m} = -6.7 \text{ J}$$

Along $\vec{s}_2$, $\vec{F}$ is perpendicular to $d\mathbf{s}_2$ and

$$W_2 = \int \vec{F} \cdot d\vec{s}_2 = 0$$

Along $\mathbf{s}_3$, $\cos \theta = 45°$, $ds_3 = (\cos \theta)^{-1} \, dx$ and we have

$$W_3 = \int \vec{F} \cdot d\vec{s}_3 = \int F \cos \theta \, ds_3$$

$$W_3 = \int_0^{1\,m} (F \cos \theta) \frac{dx}{\cos \theta}$$

$$W_3 = \int_0^{1\,m} F \, dx = -6.7 \text{ J}$$

Note $W_3 = W_1 + W_2$, i.e. the work done is the same for each complete path connecting initial and final points. This is an example of a conservative force.

Example 2

Refer to Example 6 of Chapter 6. Solve this problem using the potential energy function to calculate the work of the conservative forces.

{For reference the problem statement was: A body is released from rest at the top of a rough inclined plane of height y and angle θ. The coefficient of friction is μ. Use the work-energy relation to find the velocity of the body at the bottom of the incline in terms of m, y, μ and θ.}

Solution:

In this approach, you handle the conservative part of the total force, the gravitation, through the gravitational potential energy U = mgy. Then the work W' of all other forces is

$$W' = \Delta E = E_2 - E_1 = (K_2 + U_2) - (K_1 + U_1)$$

$$W' = \left(\frac{1}{2} mv_2^2 + mgy_2\right) - \left(\frac{1}{2} mv_1^2 + mgy_1\right)$$

$$W' = \frac{1}{2} mv_2^2 - mg\left(y_1 - y_2\right)$$

$$W' = \frac{1}{2} mv_2^2 - mgy$$

where W' is the work of all forces except the gravitational force. $W' = W_F = -\mu mgy/\tan\theta$. Thus

$$-\frac{\mu mgy}{\tan\theta} = \frac{1}{2} mv_2^2 - mgy$$

resulting in the same answer for v_2 as obtained in Example 6, Chapter 6.

Example 3

A 2 kg mass attached by a string of length L = 1 m to the ceiling is released from rest at an angle of $\theta_0 = 60°$ with the vertical. Find its maximum velocity.

Solution:

Refer to the free-body diagram and coordinate system of Fig. 7-6:

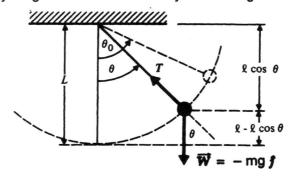

Figure 7-6

The only forces acting on the body are gravity and the tension T. T does no work ($\vec{T}$ and $d\vec{s}$ are perpendicular) so W' = 0 and mechanical energy is conserved,

$$W' = 0 = \Delta E = E_2 - E_1$$

The initial energy is

$$E_1 = K_1 + U_1 = 0 + mgy_1 = mgL(1 - \cos\theta_0)$$

The final energy at an angle θ is

$$E_2 = K_2 + U_2 = \frac{1}{2}mv_2^2 + mgy_2 = \frac{1}{2}mv_2^2 + mgL\left(1 - \cos\theta\right)$$

The conservation of energy yields

$$\frac{1}{2}mv_2^2 + mgL\left(1 - \cos\theta\right) = mgL\left(1 - \cos\theta_0\right)$$

$$v_2 = \sqrt{2gL\left(\cos\theta - \cos\theta_0\right)}$$

$$v_2 = \sqrt{2gL\left(\cos\theta - \frac{1}{2}\right)}$$

The velocity is maximum at the bottom of the swing when $\theta = 0$ and

$$v_2 = (gL)^{1/2} = (9.8 \text{ m/s}^2 \cdot 1 \text{ m})^{1/2} = 3.1 \text{ m/s}.$$

155

Example 4

In the last problem find the tension T in the string as a function of θ, and its value when T is maximum.

Solution:

In circular motion, the centripetal acceleration is v^2/R. The total force in the direction of this acceleration, toward the center, is:

$$\Sigma\, F = T - mg \cos \theta = \frac{mv^2}{R} = \frac{mv^2}{L}$$

$$T = mg \cos \theta + (m/L)(2gL)(\cos \theta - 1/2)$$

$$T = mg(3 \cos \theta - 1)$$

T is maximum when $\theta = 0$ at the bottom of its swing, where

$$T = 2\, mg = (2.2 \text{ kg}) \cdot (9.8 \text{ m/s}^2) = 39.2 \text{ N}$$

Example 5

A mass of 1 kg resting on a smooth floor, attached to a wall by a spring of constant k = 100 N/m, is pulled 0.5 m from its equilibrium position.
  (a) What is its potential energy?
  (b) If released from rest, what is its velocity when it passes through the equilibrium position?
  (c) If it is given an initial velocity of 5 m/s at x = 0.5 m, what is its velocity at x = 0?

Solution:

The system conserves total mechanical energy, as no other forces acting on the mass do work. Let x_1, v_1 be the initial position and velocity and x_2, v_2 the final position and velocity. We are given that $x_1 = 0.5$ m ; $x_2 = 0$; $v_1 = 0$; and v_2 is unknown and to be found.

At arbitrary position x and arbitrary velocity v, the mechanical energy E is:
$$E = \frac{1}{2} mv^2 + \frac{1}{2} kx^2 = K + U$$
(a) At the initial point, 0.5 m from equilibrium,
$$U_1 = \frac{1}{2} kx^2 = \frac{1}{2} (100 \text{ N/m})(0.5 \text{ m})^2 = 12.5 \text{ J}$$
(b) At this point the mass is at rest, and

$$E_1 = K_1 + U_1 = 12.5 \text{ J} \quad (K_1 = 0)$$

156

At the final point, the potential energy is zero giving

$$E_2 = K_2 + U_2 = \frac{1}{2} mv_2^{\,2} + \frac{1}{2} kx_2^{\,2} = \frac{1}{2} mv_2^{\,2}$$

and since the energies are equal, $E_1 = E_2$

$$\frac{1}{2} mv_2^{\,2} = 12.5 \text{ J}$$

$$v_2 = \sqrt{\frac{2(12.5 \text{ J})}{(1 \text{ kg})}} = 5 \text{ m/s}$$

(c)

$$E_1 = K_1 + U_1 = \frac{1}{2} mv_1^{\,2} + \frac{1}{2} kx_1^{\,2}$$

$$E_1 = \frac{1}{2} (1 \text{ kg})(5 \text{ m/s})^2 + 12.5 \text{ J} = 25 \text{ J}$$

$$E_1 = E_2 = \frac{1}{2} mv_2^{\,2} + \frac{1}{2} kx_2^{\,2} = \frac{1}{2} mv_2^{\,2}$$

$$v_2 = \sqrt{\frac{2(25 \text{ J})}{(1 \text{ kg})}} = 7.1 \text{ m/s}$$

Note it does not matter whether the initial velocity is toward or away from the equilibrium position.

Example 6

A mass of 0.5 kg hangs from a spring whose unstretched length is 1 m = 2 L. It is stretched a length L = 0.5 m. Find the spring constant.

157

Solution:

Referring to Fig. 7-7b, and the free-body diagram Fig. 7-7C, when the mass is in equilibrium we have

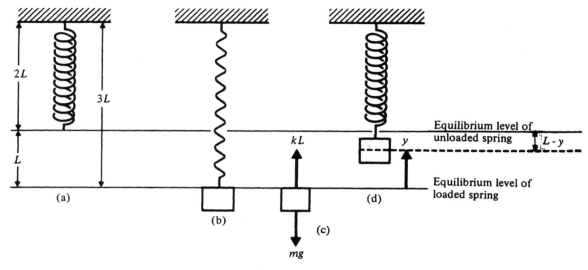

Figure 7-7

$$kL = mg \quad so \; k = \frac{mg}{L} = \frac{(0.5 \, kg)(9.8 \, m/s^2)}{(0.5 \, m)}$$

k = 9.8 N/m.

Example 7

The same mass is pulled down 0.5 m from its equilibrium position and released from rest. Find its maximum velocity and its maximum height.

Solution:

As shown in Fig. 7-7c, the coordinate y is the vertical distance from the equilibrium position of the loaded spring. The system has gravitational potential energy U_g **and** elastic potential energy U_e. The elastic potential energy is

$$U_e = \frac{1}{2} k(L-y)^2 = \frac{1}{2} k\left(\frac{mg}{k} - y\right)^2$$

because L - y is the amount the spring is stretched, where we used the kL = mg relation. Thus

$$U_e = \frac{1}{2} kL^2 - kLy + \frac{1}{2} ky^2$$

$$U_e = \frac{1}{2} kL^2 - mgy + \frac{1}{2} ky^2,$$

and

$$U_g = mgy.$$

The total potential energy is the <u>sum</u> of the elastic <u>and</u> gravitational potential energies

$$U = U_e + U_g = \frac{1}{2} kL^2 + \frac{1}{2} ky^2.$$

The constant term $(kL^2/2)$ is the elastic potential energy when the mass is in its equilibrium condition. Since only potential energy <u>differences</u> enter the work-energy relation,

$$W' = E_2 - E_1 = (K_2 - K_1) + (U_2 - U_1)$$

this constant may be dropped without changing any physical result. Alternatively if the zero of gravitational potential energy is taken to be at L/2,

$$U_g = mg\left(y - \frac{L}{2}\right) = mgy - \frac{1}{2} kL^2,$$

and

$$U = U_e + U_g = \frac{1}{2} ky^2$$

Thus when treating a spring in a gravitational field only the elastic potential energy appears in the potential function.

Here we have W' = 0 = E$_2$ - E$_1$, with

$$E_1 = \frac{1}{2} mv_1^2 + U = 0 + \frac{1}{2} ky_1^2 \text{ where } y_1 = -0.5 \text{ m.}$$

At an arbitrary final position y and velocity v

$$E_2 = \frac{1}{2} mv^2 + \frac{1}{2} ky^2 = E_1 = \frac{1}{2} ky_1^2$$

$$\frac{1}{2} mv^2 = \frac{1}{2} k\left(y_1^2 - y_2^2\right);$$

$$\frac{1}{2} m \left(v_{max} \right)^2 = \frac{1}{2} k y_1{}^2$$

The maximum velocity is when y = 0, where

$$v_{max} = \sqrt{\frac{k}{m}} \ y_1 = \sqrt{\frac{(9.8\,\text{N/m})}{(0.5\,\text{kg})}} \ (0.5\,\text{m}) = 2.2\,\text{m/s}$$

The height y as a function of v is

$$y = \pm \sqrt{\left(y_1^2 - \frac{m}{k}\,v^2 \right)}$$

and is maximum when y = + |y₁| = 0.5 m.

Example 8

For the frictionless system of Fig. 7-8a, find the velocity of the mass m when it hits the floor if it is released from rest a height L above the floor.
(m > m')

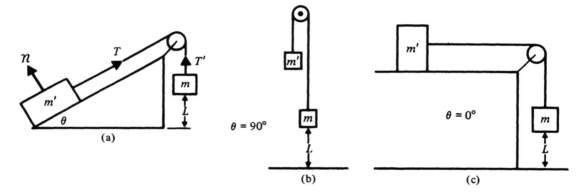

(a) $\theta = 90°$ (b) $\theta = 0°$ (c)

Figure 7-8

160

Solution:

Since n, the normal force of the plane on the mass, does no work, and the work of T is equal in magnitude but opposite in sign to the work of T', $W' = 0$ and the system is conservative, with mechanical energy conserved and given by:

$$E = \frac{1}{2} mv^2 + \frac{1}{2} m'v^2 + mgy + m'gy' = K + U$$

The initial values are

$$K_1 = 0, \quad U_1 = mgL$$

$$E_1 = K_1 + U_1 = mgL.$$

If m falls a distance L, m' rises a distance L up the plane through a vertical height $y_2' = L \sin \theta$. Thus the final energy is

$$K_2 = \frac{1}{2} (m + m')v^2 \quad \text{and } U_2 = m'gL \sin \theta$$

$$E_2 = \frac{1}{2} (m + m')v_2^2 + m'gL \sin \theta$$

The final potential energy of m is zero because its final y coordinate is zero. Energy conservation $E_2 = E_1 = mgL$ yields

$$mgL = \frac{1}{2} (m + m')v_2^2 + m'gL \sin \theta$$

The final velocity is thus

$$v_2 = \sqrt{2\frac{(m - m' \sin \theta)}{(m + m')} gL} \; .$$

As θ approaches 90° we have an Atwood's machine (Fig. 7-8b) with

$$v_2 = \sqrt{2\frac{(m - m')}{(m + m')} gL} \; .$$

As θ approaches 0 we have the table (Fig. 7-8c) with

$$v_2 = \sqrt{\frac{(2m)}{(m + m')} gL} \; .$$

Example 9

A 1 kg block attached to a spring and resting on a rough surface is given an initial velocity of v = 10 m/s. The spring, initially unstretched, has a constant k = 3 N/m. The maximum compression of the spring is L = 4 m. Find the coefficient of friction.

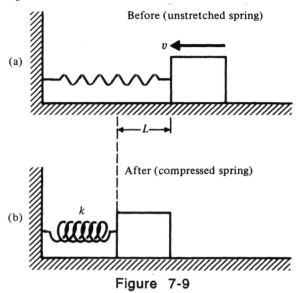

Figure 7-9

Solution:

In this case we must take into account the work of the frictional force in the work-energy relation. Referring to Fig. 7-9, we have

$$W' = E_2 - E_1; \quad W' = -fL = -\mu mgL; \quad E = \frac{1}{2}mv^2 + \frac{1}{2}kx^2$$

where W' is the work of the friction force. Evaluating the initial and final energies we have

$$E_2 = 0 + \frac{1}{2}kL^2 \quad \text{and} \quad E_1 = \frac{1}{2}mv^2 + 0$$

Thus the work-energy relation yields

$$-\mu mgL = \frac{1}{2}kL^2 - \frac{1}{2}mv^2$$

$$\mu = \frac{\left(\frac{1}{2}mv^2 - \frac{1}{2}kL^2\right)}{mgL}$$

$$\mu = \frac{\left[\frac{1}{2}(1\text{ kg})(10\text{ m/s})^2 - \frac{1}{2}(3\text{ N/m})(4\text{ m})^2\right]}{(1\text{ kg})(9.8\text{ m/s}^2)(4\text{ m})} = 0.67$$

Example 10

A 0.5 kg mass is thrown vertically at 5 m/s from the roof of a 100 m high building. Use the conservation of energy to find its velocity when it hits the ground.

Solution:

$$W' = 0 = E_2 - E_1 = \frac{1}{2} mv_2^{\,2} + mgy_2 - \frac{1}{2} mv_1^{\,2} - mgy_1$$

$$0 = \frac{1}{2} m\left(v_2^{\,2} - v_1^{\,2}\right) - mg\left(y_1 - y_2\right)$$

$v_2{}^2 = v_1{}^2 + 2g(y_1 - y_2)$

$v_1 = 5$ m/s

$y_1 - y_2 = 100$ m

$v_2 = [(5 \text{ m/s})^2 + 2(9.8 \text{ m/s}^2)100 \text{ m}]^{1/2} = 44.5$ m/s.

Example 11

A mass slides in a frictionless vertical loop-the-loop, having been released from rest a distance L above the bottom of the loop.
  (a) Find the velocity when the mass makes an angle θ with the vertical.
  (b) Find the normal force of the loop on the mass.

Solution:

  Referring to Fig. 7-10, we see that

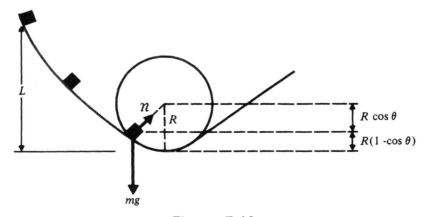

·Figure 7-10

163

the kinetic plus the gravitational potential energy is conserved because the normal force does no work,

(a)
$$W' = 0 = E_2 - E_1; \quad E = \frac{1}{2}mv^2 + mgy;$$

$y_2 = R - R \cos \theta; \; y_1 = L$

$$E_2 = \frac{1}{2}mv^2 + mgR\left(1 - \cos\theta\right)$$

$E_1 = mgL$

$v^2 = 2gL - 2gR(1 - \cos\theta)$

(b) Applying Newton's Second Law in the direction of the centripetal acceleration,
$$\sum F = n - mg \cos \theta = \frac{mv^2}{R}$$

$$n = mg \cos \theta + \frac{m}{R}\left[2gL - 2gR\left(1 - \cos\theta\right)\right]$$

$$n = 3mg \cos \theta - 2mg + 2mg \frac{L}{R}$$

The condition for the mass not to fall off the loop is n > 0. (If the mass were a bead on a wire, the wire could exert forces outward (n < 0) and the mass would never fall off). At the top of the loop when $\theta = 180°$ this condition is
$$n = -5mg + 2mg \frac{L}{R} > 0$$

or

$$L > \frac{5}{2} R$$

Example 12

Suppose in the last problem the block is released from rest a distance L = 2R above the bottom of the loop. When does the block fall off the loop?

164

Solution:

The block falls off when $n = 0$, i.e.

$$0 = 3 \, mg \cos \theta - 2 \, mg + 4 \, mg$$

$$\cos \theta = -2/3 \qquad \theta = 132°$$

Example 13

Discuss qualitatively the motion of a body whose potential energy is given in each part of Fig. 7-11.

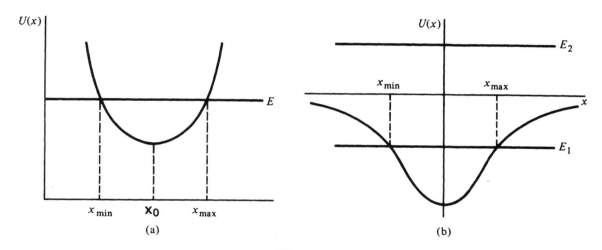

Figure 7-11

Solution:

(a) The particle is confined between x_{max} and x_{min} since $K = E - U > 0; E > U$. At these endpoints $K = 0$, $E = U$, and the velocity is zero. The body oscillates back and forth between x_{min} and x_{max}. The maximum velocity is at x_0 where U is a minimum and K is a maximum.

(b) Case E_1 is similar to (a) above. Case E_2 has no endpoints; the kinetic energy $K = E_2 - U$ is always positive but greatest at $x = 0$. If the body comes in from the far left with a kinetic energy $K = E_2 - U \approx E_2$, it speeds up as it approaches the origin, continuing to move in the positive direction, eventually far to the right with kinetic energy $K = E_2$. For $E > 0$ the body's motion is unbounded. For $E < 0$ the body's motion is bounded, that is has finite end points; it is confined to the region between x_{max} and x_{min}.

Example 14

The potential energy function of a particle is:

$$U(x) = ax + \frac{1}{2} kx^2.$$

(a) What is the force on the particle?   (b) What is its equilibrium position?

Solution:

(a) The force on the particle is:

$$F = -\frac{dU}{dx} = -a - kx = -(a + kx)$$

which represents a constant force and an "elastic force".

(b) When the force equals zero, the particle is at the equilibrium position x_0:

$$F = 0 = -a - k x_0 \quad \text{or} \quad x_0 = -(a/k).$$

Note that in terms of the displacement from equilibrium, $x' = x - x_0$,

$$F = -a - k(x' - x_0) = -a - k(x' - \frac{a}{k}) = -kx'$$

and the force is "elastic" about the equilibrium position.

QUIZ

1. A block of mass 0.5 kg is held against a spring compressed 5 cm from its relaxed length. When the block is released, it reaches a speed of 50 cm/s. What is the spring constant?

Answer: 50 N/m.

2. A cube of ice falls from the rim of a bowl down toward the bottom, 3 cm below. What is its maximum speed?

Answer: 0.77 m/s.

3. A mass weighing 10 N slides without friction around a loop-the-loop of radius R. It is released from rest a distance 3R above the bottom of the loop. If θ is the angle the mass makes with the vertical, find the force of the loop against the mass at the top of the loop, at $\theta = 90°$, and at the bottom of the loop.

Answer: 10 N, 40 N, 70 N.

4. 6 kg and 4 kg masses attached by a rope hang from a frictionless pulley. The 6 kg weight is 3 m above the ground, when the system is released from rest. Find the velocity of the system when the 6 kg mass hits the ground.

Answer: 3.4 m/s.

8

MOMENTUM, IMPULSE, AND COLLISIONS

OBJECTIVES

In this chapter the concepts of momentum, momentum conservation, and impulse are developed. Your objectives are to:

Distinguish between *internal* and *external* forces among interacting bodies.

Recognize when *momentum is conserved.*

Apply the *conservation* of *momentum* to a variety of problems involving collisions(*elastic* and *inelastic*)and other kinds of interaction between bodies.

Calculate the *impulse* of a force and relate it to the change of *momentum* of a body.

Apply the impulse-momentum relation to a variety of problems.

Calculate the coordinates of the *center of mass* of a body or a system of point masses.

REVIEW

Conservation of Momentum

When two bodies, for example two colliding billiard balls, interact with each other they exert forces on *each other.* These are *internal forces.* At the same time they may be acted upon by other forces, such as the friction on the table. These are *external forces.*

If the total force $\vec{F}^{ext}$ on the two billiard balls vanishes then the *total momentum* is conserved:

$$\vec{P}_1 = \vec{P}_2; \ \Delta\vec{P} = 0$$

The result may be generalized to any number of interacting bodies: the total momentum of a system is conserved if there is no net *external* force.

Collisions

When two bodies collide, internal forces are exerted during the short times of impact. These forces may be conservative or dissipative. If the forces are conservative, the collision is said to be *elastic. In an elastic collision, the total energy and hence the total kinetic energy is conserved*, since before and after the collision the potential energy is the same (generally zero.)

A collision is *inelastic* if total energy is not conserved.

An example of an approximately elastic collision is that between two billiard balls. No energy is lost or dissipated in permanent changes of the billiard balls; all energy is kinetic before and after the collision. During the collision kinetic energy may be transformed into elastic potential energy as the balls in contact compress each other; this potential energy is then, without loss, transformed back into kinetic energy of motion as the balls spring apart to their original shape and state.

An example of an inelastic collision is the impact of a bullet with a block of wood. Mechanical energy is not conserved; it is dissipated into heat and permanent changes of shape and state in the bullet and block.

Elastic Collision Between Two Bodies

If the initial velocities of two bodies (v_{A1} and v_{B1}) are known, then the two conditions of energy conservation and momentum conservation may be used to calculate their final velocities after a collision on a straight line. The general case is derived in the text. Note all velocities are referred to a single direction on a line (eq. positive when moving to the right, negative when moving to the left). A and B refer to the two bodies; 1 and 2 to before and after collisions. For an elastic collision between m_A and m_B with body B <u>initially at rest</u> the final velocities are

$$V_{A2} = \frac{m_A - m_B}{m_A + m_B} V_{A1} \qquad V_{B2} = \frac{2m_A}{m_A + m_B} V_{A1}$$

If $m_A \gg m_B$ (for example, baseball A hits ping-pong ball B at rest)

$$V_{A2} \cong V_{A1}; \quad V_{B2} \cong 2V_{A1}.$$

The baseball hardly changes speed; the ping-pong ball flies off with twice the initial baseball velocity. Viewed from a frame of reference moving with the baseball, the ping-pong ball approaches with a speed V_{A1} and is reflected forward upon collision with an equal but opposite velocity. The *change* of velocity is $2V_{A1}$ in both frames.

If $m_A \ll m_B$ (for example, ping-pong ball A hits baseball B at rest), then

$$V_{A2} \cong -V_{A1}; \qquad V_{B2} \cong 0.$$

The baseball hardly moves and the ping-pong ball reverses its direction, changing its velocity by $2V_{A1}$, as in the last example. If

$$m_A = m_B, \; V_{A2} = 0 \text{ and } V_{B2} = V_{A1};$$

the moving ball stops upon collision, transferring all of its momentum to the ball originally at rest.

Impulse

The impulse J of a constant force acting on a single body between times t_1 and t_2 is

$$\vec{J} = \vec{F}\Delta t;$$

Note the direction of J is the direction of F and $\Delta t = t_2 - t_1$. If the force is not constant

$$\vec{J} = \int_{t_1}^{t_2} \vec{F}\, dt; \qquad J_x = \int_{t_1}^{t_2} F_x\, dt, \text{ and similar for } J_y \text{ and } J_z$$

If the *average* force is defined as

$$\vec{F}_{av} = \frac{1}{t_2 - t_1} \int_{t_1}^{t_2} \vec{F}\, dt$$

then

$$\vec{J} = \vec{F}_{av}(t_2 - t_1)$$

Newton's second law may be used to show that the impulse of a force gives the change in *momentum*

$$\vec{J} = \int_{t_1}^{t_2} m\, \frac{d\vec{v}}{dt}\, dt = \int_{v_1}^{v_2} m\, d\vec{v} = m\vec{v}_2 - m\vec{v}_1 = \vec{p}_2 - \vec{p}_1 = \Delta\vec{p}$$

where the momentum of a single body is defined as

$$\vec{p} = m\vec{v}.$$

Note the momentum is a vector quantity (unlike energy, which is a scalar.) The impulse J in a time interval is equal to the change in momentum of the body in that interval; if F or $F_{av} = 0$, J = 0 and the momentum does not change. If the force is constant and in the x direction,

$$J_x = \int_{t_1}^{t_2} F_x \, dt = F_x \int_{t_1}^{t_2} dt = F_x \left(t_2 - t_1 \right) = m \left(v_{x2} - v_{x1} \right),$$

The last equality expresses the constant acceleration condition:

$$a_x = \frac{F_x}{m} = \frac{\left(v_{x2} - v_{x1} \right)}{\left(t_2 - t_1 \right)}$$

Center of Mass

The center of mass position vector R of a system of mass points m_1 at r_1, m_2 at r_2, etc., is given by

$$\vec{R} = \frac{m_1 \vec{r}_1 + m_2 \vec{r}_2 + m_3 \vec{r}_3 + \dots}{M} = \vec{R}(X, Y, Z)$$

where M is the sum of the individual masses ($M = m_1 + m_2 + m_3 + \dots$). The components of the center-of-mass position vector are

$$X = \frac{m_1 x_1 + m_2 x_2 + m_3 x_3 + \dots}{M}, \quad \text{and similarly for Y and Z}$$

The total momentum $\vec{P}$ of a system is the total mass times the velocity of the center of mass

$$M\vec{V} = \vec{P}$$

The rate of change of total momentum is

$$\frac{d\vec{P}}{dt} = M \frac{d\vec{V}}{dt} = M\vec{A} = \vec{F}^{\text{ext}}$$

when $\vec{A}$ is the acceleration of the center of mass. The center of mass moves as if it were a mass point at $\vec{R}$ acted upon by a total force $\vec{F}^{\text{ext}}$. If the total external force is zero, the total momentum is conserved, $\vec{A} = 0$, and the center of mass moves with uniform motion in a straight line.

PROBLEM-SOLVING STRATEGIES

Now is a good time to review the Conservation of Momentum Problem-Solving strategies of the main text. Think about these as you study the examples and solutions below. Momentum is conserved in a system if no external forces act. Even if energy is not conserved, as in an inelastic collision, momentum is constant as long as the system is isolated from external forces.

Momentum is a vector; if $\vec{P}$ is conserved, the components P_x, P_y, and P_z are constant. If all motion is along a straight line, only that component of momentum need be considered.

Make "before" and "after" diagrams. As with energy conservation, list all your initial and final momenta, indicating knowns and unknowns.

Note that components of momentum may be positive or negative depending on direction of motion.

QUESTIONS AND ANSWERS

QUESTION. Must the center-of-mass of an object be inside the object?

ANSWER. No. For example, consider a uniform ring. All the mass is concentrated around the rim but the center-of-mass is at the geometric center where there is no mass. For an "L-shaped" object, the center-of-mass does is not on either piece.

QUESTION. The same <u>horizontal</u> impulse is imparted to a ping pong ball and a base ball. Which ball reaches the ground first? Which has the greater momentum? Which has the greater horizontal range?

ANSWER. Both balls will reach the ground in the same time, neglecting air resistance since they will have the same vertical acceleration and zero initial velocity in the vertical direction. The ping pong ball and the baseball will have equal momenta in the horizontal direction since they had equal impulses but the ping pong ball will have a larger velocity than the baseball because of its smaller mass. Therefore the ping pong ball will have a greater horizontal range.

EXAMPLES AND SOLUTIONS

Example 1

A 2000 kg car moving at 30 km/hr strikes a stopped 1000 kg car and the two lock bumpers.
    (a) What is their final common velocity just after the collision?
    (b) What fraction of the initial energy is dissipated in the collision?

Solution:

The first problem-solving strategy is to decide if momentum is conserved. Just before and just after and during the collision, the only forces that are important are the internal forces between the two cars. Momentum is conserved because only internal forces act during the collision; energy is not because the two cars lock bumpers--the collision is inelastic.

We are given that:
$$m_A = 2000 \text{ kg} \qquad v_{A1} = 30 \text{ km/hr}$$

$$m_B = 1000 \text{ kg} \qquad v_{B1} = 0$$

Since B is at rest initially, then the initial momentum is carried only by A and the final momentum is carried by the wreckage (A + B). Equating the two, we find

$$m_A v_{A1} = (m_A + m_B) v_2$$

where v_2 is the common final velocity $v_{A2} = v_{B2} = v_2$.

$$v_2 = \frac{m_A}{m_A + m_B} v_{A1} = \frac{(2000 \text{ kg})}{(3000 \text{ kg})} (30 \text{ km/hr}) = 20 \text{ km/hr}$$

The final and initial mechanical energies are

$$E_2 = \frac{1}{2} m_A v_2^{\,2} + \frac{1}{2} m_B v_2^{\,2}$$

$$E_2 = \frac{1}{2} (m_A + m_B) v_2^{\,2}$$

$$E_1 = \frac{1}{2} m_A v_{A1}^{\,2}$$

$$\frac{E_2}{E_1} = \left(\frac{m_A + m_B}{m_A} \right) \left(\frac{v_2}{v_{A1}} \right)^2 = \left(\frac{m_A + m_B}{m_A} \right) \left(\frac{m_A}{m_A + m_B} \right)^2$$

$$\frac{E_2}{E_1} = \frac{m_A}{m_A + m_B} = \frac{(2000 \text{ kg})}{(3000 \text{ kg})} = \frac{2}{3}$$

2/3 of the initial energy remains kinetic; 1/3 is dissipated during the collision.

Example 2

Suppose in the last example the cars have spring loaded bumpers, so that the collision is perfectly elastic.
    (a) What is the velocity of each car after the collision?
    (b) What are the final momenta and energies?

Solution:

For this case (B initially at rest) the result of simultaneously solving the momentum conservation equation

$$m_A v_{A1} = m_A v_{A2} + m_B v_{B2} \qquad (\text{ initial momentum = final momentum})$$

and energy conservation equation

$$\frac{1}{2} m_A v_{A1}^{\,2} = \frac{1}{2} m_A v_{A2}^{\,2} + \frac{1}{2} m_B v_{B2}^{\,2} \qquad (\text{initial energy = final energy})$$

is (see text)

$$V_{A2} = \frac{m_A - m_B}{m_A + m_B} V_{A1} = \frac{(2000 \text{ kg} - 1000 \text{ kg})}{(3000 \text{ kg})} (30 \text{ km/hr}) = 10 \text{ km/hr}$$

$$V_{B2} = \frac{2m_A}{m_A + m_B} V_{A1} = \frac{2(2000 \text{ kg})}{(3000 \text{ kg})} (30 \text{ km/hr}) = 40 \text{ km/hr}$$

The energies are

$$E_{A1} = \frac{1}{2} m_A V_{A1}^2 = \frac{1}{2}(2000 \text{ kg})\left(\frac{30 \times 10^3 \text{ m}}{3600 \text{ s}}\right)^2 = 70 \times 10^3 \text{ J}$$

$$E_{A2} = \frac{1}{2} m_A V_{A2}^2 = \frac{1}{2}(2000 \text{ kg})\left(\frac{10 \times 10^3 \text{ m}}{3600 \text{ s}}\right)^2 = 8 \times 10^3 \text{ J}$$

$$E_{B2} = \frac{1}{2} m_B V_{B2}^2 = \frac{1}{2}(1000 \text{ kg})\left(\frac{40 \times 10^3 \text{ m}}{3600 \text{ s}}\right)^2 = 62 \times 10^3 \text{ J}$$

Note $E_{A1} = E_{A2} + E_{B2}$

The momenta are

$$p_{A1} = m_A V_{A1} = (2000 \text{ kg})\left(\frac{30 \times 10^3 \text{ m}}{3600 \text{ s}}\right) = 16.7 \times 10^3 \text{ kg·m/s}$$

$$p_{A2} = m_A V_{A2} = (2000 \text{ kg})\left(\frac{10 \times 10^3 \text{ m}}{3600 \text{ s}}\right) = 5.6 \times 10^3 \text{ kg·m/s}$$

$$p_{B2} = m_B V_{B2} = (1000 \text{ kg})\left(\frac{40 \times 10^3 \text{ m}}{3600 \text{ s}}\right) = 11.1 \times 10^3 \text{ kg·m/s}$$

Note

$$p_{A1} = p_{A2} + p_{B2}$$

Example 3

A 2 kg block, moving to the right on a frictionless table at 3 m/s makes a head-on collision with another 2 kg block moving 4 m/s to the left.

    (a) If the collision is completely elastic, find the final velocities of the blocks.
    (b) If the collision is completely inelastic find the final velocity of the blocks.
    (c) If half the initial energy is dissipated in the collision, find the final velocities of the blocks.

174

Solution:

Momentum and energy are conserved because the table is frictionless and you are told that the collision is "elastic". Referring to Fig. 8-1, the initial velocities before collision (subscript '1') are

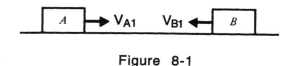

Figure 8-1

$$v_{A1} = 3 \text{ m/s}, \ v_{B1} = -4 \text{ m/s}$$

and

$$m_A = m_B = m = 2 \text{ kg}.$$

(a) The equations for conservation of momentum and energy are

$$mv_{A1} + mv_{B1} = m_A v_{A2} + mv_{B2} \quad \text{(momentum)}$$

$$v_{A1} + v_{B1} = [v_{A2} + v_{B2} = 3 \text{ m/s} - 4 \text{ m/s} = -1 \text{ m/s}]$$

$$\frac{1}{2}mv_{A1}^{2} + \frac{1}{2}mv_{B1}^{2} = \frac{1}{2}mv_{A2}^{2} + \frac{1}{2}mv_{B2}^{2} \quad (\text{energy})$$
$$(3\,\text{m/s})^{2} + (-4\,\text{m/s})^{2} = v_{A2}^{2} + v_{B2}^{2} = 25\,\text{m}^{2}/\text{s}^{2}$$

$$[v_{A2}^2 + v_{B2}^2 = 25 \text{ m}^2/\text{s}^2].$$

The two equations above, set off in square brackets [...] may be solved simultaneously for v_{A2}, v_{B2} by solving the first equation for v_{B2} and substituting the result in the second equation, producing a quadratic equation for v_{A2}, which can be factored:

$$v_{A2}^2 + (-1 - v_{A2})^2 = 25$$

$$2v_{A2}^2 + 2v_{A2} - 24 = 0$$

$$(v_{A2} + 4)(v_{A2} - 3) = 0$$

$$v_{A2} = -4 \text{ m/s}, \ 3 \text{ m/s}$$

The second root is extraneous, corresponding to the case where the bodies miss each other. Then, using the first boxed equation,

175

v_{B2} = -1 + 4 = 3 m/s.

(b) "Inelastic" indicates that mechanical energy is <u>not</u> conserved. In this case $v_{A2} = v_{B2} = v_2$ and we may only use the first equation, since only momentum is conserved:

$$2v_2 = -1 \text{ m/s} \qquad v_2 = -0.5 \text{ m/s}$$

(c) Here we use conservation of momentum,

$$v_{A2} + v_{B2} = -1 \text{ m/s}$$

and the fact that

$$E_{A2} + E_{B2} = 1/2(E_{A1} + E_{B1}) = \text{half the initial energy.}$$

$$\frac{1}{2}mv_{A2}^2 + \frac{1}{2}mv_{B2}^2 = \frac{1}{2}\left[\frac{1}{2}mv_{A1}^2 + \frac{1}{2}mv_{B1}^2\right]$$

$$v_{A2}^2 + v_{B2}^2 = 12.5 \text{ m}^2/\text{s}^2$$

Solving the last two italicized equations simultaneously,

$$v_{A2}^2 + (-1 - v_{A2})^2 = 12.5$$

$$2v_{A2}^2 + 2v_{A2} + 1 - 12.5 = 0$$

$$v_{A2}^2 + v_{A2} - 5.75 = 0$$

$$v_{A2} = \frac{-1 \pm \sqrt{1 - 4(1)(-5.75)}}{2}$$

$$v_{A2} = \frac{-1 \pm 4.9}{2} = +1.95 \text{ m}/\text{s}, \text{ or} -2.95 \text{ m}/\text{s}$$

$$v_{B2} = -1 - v_{A2} = \text{-2.95 m/s, 1.95 m/s}$$

Here the two physically relevant roots are, v_{A2} = 1.95 m/s and v_{B2} = -2.95 m/s. The others correspond to the case where the two bodies miss each other but a loss of energy is nonetheless suffered as a result of the "collision".

Example 4

A 200 g block moves to the right at a speed of 100 cm/s and meets a 400 g block moving to the left with a speed of 80 cm/s. Find the final velocity of each block if the collision is elastic.

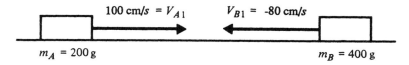

Figure 8-2

Solution:

Both momentum <u>and</u> energy are conserved because the collision is elastic. Refer to Fig. 8-2 for the initial configuration.

$$m_A = 200 \text{ g}; \qquad 100 \text{ cm/s} = v_{A1}$$

$$v_{B1} = -80 \text{ cm/s}; \qquad m_B = 400 \text{ g}$$

For an elastic collision, we have, from the text derivation,

$$v_{A2} = \frac{2m_B v_{B1} + v_{A1}(m_A - m_B)}{m_A + m_B}$$

$$v_{A2} = \frac{2(400 \text{ g})(-80 \text{ cm/s}) + (100 \text{ cm/s})(200 \text{ g} - 400 \text{ g})}{(200 \text{ g} + 400 \text{ g})}$$

$$= -140 \text{ cm/s}$$

$$v_{B2} = \frac{2m_A v_{A1} - v_{B1}(m_A - m_B)}{m_A + m_B}$$

$$v_{B2} = \frac{2(200 \text{ g})(100 \text{ cm/s}) - (-80 \text{ cm/s})(200 \text{ g} - 400 \text{ g})}{(200 \text{ g} + 400 \text{ g})}$$

$$= 40 \text{ cm/s}$$

Example 5

A 200 g block (A) moving to the right at 100 cm/s hits a 100 g block (B) moving to the right at 50 cm/s. Find each final velocity if the collision is elastic.

Solution:

Taking, as before, all velocities to the right as positive, then $m_A = 200$ g and $v_{A1} = 100$ cm/s, with $m_B = 100$ g and $v_{B1} = 50$ cm/s.

Thus we have, from the text derivation

$$v_{A2} = \frac{2m_B v_{B1} + v_{A1}\left(m_A - m_B\right)}{m_A + m_B}$$

$$v_{A2} = \frac{2\left(100 \text{ g}\right)\left(50 \text{ cm/s}\right) + \left(100 \text{ cm/s}\right)\left(200 \text{ g} - 100 \text{ g}\right)}{\left(200 \text{ g} + 100 \text{ g}\right)}$$

$$= 66.7 \text{ cm/s}$$

$$v_{B2} = \frac{2m_A v_{A1} - v_{B1}\left(m_A - m_B\right)}{m_A + m_B}$$

$$v_{B2} = \frac{2\left(200 \text{ g}\right)\left(100 \text{ cm/s}\right) - \left(50 \text{ cm/s}\right)\left(200 \text{ g} - 100 \text{ g}\right)}{\left(200 \text{ g} + 100 \text{ g}\right)}$$

$$= 116 \text{ cm/s}$$

Example 6

A 200 g body traveling at 100 cm/s hits a 500 g body at rest. Find the final velocities if the collision is elastic.

Solution

Referring to Fig. 8-3,

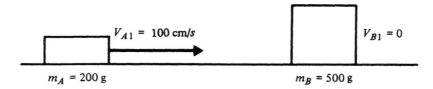

Figure 8-3

we have, for the case of $v_{B1} = 0$, again using the equations developed in the text for two-body elastic collisions,

$$v_{A2} = \frac{m_A - m_B}{m_A + m_B} v_{A1} = \frac{(200 \text{ g} - 500 \text{ g})}{(200 \text{ g} + 500 \text{ g})} (100 \text{ cm/s})$$

$$= -43 \text{ cm/s}$$

$$v_{B2} = \frac{2m_A}{m_A + m_B} v_{A1} = \frac{2(200 \text{ g})}{(200 \text{ g} + 500 \text{ g})} (100 \text{ cm/s})$$

$$= 57 \text{ cm/s}$$

Example 7

A body falls for 5 s in a gravitational field. Its initial velocity is $v_0 = 20$ m/s. Use the impulse-momentum relation to find the final velocity.

Solution:

Here momentum is not conserved because the external gravitational force acts on the body. The impulse of this force is:

$$J = \int_0^t F \, dt = \int_0^t (-mg) \, dt = -mgt.$$

and the resulting change in momentum is:

$$\Delta p = mv - mv_0$$

Since $J = \Delta p \approx -mgt = mv - mv_0$, we may solve for v:

$$v = v_0 - gt = 20 \text{ m/s} - (9.8 \text{ m/s}^2)5 \text{ s} = -29 \text{ m/s}$$

Example 8

A force acts on a 0.5 kg body along a given direction with a magnitude given by the graph of Fig. 8-4. If the body is initially at rest find its velocity at t = 5 s.

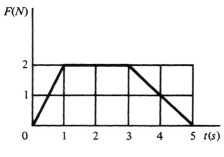

Figure 8-4

Solution:

First calculate the impulse in this interval:

$$J = \int_{0}^{5\,s} F\,dt = \text{area under the curve in Fig. } 8-4.$$

$$= 7 \times \text{area of one box} \quad \text{(Count boxes!)}$$

$$= 7 \text{ N·s}$$

Then equate that impulse to the change in momentum:

$$\Delta p = (0.5 \text{ kg})v$$

$$7 \text{ N·s} = (0.5 \text{ kg})v$$

$$v = \frac{(7 \text{ N/s})}{(0.5 \text{ kg})} = 14 \text{ m/s}$$

Example 9

A 0.15 kg baseball is pitched horizontally at 25 m/s. It is then batted at a velocity of 40 m/s in the opposite direction. The ball is in contact with the bat for 0.004 s. Find the average force exerted by the bat.

Solution:

Take the direction of the batted ball to be positive. The impulse, the product of the average force with the time interval, is positive in the direction of the batted ball:

$$J = \int_{t_1}^{t_2} F\, dt = F_{av}(\Delta t) = mv_2 - mv_1$$

The average force is the momentum change divided by Δt:

$$F_{av} = \frac{m(v_2 - v_1)}{(\Delta t)} = \frac{(0.15\ \text{kg})(40\ \text{m/s} - (-25\ \text{m/s}))}{(0.004\ \text{s})}$$

$$= 2400\ \text{N, in the direction of the batted ball.}$$

Note v_1, the pitched ball velocity, is negative because the direction of the batted ball, v_2, has been taken to be positive. Again, as always, you must be careful about sign conventions and directions!

Example 10

A box is kicked, and acquires a horizontal velocity $v_1 = 3$ m/s. It slides for 2 s along a rough floor before coming to rest. Use the impulse momentum relation to find the coefficient of friction.

Solution:

Choose the kick direction (and force) to be positive. Then after the kick, the only external force is the frictional force, pointing in the negatice direction,

$$J = \int_{t_1}^{t_2} F\, dt = \int_{t_1}^{t_2} \left(-\mu \mathcal{N}\right) dt = -\mu mg(\Delta t)$$

This impulse is equal to the change in momentum:

$$\Delta p = m(v_2 - v_1) = -mv_1; \qquad v_2 = \text{final velocity} = 0$$

$$\mu = \frac{v_1}{g(\Delta t)} = \frac{(3\ \text{m/s})}{(9.8\ \text{m/s}^2)(2\ \text{s})} = 0.15$$

Note the devil is in the details here of keeping signs and directions straight.

181

Example 11

Sand is dropped on a kitchen scale at the uniform rate of 100 g/s from a height of $h = 1$ m. Find the force on the scale and its reading if it is calibrated in kg.

Solution:

In a time Δt a mass Δm of sand hitting the scale has its velocity changed from v to zero,

$$v^2 = v_0^2 + 2gh; \qquad v_0 = 0$$

$$v = (2gh)^{1/2} \quad (= \text{velocity gained in free fall through distance h})$$

The <u>change</u> in momentum of the sand upon impact is

$$\Delta p = \Delta mv = \Delta m(2gh)^{1/2}$$

and is upward since the initial momentum is downward. This change of momentum is produced by the impulse of the upward force of the scale on the sand,

$$J = \int F dt = F\Delta t = \Delta p = \Delta m(2gh)^{1/2}$$

$$F = \left(\frac{\Delta m}{\Delta t}\right)\sqrt{2gh} = (0.1\,\text{kg/s})\sqrt{2(9.8\,\text{m/s}^2)(0.1\,\text{m})}$$

$$F = 0.44\,\text{N}$$

The scale is calibrated so that a force F is registered as its corresponding mass:

$$F = mg; \quad m = \frac{F}{g} = \frac{(0.44\,\text{N})}{(9.8\,\text{m/s}^2)} = 0.045\,\text{kg}$$

$$m = 45\,\text{g} \quad (45\,\text{grams})$$

Signs, as always, require care.

Example 12

Sand is dropped on a conveyor belt at the rate of 10 kg/s. What power is required to keep the belt moving at a constant velocity of $v = 1$ m/s?

Solution:

Consider the momentum and impulse to be positive in the direction of the belt's motion. In a time Δt an impulse $F\Delta t$ must be delivered to change the momentum of sand from $p_1 = 0$ in the belt direction to $p_2 = \Delta mv$ in the belt direction, that is

$$F\Delta t = \Delta p = p_2 - p_1 = \Delta mv$$

or

$$F = \left(\frac{\Delta m}{\Delta t}\right)v = (10\,\text{kg/s})(1\,\text{m/s}) = 10\,\text{N}.$$

$$P = Fv = 10\,\text{N}\,(1\,\text{m/s}) = 10\,\text{watts}$$

To get the sand up to velocity, friction must act and energy is dissipated. The rate at which kinetic energy is added to the sand is

$$\frac{d}{dt}\left(\frac{1}{2}\,mv^2\right) = \frac{1}{2}\frac{dm}{dt}\,v^2 = \frac{1}{2}\left(\frac{dm}{dt}\,v\right)v = \frac{1}{2}\,Pv$$

The rest of the energy is dissipated in friction.

Example 13

A 300 g ball is thrown a horizontal distance of 80 m as shown in Fig. 8-5. The initial projection angle is 458. The thrower's hand is in contact with the ball for 0.3 s. Find the average force on the ball during contact with the thrower. Neglect air resistance.

Solution:

Refer to Fig. 8-5, where two coordinate systems will be used.

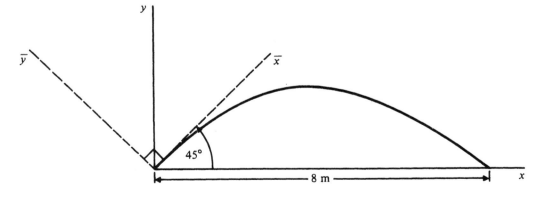

Figure 8-5

183

To discuss the motion after the ball leaves the thrower's hand, consider the conventional (untilted) coordinate system. We have

$$x = v_{0x}t = \frac{v_0 t}{\sqrt{2}} \qquad v_{0x} = v_0 \cos 45° = \frac{v_0}{\sqrt{2}}$$

$$y = v_{0y}t - \frac{1}{2} gt^2 = \frac{v_0 t}{\sqrt{2}} - \frac{1}{2} gt^2$$

Setting y = 0 we can solve for the time t of impact,

$$t = \frac{2v_0}{\sqrt{2} \; g}$$

yielding the range

$$x = \frac{v_0 t}{\sqrt{2}} = \frac{v_0}{\sqrt{2}} \frac{2v_0}{\sqrt{2} \; g} = \frac{v_0^2}{g}$$

The initial velocity is thus

$$v_0 = (gx)^{1/2}$$

Let's now zoom in on the motion before the ball leaves the thrower's hand. We now change to the rotated (tilted) coordinate system (Fig. 8-5) and consider the impulse (J). The impulse necessary to give the ball this initial velocity (in the rotated x direction) is

$$J = \int_0^t F_{\bar{x}} \, dt = \left(F_{\bar{x}} \right)_{av} \Delta t = \Delta p = mv_0$$

$$\left(F_{\bar{x}} \right)_{av} = \frac{mv_0}{\Delta t} = \frac{(0.3 \text{ kg}) \sqrt{(9.8 \text{ m/s}^2)(80 \text{ m})}}{(0.3 \text{ s})}$$

$$= 28 \text{ N}$$

where Δt is time the thrower's hand is in contact with the ball. Note this is really two problems, using two different coordinate systems. Each coordinate system is chosen to simplify the description of one specific part of the motion.

Example 14

A 0.5 kg mass of putty is hurled at a velocity of 1 m/s against a 0.5 kg mass attached to a spring as shown in Fig. 8-6. The mass attached to the spring slides across the frictionless horizontal surface, depressing the spring a maximum distance of 10 cm. Find the spring constant if the putty *sticks* to the mass on the spring.

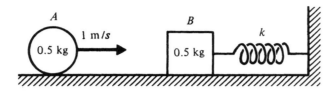

Figure 8-6

Solution:

"Sticks" indicates the collision is inelastic, but momentum is conserved during the interval from just before the collision to just after the collision, <u>before</u> the external force of the spring begins to act, i.e., before the spring is compressed appreciably. So we are dividing the motion into two intervals. The first interval is momentum conserving but non-energy conserving as the collision takes place.

$$v_{A1} = 1 \text{ m/s} \qquad\qquad v_{B1} = 0$$

$$m_A = 0.5 \text{ kg} = m_B \qquad v_{A2} = v_{B2} = v_2$$

the conservation of momentum relation is

$$m_A v_{A1} = (m_A + m_B)v_2 = 2m_A v_2$$

$$v_2 = v_{A1}/2 = 0.5 \text{ m/s}$$

Now consider the second interval. Just after the collision, mechanical energy is conserved, the dissipative forces between the putty and the mass B having ceased to act. The initial energy, just after the collision is

$$E_1 = \frac{1}{2}\left(m_A + m_B\right)v_2^{\,2}$$

$$E_1 = \frac{1}{2}\left(0.5 \text{ kg} + 0.5 \text{ kg}\right)\left(0.5 \text{ m/s}\right)^2 = 0.125 \text{ J.}$$

When the velocity of the putty stuck to the mass becomes zero and the spring is maximally compressed,

185

$$E_2 = \frac{1}{2}kx^2.$$

Conserving energy, we have

$$\frac{1}{2}kx^2 = 0.125 \text{ J}$$

$$k = \frac{2(0.125 \text{ J})}{(0.1 \text{ m})^2} = 25 \text{ J/m}^2 = 25 \text{ N/m}.$$

The key here was to recognize that an inelastic collision was followed by a conservation of energy situation.

Example 15

A 135 kg defensive lineman traveling 5 m/s makes a completely inelastic collision with an 85 kg quarterback at rest. They are observed to slide 2 m on wet grass.
 (a) What is the coefficient of friction between the grass and the players?
 (b) What energy is dissipated in the collision?

Solution:

 Letting

m_A = 135 kg	v_{A1} = 5 m/s	$v_{A2} = v_{B2} = v_2$
m_B = 85 kg	v_{B1} = 0	

the momentum conservation condition is

$$m_A v_{A1} = (m_A + m_B)v_2$$

where v_2 is the velocity of the two players after the collision. Thus we have

$$v_2 = \frac{m_A}{m_A + m_B} v_{A1} = \frac{(135 \text{ kg})}{(135 \text{ kg} + 85 \text{ kg})} (5 \text{ m/s})$$

$$= 3.07 \text{ m/s}$$

(Although the two players are not a completely isolated system, the external forces are so small compared to the forces that they exert on each other, that conservation of momentum is a reasonable approximation.)

From now on one can proceed in various ways to the coefficient of friction. We choose to use the work-energy relation, which here implies that the work of the frictional force, $-\mu$ n s, is equal to the increase in mechanical energy from the moment just after collision to the end of the slide,

$$- \mu n\, s = -\mu \left(m_A + m_B \right) gs = 0 - \frac{1}{2}\left(m_A + m_B \right) v_2^{\,2}$$

or

$$\mu = \frac{v_2^{\,2}}{2gs} = \frac{\left(3.07\,\mathrm{m/s} \right)^2}{2\left(9.8\,\mathrm{m/s^2} \right)\left(2\,\mathrm{m} \right)}$$

$$= .24$$

The energy before collision is

$$E_1 = \frac{1}{2}\, m_A v_{A1}^{\,2} = \frac{1}{2}\left(135\,\mathrm{kg} \right)\left(5\,\mathrm{m/s} \right)^2 = 1687\,\mathrm{J}$$

The energy just after collision is

$$E_2 = \frac{1}{2}\left(m_A + m_B \right) v_2^{\,2} = \frac{1}{2}\left(135\,\mathrm{kg} + 85\,\mathrm{kg} \right)\left(3.07\,\mathrm{m/s} \right)^2 = 1037\,\mathrm{J}$$

The energy dissipated in the collision is thus

$$E_1 - E_2 = 1687 - 1037 = 650\,\mathrm{J}$$

Example 16

A body of mass m = 100 g with a velocity 10 cm/s hits another identical body at rest and the two recoil as shown in Fig. 8-7.

      (a) Find the velocity of each mass after the collision.
      (b) Is the collision elastic?
      (c) Find the dissipated energy.

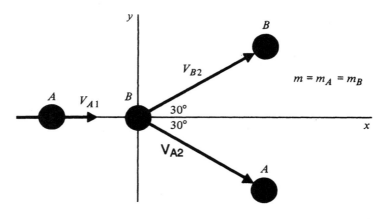

Figure 8-7

Solution:

This collision is in a plane rather than along a line as in the previous problems, so each component of momentum must be separately conserved.

The conservation of momentum yields $\vec{p}_1 = \vec{p}_2$, or in components,

    (x comp) $mv_{A1} = mv_{A2} \cos 308 + mv_{B2} \cos 308$

    (y comp) $0 = mv_{B2} \sin 308 - mv_{A2} \sin 308$

The last relation implies $v_{B2} = v_{A2}$. The first then yields

    $v_{A1} = 2v_{A2} \cos 308$

(a) $v_{A2} = \dfrac{v_{A1}}{(2 \cos 30°)} = 0.58\ v_{A1} = 5.8\ \text{cm/s}$

The initial energy is

    $E_1 = \dfrac{1}{2}\ mv_{A1}{}^2$

and the final energy is, using $v_{B2} = v_{A2}$,

$$E_2 = \frac{1}{2} mv_{A2}^2 + \frac{1}{2} mv_{B2}^2 = 2\left(\frac{1}{2} mv_{A2}^2\right)$$

$$E_2 = 2\left(\frac{1}{2} m(0.58\, v_{A1})^2\right) = 0.67\left(\frac{1}{2} mv_{A1}^2\right) = 0.67\, E_1$$

Thus the collision is inelastic and the dissipated energy is

$$E_1 - E_2 = (1 - 0.67)\left(\frac{1}{2} mv_{A1}^2\right)$$

$$E_1 - E_2 = (0.33)\left[\frac{1}{2}(0.1\text{ kg})(0.1\text{ m/s})^2\right]$$

$$= 1.65 \times 10^{-4}\text{ J}$$

Example 17

A bullet weighing .02 lb is fired with a muzzle velocity v = 2700 ft/s into a ballistic pendulum weighing 20 lbs. Find the maximum height through which it rises.

Solution:

Referring to Fig. 8-8,

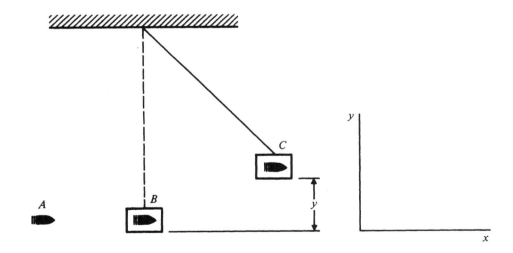

Figure 8-8

we let

m = bullet mass   M = block mass

v = velocity of bullet   V = velocity of bullet plus block just after collision

Consider the system of bullet plus block between time A when the bullet approaches the block to the time B just after collision when the system is about to start its upward swing. In this interval only internal forces act in the x direction and p_x is conserved

$$mv = (m + M)V; \text{ so that } V = \frac{m}{(m + M)} v$$

In this interval mechanical energy is not conserved; the collision of bullet and block is inelastic, energy being dissipated as the bullet drills the block.

Now consider the interval between B and C. During this time all forces are conservative (mg) or do no work (T). Thus mechanical energy is conserved,

$$E_1 = \frac{1}{2}(m + M)V^2 = (m + M)gy = E_2$$

Thus

$$y = \frac{V^2}{2g} = \frac{1}{2g}\left(\frac{m}{m + M}\right)^2 v^2$$

$$y = \frac{1}{2(32 \text{ ft/s}^2)}\left(\frac{0.02 \text{ lb/g}}{0.02 \text{ lb/g} + 20 \text{ lb/g}}\right)^2 (2700 \text{ ft/s})^2$$

$$= 0.11 \text{ ft.}$$

This is similar to Example 8-14, an inelastic collision, momentum conserving, followed by a conservation of energy problem.

Example 18

A 75 kg man stands on ice and shoots a machine gun at 120 shots per minute for a 10 s burst. The mass of each bullet is 10 g and the muzzle velocity is 800 m/s. Find the average force on the man, his velocity after the burst, and how far he has recoiled from his initial position.

Solution:

The impulse delivered to the man is,

$$J = \int F dt = F_{av} \Delta t = \Delta p,$$

which is equal in magnitude to the impulse delivered to the bullets,

$$F_{av} = \left[\frac{\Delta p}{\Delta t}\right]_{man} = \left[\frac{\Delta p}{\Delta t}\right]_{bullets}$$

In a time of 1 min there are 120 shots or 0.5 s per shot, a change $\Delta p = mv$ of momentum occurs where m is the bullet mass and v the muzzle velocity. This is approximately true when the man's velocity is small compared to the muzzle velocity. Thus

$$F_{av} = \frac{mv}{(0.5\ s)} = \frac{(0.010\ kg)(800\ m/s)}{(0.5\ s)} = 16\ N.$$

If M is the mass of the man, his acceleration is

$$a = \frac{F_{av}}{m} = \frac{(16\ N)}{(75\ kg)} = 0.21\ m/s^2$$

His velocity is thus

$$v = at = (0.21\ m/s^2)(10\ s) = 2.1\ m/s$$

and he recoils a distance

$$x = \frac{1}{2}at^2 = \frac{1}{2}(0.21\ m/s^2)(10\ s)^2 = 10.5\ m.$$

In the above solution certain approximations have been made. To make these explicit, consider the exact rocket equation

$$m\frac{dv}{dt} = v_r\frac{dm}{dt}$$

where in this case

m = mass of man + gun + unfired ammunition

v_r = velocity of the bullets with respect to the gun

If m >> mass of the ammunition, m $\cong$ M, the mass of the man and gun, and

$$F = M\frac{dv}{dt} = v_r\frac{dm}{dt} = (800\ m/s)\frac{(0.010\ kg)}{(0.5\ s)} = 16\ N$$

as before.

Example 19

A 100 kg man begins to walk at 1 m/s relative to the deck of a 200 kg boat. Their total momentum is zero. (Both are initially at rest.)
   (a) Find the velocity of the boat relative to the water.
   (b) Show that the center of mass of the man and the boat remains stationary.

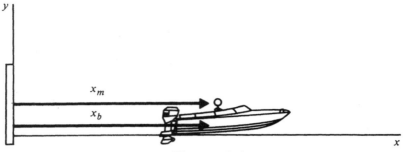

Solution: Figure 8-9

(a) Referring to Fig. 8-9, we see that the momentum of the system man plus boat is conserved if the friction between the boat and the water is neglected. Thus

$$m_B v_B + m_m v_m = 0$$

where m_B = mass of boat and m_m = mass of man. The velocity of the man relative to the boat is $(v_{mb} = v_{me} + v_{eb} = v_{me} - v_{be}$ where e = earth:)

$$v_m - v_B = 1 \text{ m/s}$$

   Substituting the last relation into the first condition, we have

$$m_B v_B + m_m(1 \text{ m/s} + v_B) = 0$$

or

$$v_b = -\frac{m_m}{m_b + m_m}(1 \text{ m/s}) = -\frac{(100 \text{ kg})}{(200 \text{ kg} + 100 \text{ kg})}(1 \text{ m/s}) = -0.33 \text{ m/s}$$

(As the man walks forward, the boat drifts backward.) If R is the center of the mass coordinate, we have

(b) $$m_m x_m + m_B x_B = R(m_B + m_m).$$

Solving for R and taking the time derivative we have

$$V = \frac{dR}{dt} = \frac{(m_b v_b + m_m v_m)}{m_b + m_m} = 0$$

by the conservation of momentum condition.

Example 20

A 5 kg mass traveling 1 m/s to the right makes a totally inelastic collision with a 3 kg mass traveling to the left at 5 m/s. Find the velocity of the center of mass of the system before and after the collision.

Solution:

Refer to Fig. 8-10:

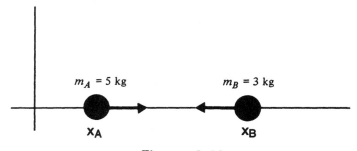

$m_A = 5$ kg $\qquad\qquad m_B = 3$ kg

x_A $\qquad\qquad\qquad\qquad\qquad x_B$

Figure 8-10

Before the collision, the center-of-mass position and velocity are

$$R_1 = \frac{\left(m_A x_A + m_B x_B\right)}{m_A + m_B}$$

$$V_1 = \frac{dR_1}{dt} = \frac{\left(m_A v_{A1} + m_B v_{B1}\right)}{m_A + m_B}$$

$$V_1 = \frac{\left[(5\text{ kg})(1\text{ m/s}) + (3\text{ kg})(-5\text{ m/s})\right]}{(5\text{ kg} + 3\text{ kg})}$$

$$= -1.25 \text{ m/s}$$

After the collision the masses stick together and their common velocity v_2 is given by
$(m_A + m_B)v_2 = m_A v_{A1} + m_B v_{B1}$

$$v_2 = \frac{\left(m_A v_{A1} + m_B v_{B1}\right)}{m_A + m_B} = v_1 = -1.25 \text{ m/s}$$

The common velocity is equal to the center of mass velocity (for a totally inelastic collision) which remains the same because no external forces act on the system.

Example 21

Find the center of mass coordinates (x,y) of the three mass points
      $m_1 = 1$ kg at (1 m, 2 m)
      $m_2 = 3$ kg at (0, 1 m)
      $m_3 = 4$ kg at (3 m, 1 m)
and plot the results.

Solution:

$$X = \frac{(m_1 x_1 + m_2 x_2 + m_3 x_3)}{m_1 + m_2 + m_3}$$

$$X = \frac{[(1\,\text{kg})(1\,\text{m}) + (3\,\text{kg})(0\,\text{m}) + (4\,\text{kg})(3\,\text{m})]}{(1\,\text{kg} + 3\,\text{kg} + 4\,\text{kg})} = 1.63\,\text{m}$$

$$Y = \frac{(m_1 y_1 + m_2 y_2 + m_3 y_3)}{m_1 + m_2 + m_3}$$

$$Y = \frac{[(1\,\text{kg})(2\,\text{m}) + (3\,\text{kg})(1\,\text{m}) + (4\,\text{kg})(1\,\text{m})]}{(1\,\text{kg} + 3\,\text{kg} + 4\,\text{kg})} = 1.13\,\text{m}$$

(See Fig. 8-11, where CM marks the center of mass)

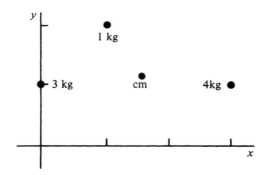

Figure 8-11

194

Example 22

Show that the coordinates of the center of mass of a composite object composed of parts A and B with masses m_A and m_B are given by

$$X_{AB} = \frac{(m_A X_A + m_B X_B)}{m_A + m_B}$$

$$Y_{AB} = \frac{(m_A Y_A + m_B Y_B)}{m_A + m_B}$$

Solution:

The sum of $m_i x_i$ over all the mass points composing the composite object may be split in the sum of $m_i x_i$ over the points in part A and the sum $m_i x_i$ over the points in part B,

$$X_{AB} = \frac{1}{M} \sum_i m_i x_i = \frac{1}{M} \sum_{Ai} m_i x_i + \frac{1}{M} \sum_{Bi} m_i x_i$$

$$X_{AB} = \frac{1}{M} (m_A X_A + m_B X_B) = \frac{(m_A X_A + m_B X_B)}{m_A + m_B}$$

with a similar derivation for Y_{AB}.

Example 23

The plate in Fig. 8-12a has a hole cut out of it at B. Find its center of mass.

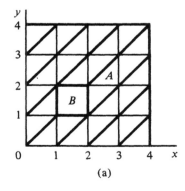 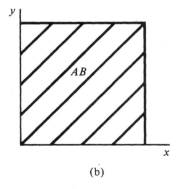 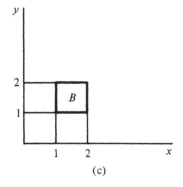

(a)   (b)   (c)

Figure 8-12

Solution:

The object A of Fig. 8-12a (slant shaded) may be considered a part of the composite object AB of Fig. 8-12b where B is the missing material of the square hole as shown in Fig. 8-12c.

$$X_{AB} = 2 \qquad Y_{AB} = 2$$

$$X_B = 1.5 \qquad Y_B = 1.5$$

From the previous problem, we find that

$$(M_A + M_B)X_{AB} = M_A X_A + M_B X_B$$

yielding

$$X_A = \frac{(M_A + M_B)}{(M_A)}(X_{AB}) - \left(\frac{M_B}{M_A}\right)(X_B)$$

$$X_A = \frac{(16)}{(15)}(2) - \left(\frac{1}{15}\right)(1.5) = 2.03$$

Similarly,

$$Y_A = 2.03$$

The center of mass of the plate is shifted slightly off center, away from the hole.

QUIZ

1. What is the momentum of a 1000 kg car travelling 60 km/hr?

Answer: 17,000 kg·m/s

2. A 1000 kg automobile travelling 60 km/hr brakes suddenly and the wheels lock as the car comes to a stop in 5 s.
    (a) What average force acted on the car?
    (b) What was the average acceleration?

Answer: (a) 3,300 N   (b) 3.3 m/s^2

3. A 7500 kg truck moving at 50 km/hr hits a stopped 3000 kg truck. They lock bumpers and skid to a stop.
    (a) What is the velocity of the two trucks just after the collision?
    (b) If the coefficient of friction is 0.25, how far do they skid?

Answer: 35 km/hr, 20 m

4. A bullet of mass 15 g strikes a ballistic pendulum of mass 3 kg. The center of mass of the pendulum rises a vertical distance of 10 cm. Find the velocity of the bullet.

Answer: 281 m/s

9

ROTATION OF RIGID BODIES

OBJECTIVES

In the next two chapters you will learn the laws of motion for rigid bodies and apply them to bodies rotating about an axis of fixed direction. This chapter concentrates on the description of motion using angular quantities and on rotational energy. Your objectives are to:

Master *radian measure* for the angle describing a body's orientation. Convert radians to revolutions.

Describe rotational motion by use of *angular velocity* and *angular acceleration*. Convert rpm to rev/s to rad/s.

Solve problems involving *constant angular acceleration*. This will be closely analogous to material of chapter 2. The methods will be the same. The difference is that you will be dealing with the angular quantities θ, $d\theta/dt = \omega$, and $d\omega/dt = \alpha$ instead of the linear quantities x, dx/dt = v, and dv/dt = a.

Relate *linear quantities to angular quantities*. Here you will apply some of the circular motion material of chapters 3 and 6.

Solve problems involving *rotational energy* and the *conservation* of rotational energy plus translational energy for simple coupled systems.

Calculate the *moment of inertia* I of a rigid body by use of integration and the *parallel axis theorem*, for simple geometries such as square plates and cylinders.

REVIEW

The orientation of a rigid body rotating about an axis of fixed direction is described by a single angle θ. It will be crucial in some of the equations of this chapter that θ be measured in radians (rad),

$$\theta = \frac{s}{r}$$

where s is the arc subtended by θ on a circle of radius r. A complete revolution (rev) is given by an arc of length $2\pi r$. Thus

$$1 \text{ rev} = 360° = 2\pi \text{ rad}$$

$$1 \text{ rad} = \frac{360°}{2\pi} = 57.3° = \frac{\text{rev}}{2\pi}$$

The angular velocity is

$$\omega = \frac{d\theta}{dt}$$

and may be given in rad/s, deg/s, rev/s, or rpm (= rev/min). Useful conversions are

$$1 \text{ rev/s} = 2\pi \text{ rad/s}$$

$$1 \text{ rpm} = \frac{\text{rev}}{60 \text{ s}} = \frac{2\pi \text{ rad/s}}{60 \text{ s}}$$

The angular acceleration is given by

$$\alpha = \frac{d\omega}{dt} = \omega\frac{d\omega}{d\theta} = \frac{d^2\theta}{dt^2}$$

and has units rev/s^2 = 2π rad/s^2, etc.

If the *acceleration α is constant*, the basic equations of rotational motion are

$$\theta = \theta_0 + \omega_0 t + \frac{1}{2}\alpha t^2 \qquad \theta(t = 0) = \theta_0$$

$$\omega = \omega_0 + \alpha t \qquad \omega(t = 0) = \omega_0$$

$$\omega^2 = \omega_0^2 + 2\alpha(\theta - \theta_0)$$

The relation between linear and angular quantities is

$$s = r\theta \qquad (\theta \text{ in rad})$$

where s is the arc subtended by a point in a rigid body a distance r from the axis as the body rotates through θ. This point has speed

$$v = \frac{ds}{dt} = r\frac{d\theta}{dt} = r\omega \qquad \left(\omega \text{ in rad/s}\right)$$

Its component of acceleration, a_t, along the arc is

$$a_t = \frac{dv}{dt} = r\frac{d^2\theta}{dt^2} = r\alpha \qquad \left(\alpha \text{ in rad/s}^2\right)$$

Note radian measure must be used in the last three formulas. From the results of chapter 5, the centripetal acceleration toward the axis is

$$a_{rad} = \frac{v^2}{r} = \omega^2 r$$

The rotational counterpart to mass is the moment of inertia I,

$$I = \Sigma \, mr^2 = \int (r^2)dm = \int (r^2)\rho \, dV$$

where the sum is over all points in the body and r is the perpendicular distance to the axis of rotation. I depends on the axis of rotation. If the body is continuous, I must be evaluated by integration. If the body has uniform density,

$$I = \rho \int \left(r^2\right) dV = \frac{M}{V} \int \left(r^2\right) dV$$

Moments of inertia of various simple solids are given in the text.

The parallel axis theorem helps in calculating moments of inertia. If I_{cm} is the moment of inertia about the cm and I_p is the moment about a parallel axis a distance d from the cm,

$$I_p = I_{cm} + Md^2$$

where M is the body's mass.

PROBLEM-SOLVING STRATEGIES

Review the problem-solving strategies of the main text. These stress that all the old rules and hints still apply. The problems of this chapter are very similar to those of chapters 2-8. The difference is that they involve angular rather than linear quantities. As before, decide on a positive direction and stick to it. List all pertinent known and unknown quantities. Typical problems encountered involve:

Radian measure, conversions, and the relation between angular and linear quantities. See examples 1 and 2 below.

Equations of constant angular acceleration. See examples 3 and 4 below.

Finding the radial and tangential acceleration of a point in a rigid body. See examples 5 and 6.

Calculation of moments of inertia. See examples 7 through 9 below.

Rotational work and power, and the conservation of energy, including rotational energy for conservative systems. See examples 10 and 11.

Application of the parallel axis theorem. See Example 12.

QUESTIONS AND ANSWERS

QUESTION. A rigid body rotates about a fixed axis. Is the speed of any point proportional to the point's perpendicular distance to the axis? Explain.

ANSWER. Yes. The rigid body has a single angular velocity (v) so that the speed (magnitude of velocity) of <u>any</u> point P is equal to the product of R_P v, where R_P is the perpendicular distance from P to the axis of rotation.

QUESTION. What is the approximate speed of a point at the Equator on the Earth's surface?

ANSWER. Taking the radius to be R_E = 4000 miles and the period of rotation to be T_E = 24 hours, the speed at the Equator is approximately v = $(2\pi R_E)/T_E$ = 24000 miles/24 hours = 1000 MPH!

EXAMPLES AND SOLUTIONS

Example 1

Referring to Fig. 9-1, find the angle θ in degrees and radians, and the arc length s'.

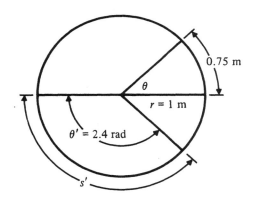

Figure 9-1

Solution:
This is basic for rotation--you need to talk the language of radians, degrees, and translate from angular to linear quantities!

$$\theta = \frac{s}{r} = \frac{(0.75\,\text{m})}{(1\,\text{m})} = 0.75\,\text{rad}$$

$$\theta = (0.75 \text{ rad})\frac{(360°)}{(2\pi \text{ rad})} = 43°$$

s' = θ'r = 2.4(1 m) = 2.4 m

The equation s = θr is true only when θ is measured in radians.

Example 2

A phonograph record rotates at 33(1/3) rpm (revolutions per minute). Find its angular velocity in rad/s and the speed of a point 5 cm from the center. (A phonograph record, formerly used for recording music, is a plastic circular disk that rotates about a fixed perpendicular axis.)

Solution:

1 rev = 2π rad

$$33\frac{1}{3} \text{ rpm} = \omega = \left(33\frac{1}{3} \text{ rev/min}\right)\left(\frac{1 \text{ min}}{60 \text{ s}}\right)\left(\frac{2\pi \text{ rad}}{\text{rev}}\right)$$

ω = 3.5 rad/s = 3.5 s⁻¹

$v = \omega r = (3.5 \text{ s}^{-1})(5 \text{ cm}) = 17.5 \text{ cm/s}.$

Example 3

The record in the previous problem starts from rest and accelerates with constant angular acceleration to 33(1/3) rpm in 2 s. Find the angular acceleration and the angle turned through, in degrees and revs.

Solution:

For constant angular acceleration we have:
$$\theta = \theta_0 + \omega_0 t + \frac{1}{2}\alpha t^2$$

$\omega = \omega_0 + \alpha t$

In this problem, the initial parameters are: θ = 0 and ω = 0. Thus we have

$$\alpha = \frac{\omega - \omega_0}{t} = \frac{3.5 \text{ rad/s}}{2 \text{ s}} = 1.75 \text{ rad/s}^2$$

$$\theta = \frac{1}{2} \alpha t^2 = \frac{1}{2} (1.75 \text{ rad/s}^2)(2 \text{ s})^2 = 3.5 \text{ rad}$$

$$\theta = (3.5 \text{ rad}) \frac{(360^\circ)}{(2\pi \text{ rad})} = 200^\circ$$

$$\theta = (3.5 \text{ rad}) \frac{(1 \text{ rev})}{(2\pi \text{ rad})} = 0.56 \text{ rev}$$

Example 4

The record player in the previous problem is shut off and the record decelerates uniformly from 33(1/3) rpm to a stop in 2 complete revolutions. Find the angular acceleration and the time it takes to come to a stop.

Solution:

$$\theta = \theta_0 + \omega_0 t + \frac{1}{2} \alpha t^2 \qquad \theta_0 = 0$$

$$\omega = \omega_0 + \alpha t \qquad \omega_0 = 33\frac{1}{3} \text{ rpm}$$

$$\omega^2 = \omega_0^2 + 2\alpha(\theta - \theta_0) \qquad \omega = 0 \text{ at } \theta = 2 \text{ rev}$$

The last equation is most convenient to use because all terms but α are known:

$$\alpha = \frac{\omega^2 - \omega_0^2}{2(\theta - \theta_0)} = \frac{-\omega_0^2}{2\theta} = \frac{-\left(33\frac{1}{3} \text{ rpm}\right)^2}{2(2 \text{ rev})}$$

$$\alpha = -278 \text{ rev/min}^2 = -(278 \text{ rev/min}^2)\left[\left(\frac{2\pi \text{ rad}}{\text{rev}}\right)\left(\frac{1 \text{ min}}{60 \text{ s}}\right)^2\right]$$

$$= -0.485 \text{ rad/s}^2 = -0.077 \text{ rev/s}^2$$

From the middle equation

$$t = \frac{\omega - \omega_0}{\alpha} = \frac{-\omega_0}{\alpha} = \frac{-\left(33\frac{1}{3} \text{ rpm}\right)}{(-278 \text{ rev/min}^2)}$$

$$= 0.12 \text{ min} = 7.2 \text{ s}$$

Example 5

Find, in Example 3, the components of the linear acceleration of a point on the edge of a 12 inch record

    (a) 1 s after the turntable is turned on, and

    (b) 3 s after the turntable is turned on. (α = 0 after t = 2s)

Solution:

The linear acceleration has two components, tangential a_t and radial, a_{rad}.

(a) $\quad \omega = \omega_0 + \alpha t \qquad\qquad \omega_0 = 0$

$\quad\quad a_t = \alpha r \qquad\qquad \alpha = 1.75$ rad/s²

$\quad\quad a_{rad} = \dfrac{v^2}{r} \qquad\quad v^2 = \omega^2 r^2 = (\alpha t r)^2$

$\quad\quad r = 6$ in

Thus

$\quad\quad a_t = (1.75 \text{ rad/s}^2)\cdot(6 \text{ in}) = 10.5 \text{ in/s}^2.$

$$a_{rad} = \frac{(\alpha t r)^2}{r} = (1.75 \text{ rad}/s^2)^2 (1\text{ s})^2 (6\text{ in})$$

$$= 18.4 \text{ in/s}^2$$

(b) For t > 2 s, the angular acceleration is zero, a_t = 0, and

$$a_{rad} = \frac{v^2}{r}$$

The angular velocity at 3 s is the same as the velocity at t = 2 s when 33(1/3) rpm was reached,

$$\omega = \alpha t = (1.75 \text{ rad/s}^2)(2 \text{ s}) = 3.5 \text{ rad/s}$$

$$v = r\omega = (6 \text{ in})(3.5 \text{ rad/s}) = 21 \text{ in/s}$$

$$a_{rad} = \frac{(21 \text{ in}/s)^2}{(6\text{ in})} = 73 \text{ in}/s^2$$

Note the relations mixing angular and linear quantities, $a_t = \alpha r$, $v = \omega r$, $s = \theta r$, involve angular quantities using radian measure.

Example 6

A flywheel is a disk-shaped mass rotating about a central perpendicular axis. Such a "flywheel", 1.0 m in diameter rotating with an initial velocity of 500 rpm is rotating at 1000 rpm after 20 s. Assuming constant acceleration find
     (a) the angular acceleration,
     (b) the angle through which the flywheel rotates between 500 and 1000 rpm, and
     (c) the acceleration of a point on the rim.

Solution:

$$\theta = \theta_0 + \omega_0 t + \frac{1}{2}\alpha t^2 \qquad \omega_0 = 500 \text{ rpm}$$

$$\omega = \omega_0 + \alpha t \qquad \omega = 1000 \text{ rpm}$$

$$\omega^2 = \omega_0^2 + 2\alpha(\theta - \theta_0)$$

(a) From the middle equation, we find

$$\alpha = \frac{\omega - \omega_0}{t} = \frac{(1000 \text{ rpm} - 500 \text{ rpm})}{(20 \text{ s})}$$

$$\alpha = \frac{(500 \text{ rev/min})}{(20 \text{ s})}\left[\frac{1 \text{ min}}{60 \text{ s}}\right] = 0.42 \text{ rev/s}^2$$

$$\alpha = \left(0.42 \text{ rev/s}^2\right)\left[\frac{60 \text{ s}}{1 \text{ min}}\right]^2$$

$$= 1{,}510 \text{ rev/min}^2$$

(b) Using this last result, we find

$$\theta - \theta_0 = \frac{\omega^2 - \omega_0^2}{2\alpha} = \frac{(1000 \text{ rev/min})^2 - (500 \text{ rev/min})^2}{2\left(1510 \text{ rev/min}^2\right)}$$

$$= 248 \text{ rev}$$

(c) $a = \alpha r = (0.42)(2\pi \text{ rad/s}^2)(0.5 \text{ m})$

$$= 1.32 \text{ m/s}^2.$$

Note again in the forms $a = \alpha r$ and $v = \omega r$, α must be in rad/s^2 and ω must be in rad/s.

Example 7

Four small bodies of mass 0.5 kg are connected by light rods and form the shape of a square 1 m on a side. Find the moment of inertia of the system about an axis
    (a) Perpendicular to the plane of the square and passing through one mass.
    (b) Passing through two adjacent masses.
    (c) Passing through two opposite masses.

Solution:

Referring to Fig. 9-2,

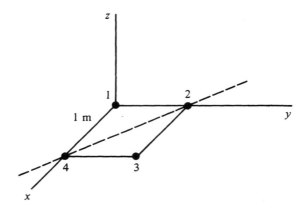

Figure 9-2

we first consider case (a) and take the axis to be the z-axis, finding

$$I_a = \Sigma \; mr^2 = m(r_1^2 + r_2^2 + r_3^2 + r_4^2)$$

where r is the *perpendicular distance* of each mass to the z-axis:

$$r_1 = 0; \qquad\qquad r_2 = \text{side} = 1 \text{ m};$$

$$r = \text{diagonal} = 1.41 \text{ m}; \qquad r_4 = \text{side} = 1 \text{ m}.$$

Thus we have

$$I_a = (0.5 \text{ kg})[(1 \text{ m})^2 + (1.4 \text{ m})^2 + (1 \text{ m})^2]$$

$$= 2 \text{ kg·m}^2.$$

For case (b) we take the axis to be the x-axis; then

$r_1 = 0;$ $r_2 = 1$ m; $r_3 = 1$ m; $r_4 = 0.$

$$I_b = (0.5 \text{ kg})[(1 \text{ m})^2 + (1 \text{ m})^2]$$

$$= 1 \text{ kg·m}^2.$$

For case (c) the axis is the dashed line in Fig. 9-2:

$$r_1 = (1/2)\text{diagonal} = 0.71 \text{ m} = r_3$$

$$r_2 = r_4 = 0$$

$$I_c = (0.5 \text{ kg})[(0.71 \text{ m})^2 + (0.71 \text{ m})^2]$$

$$= 0.5 \text{ kg·m}^2.$$

Example 8

Find the moment of inertia of a thin rectangular uniform sheet of metal, of mass M and dimension L = length by w = width, about the x axis in Fig. 9-3.

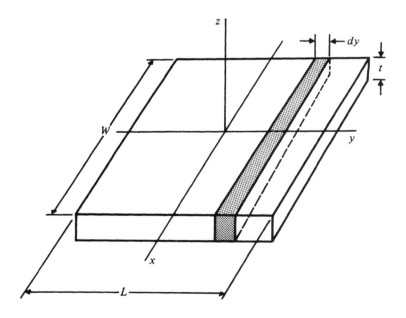

Figure 9-3

Solution:

Referring to Fig. 9-3, we sub-divide the plate into thin strips all parts of which have the same perpendicular distance r = y to the x axis. If the plate has thickness t,

$$\rho = \frac{M}{wLt} = \text{mass per unit volume}$$

dV = volume of strips = wt dy

$$I = \rho \int (r^2) \, dV = \rho \int wt (y^2) \, dy$$

$$I = \frac{M}{wLt} \, wt \int_{-L/2}^{+L/2} (y^2) \, dy = \frac{M}{L} \left[\frac{y^3}{3} \right]_{-L/2}^{+L/2} = \frac{ML^2}{12}$$

Example 9

A rod of mass M = 1 kg and length L = 1 m connects 2 small, m = 1 kg masses. Find the moment of inertia of the composite system about an axis through the center, perpendicular to the rod.

Solution:

According to the text the moment of inertia of the *rod* is

$$I_{rod} = \frac{ML^2}{12}$$

(This result could have been inferred from Example 8.) To find the total moment we must add the moment of inertia of the masses at the ends,

$$I_{masses} = m\left(\frac{L}{2}\right)^2 + m\left(\frac{L}{2}\right)^2 = 2m\left(\frac{L}{2}\right)^2 = \frac{mL^2}{2}$$

$$I_{system} = I_{rod} + I_{masses} = \frac{ML^2}{12} + \frac{mL^2}{2}$$

Since M = m in this example,

$$I_{system} = mL^2\left(\frac{1}{12} + \frac{1}{2}\right) = \frac{7}{12} \, mL^2 = 0.583 \, mL^2$$

$$= (0.583)(1 \text{ kg})(1 \text{ m})^2$$

$$= 0.583 \text{ kg·m}^2$$

Example 10

How much energy is dissipated when a 2 kg grinding wheel of radius 0.1 m is brought to rest from an initial velocity of 3000 rpm?

Solution:

The energy dissipated is the kinetic energy of the wheel,

$$K = \frac{1}{2} I \omega^2 \qquad I = \frac{1}{2} MR^2 \quad (\text{solid disk})$$

$$\omega = 3000 \text{ rpm} = (3000 \text{ rev/min}) \left[\frac{2\pi \text{ rad}}{\text{rev}} \right] \left[\frac{1 \text{ min}}{60 \text{ s}} \right];$$

Substituting the given values for M and R;

$$K = \frac{1}{2} \left[\frac{1}{2} (2 \text{ kg})(0.1 \text{ m})^2 \left(\frac{3000 \cdot 2\pi \text{ rad}}{60 \text{ s}} \right)^2 \right]$$

$$K = 493 \text{ J}.$$

Example 11

What is the average power dissipated in the last example if the grinding wheel is brought to rest in 10 rev? Assume constant angular acceleration.

Solution:

The average power dissipated is

$$P_{av} = \frac{\Delta E}{\Delta t} = \frac{\Delta K}{\Delta t}$$

and we must calculate the time it takes the wheel to stop. First, from the initial data we calculate the acceleration:

$$\omega^2 = \omega_0{}^2 + 2(\theta - \theta_0)\alpha; \qquad \omega = 0$$

$$\theta - \theta_0 = 10 \text{ rev}; \qquad \omega_0 = 3000 \text{ rpm}$$

$$\alpha = \frac{-\omega_0{}^2}{2(\theta - \theta_0)} = \frac{-(3000 \text{ rpm})^2}{2(10 \text{ rev})} = -4.5 \times 10^5 \text{ rev/min}^2$$

Then, knowing the acceleration and the initial and final velocities, we calculate the stopping time,

$$\omega = \omega_0 + \alpha t$$

$$t = \frac{\omega - \omega_0}{\alpha} = \frac{-\omega_0}{\alpha} = \frac{-(3000 \text{ rpm})}{(-4.5 \times 10^5 \text{ rev/min}^2)}$$

$$= 6.7 \times 10^{-3} \text{ min} = 0.40 \text{ s}$$

The average power dissipated is

$$\frac{\Delta K}{\Delta t} = \frac{493 \text{ J}}{0.40 \text{ s}} = 1.23 \text{ kW}$$

Example 12

The moment of inertia of a uniform circular disk about an axis through its center and perpendicular to the plane of disk is $I = 1/2(MR^2)$, where M is the mass and R the radius of the disk. Find the moment of inertia of the disk about an axis perpendicular to the disk through an edge of the disk (a distance R from the center of the disk.)

Solution:

By the parallel axis theorem, we have

$$I_p = I_{cm} + Md^2$$

where in this case d = R. Thus

$$I_p = \frac{1}{2} MR^2 + MR^2 = \frac{3}{2} MR^2$$

QUIZ

1. What is the angular speed of the earth about its axis in rad/s?

Answer: 7.3×10^{-5} rad/s.

2. A 33 and 1/3 rpm phonograph record takes 3 s to get up to speed. Find the average angular acceleration in rad/s^2.

Answer: 1.16 rad/s^2

3. A flywheel in the shape of a solid disk has a mass of 5 kg and a radius of 5 cm. Find its kinetic energy when it rotates at 100 rpm.

Answer: 0.34 J

4. A mass of 1 kg hangs by a rope wrapped around a pulley of moment of inertia 2 kg·m^2 and radius 50 cm. If the mass starts from rest, find its acceleration and how far it falls after a time of 1 s.

Answer: 1.1 m/s^2, 0.54 m

10
DYNAMICS OF ROTATIONAL MOTION

OBJECTIVES

In this second chapter on rotation, you will learn the laws of motion for rigid bodies and apply them to bodies rotating about an axis of fixed direction. This chapter is a direct application to rotational motion of methods you developed in chapters 2 through 8. Your objectives are to:

Develop the concept of torque and obtain the equations of motion for a rigid body rotating about a fixed axis.

Solve problems involving *torques* acting on bodies rotating about an axis of fixed direction.

Develop the ideas of angular momentum, the conservation of angular momentum, and impulse.

Solve problems involving *angular impulse* and *angular momentum*, using methods similar to those of chapter 8.

REVIEW

As shown in Fig. 10-1, a force $\vec{F}$ on a body has a *point of application* and a *line of action.* The *moment* or *torque* of $\vec{F}$ about an arbitrary point 0 is

$$\Gamma_0 = FL = Fr \sin \theta = F_n r$$

where $\dot{F}_n$ is the projection of $\vec{F}$ perpendicular to $\vec{r}$ (see Fig. 10-1).

L is called the lever or moment arm of $\vec{F}$. The torque Γ_0 is counted *positive* if it tends to make the body rotate counterclockwise, *negative* otherwise. The torque of $\vec{F}$ in Fig. 10-1 is positive.

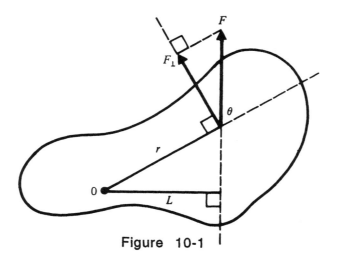

Figure 10-1

A more general definition of a moment is given by the vector

$$\vec{\Gamma}_0 = \vec{r} \times \vec{F}$$

whose magnitude is

$$\Gamma_0 = rF \sin \theta = F_n r$$

The direction of Γ_0 is perpendicular to $\vec{r}$ and $\vec{F}$. By *the right hand rule*, it points out of the paper for the example in Fig. 10-1. (Curl the fingers of your *right* hand in the direction from $\vec{r}$ to $\vec{F}$, the shorter way; your thumb points in the direction of $\vec{r} \times \vec{F} = \Gamma_0$.)

The work done by a torque Γ moving a body through an angle $\Delta\theta$ is

$$W = \Gamma\Delta\theta$$

and the power is

$$P = \Gamma\omega$$

Newton's Laws applied to rigid body motion about a fixed axis imply that

$$\Sigma \, \Gamma = I \, \alpha$$

where the sum over Γ is the total torque about the fixed axis, I is the moment of inertia, and α the angular acceleration.

The angular momentum L of a point mass moving in a circle of radius r with velocity v is

$$L = mvr = m\omega r^2.$$

If these are added up for the points in a rotating rigid body, we find the total angular momentum is

$$L = I\omega$$

The angular impulse of a torque is

$$J_\theta = \int_{t_1}^{t_2} \Gamma\, dt = \int_{t_1}^{t_2} I\alpha\, dt = \int_{t_1}^{t_2} I \frac{d\omega}{dt}\, dt = \int_{\omega_1}^{\omega_2} I\, d\omega$$

$$= I\omega_2 - I\omega_1 = L_2 - L_1 = \Delta L$$

and yields the change in angular momentum. In differential form we find

$$\frac{dL}{dt} = I\alpha$$

or

$$\sum \Gamma = I\alpha = \frac{dL}{dt} = I\frac{d\omega}{dt} = I\frac{d^2\theta}{dt^2}$$

If the total external torque on a system is zero, $\sum \vec{\Gamma} = 0$, the total angular momentum is conserved, $\Delta \vec{L} = 0$.

PROBLEM-SOLVING STRATEGIES

Refer to the problem-solving strategies of the main text. They stress again that all the old rules continue to apply; sketches, coordinate systems, and geometrical relations are still vital parts of problem solving. Again, the problems of this chapter(as in Chapter 9) are very similar to those of chapters 2-8. The difference is that they involve angular rather than linear quantities. As before, decide on a positive direction and stick to it. List all pertinent known and unknown quantities. Typical problems encountered involve:

Finding torques of forces. See examples 1 and 2.

Applications of $\Gamma = I\alpha$ to simple and coupled systems. Conservation of energy is also useful here. In the coupled systems there is often a relation between the linear coordinate of one component and the angular coordinate of the other component. See Examples 3 through 8 below.

Angular impulse and the conservation of angular momentum. See Examples 9 through 11.

QUESTIONS AND ANSWERS

QUESTION. A mechanical scale balance is off center because one arm is 1% shorter than the other but it hangs in equilibrium because one cup is heavier than the other. Will the scale give accurate weight measure when used with a set of accurately calibrated weights?

ANSWER. No. The torques will be unequal even when equal weights are placed in the two cups.

QUESTION. Pull horizontally on the end of a thread wrapped around a spool lying on the table, with the thread coming from the underside of the string. Does the spool tend to move toward or away from the pull? Why?

ANSWER. The spool of thread will move in the direction of the pull. The center-of-mass goes in the direction of the applied force.

EXAMPLES AND SOLUTIONS

Example 1

Find the sum of the moments, $\Sigma \Gamma_0$, of the forces $\vec{F}$ and $\vec{F}'$ shown in Fig. 10-2a about the point 0, in terms of the lever arms h and h'.

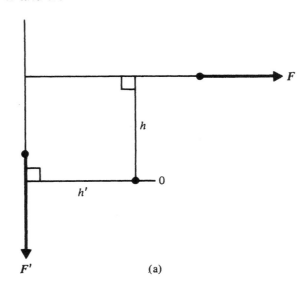

(a)

Figure 10-2a

215

Solution:

This basic question tests to see if you can find the magnitude and direction of the torque due to simple forces.

$$\Sigma \Gamma_0 = -Fh + h'F'$$

The moment of F about 0 is negative because F tends to produce a clockwise rotation about 0; correspondingly the moment of F' about 0 is positive because it tends to cause a counterclockwise rotation about 0. In this case, it was most convenient to calculate the moments in terms of the lever arms h and h' because they were directly given.

Example 2

Find the sum of the moments about 0 of the forces $\vec{F}$ and $\vec{F}'$ shown in Fig. 10-2b.

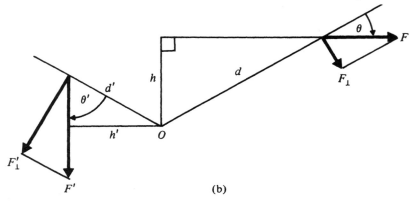

(b)

Figure 10-2b

Solution:

This is a slightly more complicated version of Example 1 because the forces have a more complex geometry.

$$\Sigma \Gamma_0 = -F (\sin \theta) d + F'(\sin \theta')d'$$

$$= -F_\perp d + F_\perp 'd' = -Fh + F'h'$$

In this example the same forces as in Example 1 were oriented by giving the distance between the points of application and the origin, and the angle between this line and the force. The convenient way to calculate the moment is to find the component of each force perpendicular ($F_\perp$) to the line d' and d. Since $h = d \sin \theta$ and $h' = d'\sin \theta'$, the result is the same. To make further progress in rotational dynamics problems, make sure you can calculate the torque of forces as in Examples 1 and 2.

216

Example 3

In example 11 of Chapter 9, what torque is necessary to produce the calculated acceleration of the grinding wheel?

Solution:

$$\Gamma = I\alpha; \quad \text{and } I = \frac{1}{2} MR^2$$

From Example 11 (Chapter 9),

$$\alpha = -4.5 \times 10^5 \text{ rev/min}^2 = -\left(4.5 \times 10^5\right)\left(2\pi \text{ rad}\right)\left(60 \text{ s}\right)^{-2}$$

$$= -785 \text{ rad/s}^2$$

$$\Gamma = \frac{1}{2} MR^2 \alpha = -\frac{1}{2}(2 \text{ kg})(0.1 \text{ m})^2\left(785 / s^2\right)$$

$$= -7.85 \text{ N·m.}$$

The negative sign indicates that the torque opposes the flywheel's motion, tending to slow it down.

Example 4

In the last example, what work is done by the torque Γ as the grinding wheel is brought to rest?

Solution:

$$\theta = 10 \text{ rev} = 10(2\pi) \text{ rad}$$

$$W = \Gamma\theta = -7.85 \text{ N·m } (10 \cdot 2\pi \text{ rad})$$

$$= -493 \text{ J.}$$

The work of Γ is negative because the angular displacement is opposite to the direction of Γ. The work done by Γ is equal to the *increase* in kinetic energy of the wheel; since this energy is decreased, W is negative.

Example 5

In the last example, what is the instantaneous power dissipated at t = 0 as the grinding wheel starts to slow down?

Solution:

The initial velocity is 3000 rpm. Thus

$$P = \Gamma\omega = \Gamma\omega_0 = -(7.85 \text{ N} \cdot \text{m})\left[\frac{(3000 \text{ rev}/\text{min})(2\pi \text{ rad}/\text{rev})}{(60 \text{ s}/\text{min})}\right]$$

$$P = -2.46 \text{ kW}$$

The initial power *dissipated* is 2.46 kW, twice the average power dissipated as calculated in Example 11.

Example 6

In the last example, what tangential force must be applied to the rim of the grindstone to yield the torque of Example 3?

Solution:

$$\Gamma = Fr$$

$$F = \frac{\Gamma}{r} = -\frac{(7.85 \text{ N} \cdot \text{m})}{(0.1 \text{ m})} = -78.5 \text{ N}$$

Example 7

A rod pivoted at an end has mass M = 1 kg and length L = 0.5 m. Neglect the friction at the pivot. It is released from rest in a horizontal position. What is its angular velocity when it is vertical?

Solution:

The system is conservative and mechanical energy is conserved. The potential and kinetic energies are, referring to Fig. 10-3, and taking U = 0 at the initial horizontal position,

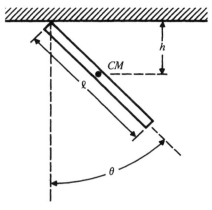

Figure 10-3

$$U = Mgh = -Mg\left(\frac{L}{2}\cos\theta\right) = \text{potential energy of CM.}$$

$$K = \frac{1}{2}I\omega^2 \qquad I = \frac{1}{3}ML^2$$

The initial energy E_i is

$$E_i = K_i + U_i = 0 \qquad \left(\theta_i = \frac{\pi}{2}; \omega_i = 0\right)$$

and the final energy E_f is

$$E_f = K_f + U_f = \frac{1}{2}I\omega_f^2 - \frac{MgL}{2} = E_i = 0$$

The final angular velocity is thus

$$\omega_f = \sqrt{\frac{MgL}{I}} = \sqrt{\frac{3g}{L}} = 7.67 \text{ rad/s}$$

Example 8

A cord is wrapped around the rim of a uniform flywheel of radius r = 0.2 m and mass M = 10 kg. A mass of m = 10 kg is suspended from the cord 10 m above the floor.
   (a) How much time does it take the mass to hit the floor?
   (b) What is the velocity when the mass hits the floor?
   (c) What is the tension in the rope?

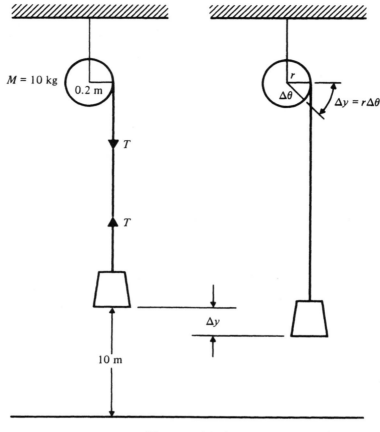

Figure 10-4

Solution:

Referring to Fig. 10-4, we first solve the problem for the general case m ≠ M.

The equation of motion of the flywheel is

$$\Gamma = I\alpha \qquad I = \frac{1}{2}Mr^2 \qquad \Gamma = Tr$$

and the equation of motion of the mass m is

mg - T = ma

where a is the downward acceleration of the mass. (Note T ≠ mg. The equality is true only in equilibrium.)

The key to solving this problem is to recognize that there is a relation between the linear acceleration a of the mass and the angular acceleration α of the flywheel. If a length Δy of cord is pulled from the flywheel it moves through an angle

$$\Delta\theta = \frac{\Delta y}{r}$$

while the mass drops a distance Δy. Thus

$$\frac{\Delta\theta}{\Delta t} = \frac{1}{r}\frac{\Delta y}{\Delta t} \quad \text{or } \omega = \frac{v}{r}$$

implying

$$\Delta\omega = \frac{\Delta v}{r} \quad \text{or} \quad \frac{\Delta\omega}{\Delta t} = \frac{1}{r}\frac{\Delta v}{\Delta t} = \frac{a}{r} = \alpha$$

yielding

$$\alpha = \frac{a}{r}$$

The equation of motion for the flywheel tells us

$$\alpha = \frac{\Gamma}{I} = \frac{(Tr)}{(Mr^2/2)} = \frac{2T}{Mr} = \frac{a}{r}$$

or

$$T = \frac{Ma}{2}$$

The equation of motion for the weight then may be solved for the acceleration, a, by substituting the above value of T in the equation of motion form, $mg - T = ma$,

$$mg - \frac{Ma}{2} = ma$$

$$a = \frac{g}{1 + \dfrac{M}{2m}}$$

When m = M, we have a = 2/3 g. In this case we have for the falling mass,

$$y = y_0 + v_0 t + \frac{1}{2}at^2 \quad \text{where} \quad y_0 = 0 \quad \text{and} \quad v_0 = 0$$

$$y = \frac{1}{2}at^2$$

$$v = \frac{dy}{dt} = at$$

$$v^2 - v_0^2 = 2ay \qquad v_0 = 0$$

From the last equation we have

$$v = \sqrt{2\,ay} = \sqrt{2\left(\frac{2}{3}\right)(9.8 \text{ m/s}^2)(10 \text{ m})} = 11.43 \text{ m/s}$$

The time to hit the floor is

$$t = \frac{v}{a} = \frac{v}{(2g/3)} = \frac{3(11.43 \text{ m/s})}{2(9.8 \text{ m/s}^2)} = 1.74 \text{ s}$$

Finally, the tension in the rope is

$$T = \frac{Ma}{2} = \left(\frac{M}{2}\right)\left(\frac{2g}{3}\right) = \left(\frac{Mg}{3}\right) = \frac{(10 \text{ kg})(9.8 \text{ m/s}^2)}{3}$$

$$= 32.7 \text{ N}$$

Another approach to this problem is through energy conservation. The initial energy, E_i, is

$$E_i = K_i + U_i$$

$$E_i = \frac{1}{2}I\omega_i^2 + \frac{1}{2}mv_i^2 + mgh_i$$

$$= \text{rotational plus translational energy} = mgh_i \; ; \; h_i = 10 \text{ m}, \; v_i = 0, \; \omega_i = 0.$$

The final energy is

$$E_f = \frac{1}{2}I\omega_f^2 + \frac{1}{2}mv_f^2 + mgh_f$$

$$E_f = \frac{1}{2}I\left(\frac{v_f}{r}\right)^2 + \frac{1}{2}mv_f^2 + mgh_f; \quad h_f = 0$$

$$E_f = \frac{1}{2}\left(\frac{1}{2}Mr^2\right)\left(\frac{v_f}{r}\right)^2 + \frac{1}{2}mv_f^2$$

$$E_f = \frac{1}{4}Mv_f^2 + \frac{1}{2}mv_f^2 = E_i = mgh_i$$

Taking M and m numerically equal, we have

$$v_f = \sqrt{4\frac{gh_i}{3}} = \sqrt{4\frac{(9.8 \text{ m/s}^2)(10 \text{ m})}{3}} = 11.4 \text{ m/s}$$

Example 9

A baggage carousel has a mass of M = 500 kg and is approximately a disk of radius R = 2 m. It is freely rotating at an angular velocity ω = 1 rad/s when 10 pieces of baggage with masses of M' = 20 kg a piece are dropped on the carousel a distance R = 2 m from the axis of rotation.

(a) Assuming no external torques act on the system of carousel plus baggage, what is the final angular velocity?

(b) What is the energy of system before and after the 10 pieces of baggage are added?

Solution:

Angular momentum is conserved because only internal forces act. The initial angular momentum is

$$L_i = I_c \omega_{ci} = \left(\frac{1}{2} MR^2\right)\omega_{ci}$$

where M is the carousel mass I_c its moment of inertia, and ω_{ci} the initial carousel velocity.

The final angular momentum is

$$L_f = I_c \omega_{cf} + 10 \, M'(R)^2 \omega_{cf}$$

Note this is the angular momentum of carousel plus baggage where ω_{cf} is the final angular velocity.

By the conservation of angular momentum,

$$L_i = L_f$$

$$I_c \omega_{ci} = I_c \omega_{cf} + 10M'(R)^2 \omega_{cf}$$

$$\left(\frac{1}{2} MR^2\right)\omega_{ci} = \left(\frac{1}{2} MR^2\right)\omega_{cf} + \left(10 \, M'R^2\right)\omega_{cf}$$

The final angular velocity is

$$\omega_{cf}\left[\left(\frac{1}{2} MR^2\right) + \left(10 \, M'R^2\right)\right] = \left(\frac{1}{2} MR^2\right)\omega_{ci}$$

$$\omega_{cf} = \left(\frac{1}{1 + 20 \, M'/m}\right)\omega_{ci}$$

$$\omega_{cf} = \left(\frac{1}{1 + 20\,(20/500)}\right)\omega_{ci} = 0.56 \, \text{rad}/s$$

(b) The initial energy is

$$E_i = K_i = \frac{1}{2} I_c \omega_{ci}^2 = \frac{1}{2} \left(\frac{1}{2} MR^2 \right) \omega_{ci}^2$$

$$E_i = \left(\frac{1}{4} MR^2 \right) \omega_{ci}^2 = \frac{1}{4} (500 \text{ kg})(2 \text{ m})^2 (1 \text{ rad/s})^2 = 500 \text{ J}$$

The final energy is

$$E_f = K_f = \frac{1}{2} I_c \omega_{cf}^2 + \frac{1}{2} I_b \omega_{cf}^2$$

with I_b the moment of inertia of the baggage,

$$I_b = 10 \text{ M'R2}$$

Thus

$$E_f = \frac{1}{2} I_c \omega_{cf}^2 + \frac{1}{2} \left(10 \text{ M'R}^2 \right) \omega_{cf}^2$$

$$E_f = \frac{1}{2} \left(\frac{1}{2} MR^2 \right) \omega_{cf}^2 + \frac{1}{2} \left(10 \text{ M'R}^2 \right) \omega_{cf}^2$$

$$E_f = \left[\frac{1}{4} (500 \text{ kg})(2 \text{ m})^2 + 5 (20 \text{ kg})(2 \text{ m})^2 \right] (0.56 \text{ rad/s})^2$$

$$E_f = 282 \text{ J}$$

The collision did not conserve energy; as the baggage dropped on the carousel, dissipation occurred.

Example 10

A block of mass 0.1 kg is attached to a cord passing through a hole in a horizontal frictionless surface. The block is originally rotating in a circle at a distance 0.2 m from the hole with an angular velocity 7 rad/s. A force P pulls on the cord and it is shortened to 0.1 m.

    (a) What is the new angular velocity?
    (b) What is the work done by the force P?

Solution:

Here we ask you to provide the sketch--it's good practice.

    (a) Since there is no external torque (P has no moment about the hole) angular momentum is conserved. The initial angular momentum is

$$L_i = I_i\omega_i = MR_i^2\omega_i$$

The final angular momentum is

$$L_f = I_f\omega_f = MR_f^2\omega_f$$

By conservation of momentum,

$$MR_i^2\omega_i = MR_f^2\omega_f$$

$$\omega_f = \left(\frac{R_i}{R_f}\right)^2 \omega_i = \left(\frac{0.2\,\text{m}}{0.1\,\text{m}}\right)^2 (7\,\text{rad}/\text{s})$$

$$= 28\,\text{rad/s}$$

    (b) The work W' done by the external force, P acting through the distance s, is equal to the increase in kinetic energy,

$$W' = K_f - K_i; \qquad s = 0.1\,\text{m}$$

$$W' = \frac{1}{2}I_f\omega_f^2 - \frac{1}{2}I_i\omega_i^2$$

$$W' = \frac{1}{2}\left(MR_f^2\right)\omega_f^2 - \frac{1}{2}\left(MR_i^2\right)\omega_i^2$$

Note however that

$$R_f^2 = R_i^2 \frac{\omega_i}{\omega_f}$$

and thus

225

$$W' = \frac{1}{2}\left(MR_i^2 \frac{\omega_i}{\omega_f}\right)\omega_f^2 - \frac{1}{2}\left(MR_i^2\right)\omega_i^2$$

$$W' = \frac{1}{2}\left(MR_i^2\right)\left(\frac{\omega_f}{\omega_i} - 1\right)\omega_i^2$$

$$W' = \frac{1}{2}(0.1 \text{ kg})(0.2 \text{ m})^2\left(\frac{28}{7} - 1\right)(7 \text{ rad/s})^2 = 0.29 \text{ J}$$

Example 11

A turntable with moment of inertia I = 2000 kg•m2 makes one revolution every 5 s. A man of mass M = 100 kg standing at the center of the turntable runs out along a radius fixed in the turntable.

    (a) What is the angular velocity when he is R = 3 m from the center?

    (b) How much work does he do during the run?

Solution:

Since there are no external torques, total angular momentum is conserved. The initial angular momentum is ($\omega_i = 2\pi$ rad / 5 s)

$$L_i = I_i \omega_i = (2000 \text{ kg} \cdot \text{m}^2)\left(\frac{2\pi \text{ rad}}{5 \text{ s}}\right)$$

$$= 2513 \text{ kg·m2/s}$$

The final angular momentum is

$$L_f = \left(I_i + MR^2\right)\omega_f = \left(2000 \text{ kg} \cdot \text{m}^2 + 100 \text{ kg}(3 \text{ m})^2\right)\omega_f$$

$$= 2900 \text{ kg·m2 } \omega_f$$

The conservation of angular momentum yields, $L_i = L_f$, or

$$\omega_f = \frac{(2513 \text{ kg} \cdot \text{m}^2/\text{s})}{(2900 \text{ kg} \cdot \text{m}^2)} = 0.87 \text{ rad/s}$$

$$\omega_f = (0.87 \text{ rad/s})\frac{(1 \text{ rev})}{(2\pi \text{ rad})} = 0.14 \text{ rav/s}$$

$$= 1 \text{ rev every 7.25 s.}$$

The work done *by* the man, W', is equal to the *increase* in kinetic energy,

$$W' = K_f - K_i = \frac{1}{2} I_f \omega_f^2 - \frac{1}{2} I_i \omega_i^2$$

$$W' = \frac{1}{2}\left[(2000 \text{ kg} \cdot \text{m}^2) + (100 \text{ kg})(3 \text{ m})^2\right](0.87 \text{ rad/s})^2 - \frac{1}{2}(2000 \text{ kg} \cdot \text{m}^2)\left(\frac{2\pi}{5 \text{ s}}\right)^2$$

$$W' = \frac{1}{2}\left[(2000 \text{ kg} \cdot \text{m}^2) + (100 \text{ kg})(3 \text{ m})^2\right](0.87 \text{ s}^{-1})^2 - \frac{1}{2}\left[(2000 \text{ kg} \cdot \text{m}^2)\right]\left(\frac{2\pi}{5 \text{ s}}\right)^2$$

W'= 1097 J - 1579 J

= -482 J

Since the work is negative, work is done *on* the man as he runs out; if he tries to run in he will have to do work to get closer to the center.

QUIZ

1. A block of mass 0.5 kg is attached to a cord passing through a hole in a horizontal frictionless surface. The block originally rotates in a circle of radius 0.4 m with an angular velocity of 14 rad/s. The cord is then pulled so that the new radius is 0.2 m.
    (a) What is the new angular velocity?
    (b) What work is done when the cord is pulled?

Answer: 56 rad/s, 23 J.

2. A penny on edge is released into a conical bowl with an angular speed ω_0. It spirals down.
(a) Is angular momentum conserved? (b) Is energy conserved? (c) Estimate the angular speed when the penny's radius (distance from the center of the bowl) has halved.

Answer: (a) Yes, (b) No,(c) 4 ω_0.

3. Two identical disks are rotating on the same vertical shaft, one at angular speed ω_0 and the other at $2\omega_0$. The top disk is suddenly released and falls on the bottom disk and they stick together. (a) What is their final common angular speed? (b) Is energy conserved?

Answer: (a) $\omega_0/2$, (b) No.

11

EQUILIBRIUM AND ELASTICITY

OBJECTIVES

The first condition of equilibrium of a rigid body is that the total force acting on the body is zero:

$$\Sigma \vec{F} = 0.$$

This guarantees *translational equilibrium. The second condition of equilibrium* guarantees that there is no tendency to rotate. This is guaranteed by the torque condition (see below.) When forces and torques are applied to bodies they change their size and shape. Your objectives in this chapter are to:

Continue to gain proficiency at calculating the *moment* or *torque* of a force.

Apply the second condition of equilibrium to a variety of problems involving *static structures* and bodies in *rotational equilibrium*, such as see-saws or teeter-totters, booms supported by guy wires, and ladders against walls.

Calculate the position of the *center-of-gravity* of a rigid body.

Define *stress* and *strain* and evaluate their magnitudes for various applied force.

Define *pressure.*

Recognize elastic behavior and define the *elastic moduli* associated with elastic behavior.

Solve problems involving *Young's modulus* Y, *Poisson's ratio* σ, the *bulk modulus* B, and the *shear modulus* S.

REVIEW

Second Condition of Equilibrium

To insure static equilibrium of a rigid body, the total torque must vanish about any axis: 0, 0', etc.,

$$\Sigma G_0 = 0, \qquad \Sigma G_{0'} = 0, \text{ etc.}$$

This is a good time to review how to calculate a torque: see Examples 12 and 13 of Chapter 9.

Center-of-Gravity

The moment of the gravitational force is the sum of the moments of gravity acting on the individual mass points,

$$\vec{\Gamma} = \sum_i \vec{r}_i \times \vec{F}_i = \sum_i r_i \times m_i \vec{g}$$

where $\vec{g}$ is a vector pointing down, of magnitude g.

Thus

$$\vec{\Gamma} = \Sigma_i m_i \vec{r}_i \times \vec{g} = M\vec{R} \times \vec{g} = \vec{R} \times \vec{W}$$

$$= \vec{R} \times \vec{W}$$

where R is the center-of-mass position vector previously defined, and

$$\vec{W} = \Sigma_i \vec{w}_i = M\vec{g}$$

is the total weight of the body.

If a body is considered as composed of two parts, *a stress is the force per unit area* of one part of the body on the other part, at the boundary layer, a plane. The force may be normal to the plane, in which case the stress is

$$\text{Normal stress} = \frac{F_n}{A} \ ,$$

or parallel to the plane, in which case the stress is

$$\text{Shear stress} = \frac{F_t}{A} \ .$$

If the body is being squeezed together at the plane, the stress is *compressive*. If the body is being pulled apart, the stress is *tensile.* The units of stress are the units of pressure, which we discuss next: force per unit area.

A body at rest and submerged in a fluid at rest feels a normal stress of compression, independent of direction, called hydrostatic pressure,

$$p = \frac{F}{A} \ ,$$

measured in units of $N/m^2 = Pa$. The average pressure of the earth's atmosphere at sea level is equal to one atmosphere,

$$1 \text{ atm} = 1.013 \times 10^5 \text{ Pa} = 14.7 \text{ lb/in}^2.$$

The strain of a body is the relative change of the dimensions of the body when subjected to stress. It is always a dimensionless quantity.

If ΔL is the change of length of a wire of initial length L_0 subjected to tensile stress, the strain is

$$\text{Tensile strain} = \frac{\Delta L}{L_0} .$$

If a body is subjected to shear stress, it tends to change its shape by altering an angle by a small amount ϕ, and the strain is

$$\text{Shear strain} = \phi \text{ (radians)}.$$

The forces responsible for hydrostatic pressure produce a volume change ΔV of a submerged object of volume V,

$$\text{Volume strain} = \frac{\Delta V}{V} .$$

If stress (force per unit area producing strain) is proportional to strain (deformation) we are in the *elastic* regime where Hooke's law is valid and *elastic moduli* are useful:

$$\text{Young's modulus} = Y = \frac{\text{tensile stress}}{\text{tensile strain}} = \frac{\text{compressive stress}}{\text{compressive strain}} = \frac{F_n/A}{\Delta L/L_0}$$

When a body stretches it also becomes smaller in width w. The fractional change in width is proportional to the fractional change in length:

$$\frac{\Delta w}{w} = -\sigma \frac{\Delta L}{L_0} ,$$

where σ is called Poisson's ratio.

$$\text{Shear modulus} = S = \frac{\text{shear stress}}{\text{shear strain}} = \frac{F_t/A}{\phi}$$

$$\text{Bulk modulus} = B = -\frac{\Delta p}{\Delta V/V_0} = -V_0 \left(\frac{\Delta p}{\Delta V} \right).$$

The compressibility is the inverse bulk modulus, $k = (1/B) = B^{-1}$.

In the elastic range the force necessary to stretch a spring a distance x is $F = kx$. The equal and opposite force of the spring is $F = -kx$. k is the spring constant.

PROBLEM-SOLVING STRATEGIES

Review the problem-solving strategies of the main text, we will constantly be referring to them in the examples below. The equilibrium problems of Chapter 5 included extended bodies but were chosen so that the moment condition was automatically satisfied. Referring to the problem-solving strategy section of the study guide, *add the condition that the total moment about any point must vanish*. After writing down the first conditions, write down the second condition *about as many points* as you wish until you have as many independent equations as there are unknowns.

As illustrated below, the equilibrium equations are not always independent; there will often be a choice as to what combinations of the two components of the first equilibrium condition and the various second equilibrium conditions you use to find a solution.

As always, painstaking care must be exercised with respect to the *signs of the moments of the forces*: In Fig. 11-1, for illustration, forces A and B have positive moments about 0; C and D have negative moments about 0. What are the signs of the moments of the forces A, B, C, D about 0'? (Negative, positive, positive, positive, respectively.)

A choice of coordinate system and a free-body diagram are always useful. If a force is resolved into components, then each component may have a torque about a given point.

A convenient choice of the point O about which moments are taken will often simplify the moment condition equation.

Refer to the text, Tables 11-1, 11-2, and 11-3 for useful elastic constants.

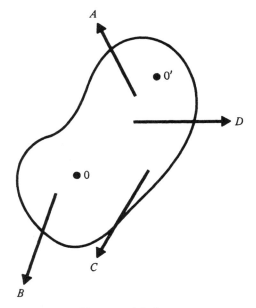

Figure 11-1

QUESTIONS AND ANSWERS

QUESTION. A body satisfies the first and second conditions of equilibrium. Is it necessarily at rest?

ANSWER. No. It could be in a state of translational motion with a constant (but non-zero) center-of-mass velocity <u>and</u> in a state of rotational motion with constant angular velocity and still satisfy the first and second conditions of equilibrium.

QUESTION. A ladder is placed against a rough wall with its foot on a frictionless floor. When released, can it remain at rest if the coefficient of friction between the ladder and the wall is large enough?

ANSWER. No. If there is no friction with the floor, there can be no "reaction" force at the wall to serve as a normal force for a vertical friction force along the wall.

EXAMPLES AND SOLUTIONS

Example 1

Show that if a body is in equilibrium under the action of three coplanar forces, no two of which are parallel, their lines of action must meet at a point.

Solution:

This is an important condition, often useful in many problems. Referring to Fig. 11-2, we write the first condition of equilibrium as $\vec{A} + \vec{B} + \vec{C} = 0$. The lines of action of any two of the three forces, say $\vec{B}$ and $\vec{C}$, must meet at a point 0, as shown in the Fig. 11-2:

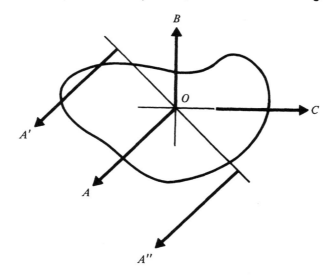

Figure 11-2

For translational equilibrium, the vector $\vec{A}$ must be equal to $- \vec{B} - \vec{C}$ but its line of action is as yet unknown, as indicated by the parallel vectors of equal magnitude, $\vec{A}, \vec{A}', \vec{A}''$. The second condition requires that all moments vanish, in particular those about 0, $\Sigma \vec{\Gamma}_0 = 0$. $\vec{B}$ and $\vec{C}$ have zero moment about 0; thus the moment of A about 0 must vanish. This is possible only if its lever arm vanishes, i.e., if its line of action passes through 0.

Note the importance of the sketch in this problem and the indication of the point about which torques are computed.

Example 2

The case excluded above, when the forces are parallel is illustrated in the classic playground question: A 20 kg boy sits on one end of a 3 m see-saw (teeter-totter). His 30 kg sister wishes to be in balance with him. Where should she adjust the pivot?

Solution:

Following the problem-solving strategy of the main text, the first step is to draw a free-body diagram showing all the forces acting on the see-saw, and choose a convenient coordinate system. Referring to Fig. 11-3, we see that

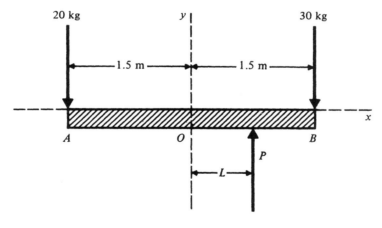

Figure 11-3

the first condition, $\Sigma F_y = 0$, implies

$$P = A + B$$

The second condition takes various forms, according to the point about which moments are calculated. When the sum of all the forces acting on a body is zero, the torque is the same for any point. You can choose any point.

About A, we have

$$0 = \Sigma\ G_A = (1.5\ m + L)P = (3\ m)B.$$

About 0, we have

$$0 = \Sigma\ G_0 = (1.5\ m)\ A + LP - (1.5\ m)\ B.$$

About P, we have

$$0 = \Sigma\ G_P = (1.5\ m + L)A - (1.5\ m - L)B.$$

About B, we have

$$0 = \Sigma\ G_B = (3\ m)\ A - (1.5\ m - L)P.$$

Note that A, P and B were convenient points because only two of the three forces have moments there. Any two of the moment equations may be used to solve for the unknowns P and L. For example, the third yields

$$\frac{1.5\ m + L}{1.5\ m - L} = \frac{B}{A} = \frac{3}{2}\ ; \quad L = 0.3\ m$$

Then the first yields

$$P = \frac{(3\ m)\ B}{1.5\ m + L} = \frac{3}{1.8}\ (30\ kg) = 50\ kg$$

and the second yields

$$P = \frac{(1.5\ m)\ B - (1.5\ m)\ A}{L} = \frac{(1.5\ m)(30\ kg) - (1.5\ m)(20\ kg)}{(0.3\ m)}$$

$$= 50\ kg$$

Thus of the four moment equations plus the $\Sigma\ F_y = 0$ condition, only two are independent. One could as well choose any other independent two to find the unknowns P and L. Again, refer to the problem-solving strategies in the main text and note how we have applied them here.

Example 3

A 150 kg uniform diving board 4 m long is mounted as in Fig. 11-4a. Find the forces acting on the board when a 100 kg man is standing on its end.

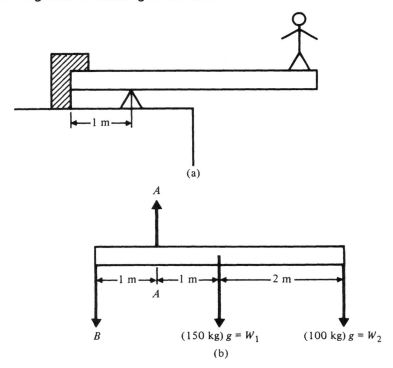

(a)

(b)

Figure 11-4

Solution

Referring to the free-body diagram of Fig. 11-4b, we see that the first condition of equilibrium yields

$$0 = \Sigma F_y = A - B - W_1 - W_2$$

Taking moments about B, we have

$$0 = \Sigma G_B = (1 \text{ m})A - (2 \text{ m})W_1 - (4 \text{ m})W_2$$

yielding

$$A = 2W_1 + 4W_2 = 2(150 \text{ kg})(9.8 \text{ m/s}^2) + 4(100 \text{ kg})(9.8 \text{ m/s}^2)$$

235

$$A = 6860 \text{ N}$$

Combining this result with the first condition yields

$$B = A - W_1 - W_2$$

$$= 6860 \text{ N} - (150 \text{ kg})(9.8 \text{ m/s}^2) - (100 \text{ kg})(9.8 \text{ m/s}^2)$$

$$= 4410 \text{ N}$$

The result can be checked by calculating the total moment about A,

$$\Sigma G_A = (1 \text{ m})B - (1 \text{ m})W_1 - (3 \text{ m})W_2$$

$$= (1 \text{ m})(4410 \text{ N}) - (1 \text{ m})(150 \text{ kg})(9.8 \text{ m/s}^2) - (3 \text{ m})(100 \text{ kg})(9.8 \text{ m/s}^2)$$

$$= (4410 - 1470 - 2940)\text{N·m}$$

$$= 0,$$

verifying that it vanishes.

Example 4

The weightless horizontal strut in Fig. 11-5a supports a 300 N weight as shown. Find the tension in the supporting cable and the force of the wall against the strut.

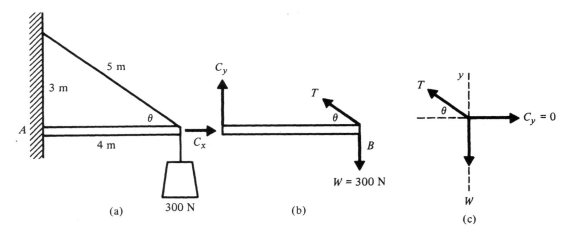

Figure 11-5

236

Solution:

The first step is to draw a free-body diagram showing all forces acting on the strut as illustrated in Fig. 11-5B, and refer them to a convenient coordinate system, Fig. 11-5c. C_x and C_y are the rectangular components of the force *of* the wall *on* the strut. Since only three forces act on the strut (C, T, W) all their lines of action must meet at a single point, as proved in Example 1. Thus the line of action of C passes through the end of the strut where T and W act, implying $C_y = 0$. Alternatively, taking moments about B, we have

$$0 = \Sigma \ G_B = - C_y \cdot 4m \qquad C_y = 0$$

leading to the same result. The first condition of equilibrium then yields (see Fig. 11-5c)

$$0 = \Sigma \ F_x = C_x - T \cos \theta = \ C_x - \frac{4}{5} T$$

$$0 = \Sigma \ F_y = T \sin \theta - W = \frac{3}{5} \ T - W,$$

with the solutions,

$$T = \frac{5}{3} \ W = \frac{5}{3} \ (300 \ N) \ = 500 \ N$$

$$C_x = \frac{4}{5} \ T = \frac{4}{5} \ (\ 500 \ N) \ = 400 \ N$$

Example 5

Find the magnitude and direction of the force exerted by the wall on the uniform strut in Fig. 11-6a if the strut weighs 100 N.

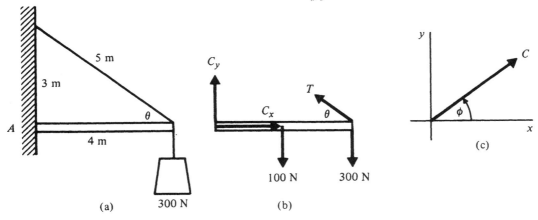

Figure 11-6

Solution:

237

In this case there are more than three forces acting on the strut and they need not meet at a single point. After drawing a free body diagram and coordinate axes (see Fig. 11-6b) we write down the first conditions of equilibrium,

$$0 = \sum F_x = C_x - T\cos\theta = C_x - \frac{4}{5}T$$

$$0 = \sum F_y = C_y + T\sin\theta - (300\text{ N}) - (100\text{ N})$$

$$0 = C_y + \frac{3}{5}T - (400\text{ N})$$

A convenient point to take moments about for the second condition of equilibrium is A because only one unknown, T, will be involved,

$$0 = \sum G_A = 4\text{ m}\cdot T\sin\theta - 2\text{ m}\cdot 100\text{ N} - 4\text{ m}\cdot 300\text{ N}$$

yielding (sin θ = 3/5)

$$T = \frac{200\text{ N}\cdot\text{m} + 1200\text{ N}\cdot\text{m}}{(4\text{ m})(3/5)} = 583\text{ N}$$

The first conditions then yield

$$C_x = \frac{4}{5}T = \frac{4}{5}(583\text{ N}) = 467\text{ N}$$

$$C_y = (400\text{ N}) - \frac{3}{5}T = (400\text{ N}) - \frac{3}{5}(583\text{ N}) = 50\text{ N}$$

In the free-body sketch of Fig. 11-6b we evidently guessed correctly the directions of C_x, C_y. Had we reversed these directions in Fig. 11-6b and calculated the equilibrium conditions accordingly, our answers would have come out negative for C_x and C_y, signaling that their directions are opposite to those originally assumed.

It is often wise to check your results by taking moments about another point: about B we have

$$\sum G_B = -4\text{ m}\cdot C_y + 2\text{ m}\cdot(100\text{ N}) = -4\text{ m}\cdot 50\text{ N} + 2\text{ m}\cdot(100\text{ N}) = 0$$

Note C, T and the two weights are not concurrent. The angle θ between C and the horizontal is

$$\phi = \arctan\frac{C_y}{C_x} = \arctan\left(\frac{50\text{ N}}{467\text{ N}}\right) = 6.1°$$

Example 6

A ladder of mass M = 25 kg rests against a frictionless wall as shown in Fig. 11-7a. Find all the forces acting on the ladder.

Solution:

The first step is to draw the free-body diagram, as shown in Fig. 11-7b.

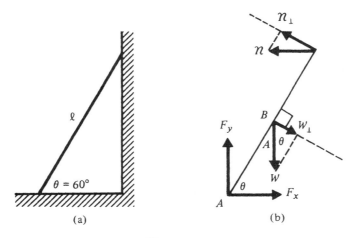

Figure 11-7

The unknown force F of the ground against the foot of the ladder has been resolved into its horizontal and vertical components. *Since the wall is frictionless, it only exerts a force on the ladder perpendicular to the wall, $\vec{n}$.* Also indicated on Fig. 11-7b is W, the projection of $W_\perp$ perpendicular to the lever arm or ladder. Note $W_\perp$ and $n_\perp$ are *not* extra forces; they are components of $\vec{W}$ and $\vec{n}$ useful for calculating moments.

The next step is to write down the conditions of equilibrium. The first conditions are

$$0 = \Sigma F_x = F_x - n$$

$$0 = \Sigma F_y = F_y - W$$

About the point A, the second condition of equilibrium is

$$0 = \Sigma \Gamma_A = \left(-\frac{L}{2} W_\perp\right) + Ln_\perp$$

$$0 = \Sigma \Gamma_A = \left(-\frac{L}{2} W \cos \theta\right) + Ln \sin \theta$$

Regarding F_x, F_y, and n as unknown in the three equations above, we solve the second one for F_y,

$$F_y = W = Mg = (25\ kg)\cdot(9.8\ m/s^2) = 245\ N,$$

the last one for n,

$$n = \frac{W \cos \theta}{2 \sin \theta} = \frac{W}{2 \tan \theta} = \frac{245\ N}{2(1.73)} = 71\ N$$

and finally the first for F_x,

$$F_x = 71\ N$$

To check these results, calculate the moments about C,

$$\sum \Gamma_C = F_x (\sin \theta) L - F_y (\cos \theta) L + W (\cos \theta)\left(\frac{L}{2}\right)$$

$$\sum \Gamma_C = L\left[(71\ N)(0.87) - (245\ N)(0.50) + (245\ N)(0.50)\left(\frac{1}{2}\right)\right] = 0$$

Note the length scale L is irrelevant in this problem.

Example 7

In the previous problem, what is the minimum coefficient of friction between the ladder and the ground which allows the ladder to stand without slipping?

Solution:

The force of friction is $f = F_x$ and the normal force at the ground is $n = F_y$. The condition

$$f \leq \mu_s\, n$$

is thus

$$F_x \leq \mu_s F_y$$

or

$$\mu_s \geq \frac{F_x}{F_y} = \frac{71\ N}{245\ N} = 0.29$$

Example 8

A boom of length L supported by a horizontal guy wire supports a weight of W = 200 N as shown in Fig. 11-8a. The boom also weighs W. Find the tension in the guy wire and the force exerted by the ground on the foot of the boom.

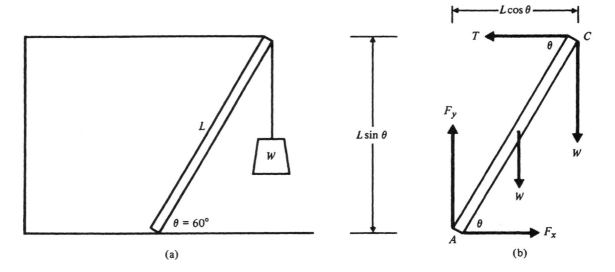

Figure 11-8a,b

Solution:

After sketching the free body diagram of Fig. 11-8b, accounting for all forces acting on the boom, we write down the first conditions of equilibrium,

$$0 = \Sigma\, F_x = F_x - T$$

$$0 = \Sigma\, F_y = F_y - W - W = F_y - 2W$$

and the second condition of equilibrium, taking moments about A,

$$0 = \Sigma\, \Gamma_A = LT \sin\theta - LW \cos\theta - \frac{1}{2} LW \cos\theta$$

$$0 = L\left(T \sin\theta - \frac{3}{2} W \cos\theta\right)$$

At this point it is perhaps well to pause, and review techniques for calculating moments. Indicated on Fig. 11-8b are the components of the ladder in the vertical and horizontal direction. The moment of T is positive because it tends to make the boom rotate

241

counterclockwise about A; its magnitude is given by

$$T_\perp L = (T\sin\theta)L = T(L\sin\theta) = TL_\perp$$

as indicated in Fig. 11-8c

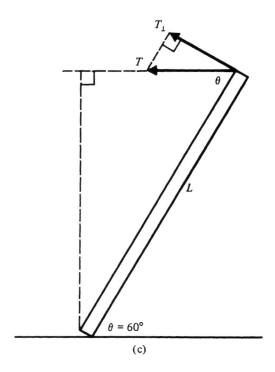

(c)

Figure 11-8c

Returning to the moment condition we have

$$T = \frac{3}{2}\frac{W\cos\theta}{\sin\theta} = \frac{3}{2}\frac{W}{\tan\theta} = \frac{3}{2}\frac{(200\text{ N})}{(1.73)} = 173\text{ N}$$

The first conditions then imply

$$F_x = 173\text{ N}$$

$$F_y = 2W = 400\text{ N}$$

To check that the solution is correct, verify that $0 = \Sigma\ G_B$.

Example 9

Find the minimum coefficient of friction between the weightless horizontal strut and the wall of the system in Fig. 11-9, if the strut is not to slip.

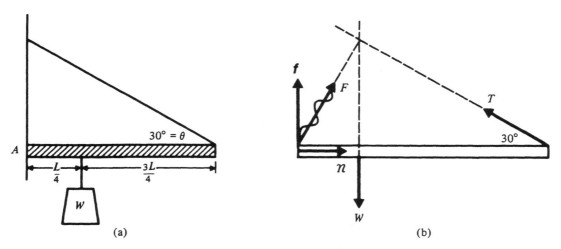

Figure 11-9

Solution:

f and n, as indicated on the free-body diagram of Fig. 11-9b, are the components of the force $\vec{F}$ of the wall on the strut. Since there are only three coplanar forces acting on the strut $(\vec{F}, \vec{W}, \vec{T})$ they must meet at a point as shown. We will not explicitly use that fact in the solution below, but the observation is useful in indicating the correct directions of $\vec{F}$ and $\vec{n}$.

The first conditions of equilibrium are

$$0 = \Sigma\ F_x = n - T \cos 30°$$

$$0 = \Sigma\ F_y = f + T \sin 30° - W$$

The second condition, taking moments about A, is

$$0 = \sum \Gamma_A = T(\sin 30°)L - \frac{1}{4} LW$$

$$T = \frac{1}{4}\ \frac{W}{(\sin 30°)}$$

From the first condition we then have

$$n = T \cos 30° = \frac{1}{4}\ \frac{W}{(\tan 30°)}$$

$$f = W - T \sin 30° = W - \frac{1}{4} W = \frac{3}{4} W$$

For no slipping we require

$$f < \mu_s n$$
$$\frac{3}{4} W < \mu_s \left(\frac{W}{4 \tan 30°} \right)$$
$$\mu_s > 3 \tan 30° = 1.73$$

Solve this problem when W hangs a distance λ from the wall, with $0 < \lambda < 1$. The answer is

$$\mu_s > \frac{1-\lambda}{\lambda} \tan 30°$$

___

Example 10

The guy wire on the hinged door of Fig. 11-10 is adjusted so that the top hinge exerts only a vertical force on the door. The hinges are set so that they share the weight of the 300 N door equally. Find the force exerted by each hinge. The hinges are 0.5 m from the top and bottom of the 2 m high door.

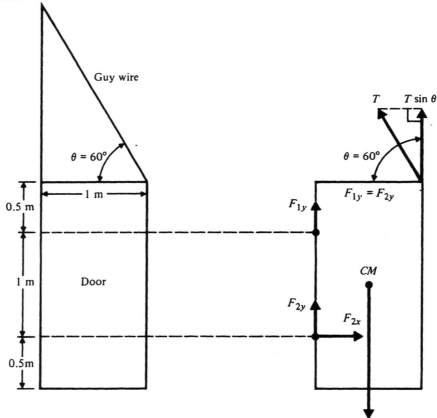

Figure 11-10

Solution:

Indicated on the free body diagram of Fig. 11-10b are the problem conditions that $F_{1y} = F_{2y}$ (equal sharing of weight by the hinges) and $F_{1x} = 0$ (no horizontal component of upper hinge.)

The first conditions of equilibrium are

$$0 = \Sigma F_x = F_{2x} - T \cos \theta$$

$$0 = \Sigma F_y = T \sin \theta + F_{1y} + F_{2y} - W$$

$$0 = T \sin \theta + 2F_{2y} - W$$

The second condition is, taking moments about the center-of-mass (CM),

$$0 = \Sigma G_{CM}$$

$$= T \sin \theta(0.5 \text{ m}) + T \cos \theta(1 \text{ m}) - (F_{1y} + F_{2y})(0.5 \text{ m}) + F_{2x}(0.5 \text{ m})$$

or since $F_{1y} = F_{2y}$,

$$0 = T \sin \theta + 2T \cos \theta - 2F_{2y} + F_{2x}$$

The underlined equations are three simultaneous equations in the unknowns T, F_{2x}, and F_{2y}. Solving the first for T and substituting the result in the other two yields [$T = F_{2x}/(\cos \theta)$]

$$0 = F_{2x} \tan \theta + 2F_{2y} - W$$

$$0 = F_{2x} \tan \theta + 2 F_{2x} - 2F_{2y} + F_{2x} = F_{2x} \tan \theta - 2F_{2y} + 3F_{2x}$$

These are two simultaneous linear equations in the unknowns F_{2x}, F_{2y}. Adding them yields

$$F_{2x} = \frac{W}{2 \tan \theta + 3} = \frac{(300 \text{ N})}{2 \tan 60° + 3} = 46.4 \text{ N}$$

Subtracting them yields

$$0 = 4F_{2y} - W - 3F_{2x}$$

$$F_{2y} = \frac{W + 3 F_{2x}}{4} = \frac{(300 \text{ N}) + 3(46.4 \text{ N})}{4} = 110 \text{ N}$$

245

Finally, the first underlined equation yields

$$T = \frac{F_{2x}}{\cos \theta} = \frac{(46.4 \text{ N})}{\cos 60°} = 93 \text{ N}$$

As a check, calculate the moment about the upper left corner and verify that it vanishes.

Example 11

Find the center-of-gravity of the uniform boomerang-like object in Fig. 11-11.

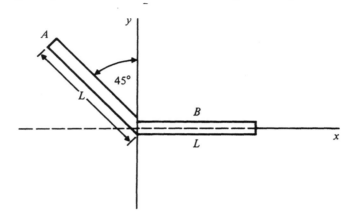

Figure 11-11

Solution:

Dividing the object into the pieces A and B, we have

$$M_A = M_B$$

$$X_A = -\frac{L}{2} \sin 45° = -\frac{L}{2} \frac{\sqrt{2}}{2}$$

$$Y_A = \frac{L}{2} \cos 45° = \frac{L}{2} \frac{\sqrt{2}}{2}$$

$$X_B = \frac{L}{2}$$

$$Y_B = 0$$

where (X_A, Y_A) and (X_B, Y_B) are the coordinates of the center-of-gravity of the two pieces. The coordinates (X, Y) of the whole object is

246

$$X = \frac{M_A X_A + M_B X_B}{M_A + M_B} = \frac{1}{2}\left(X_A + X_B\right) = \frac{1}{2}\left(-\frac{\sqrt{2}\,L}{4} + \frac{L}{2}\right)$$

$$X = \frac{L}{4}\left(1 - \frac{1}{\sqrt{2}}\right) = 0.07\,L$$

$$Y = \frac{M_A Y_A + M_B Y_B}{M_A + M_B} = \frac{1}{2}\left(Y_A + Y_B\right) = \frac{1}{2}\left(\frac{\sqrt{2}\,L}{4} + 0\right)$$

$$Y = \frac{1}{2}\left(\frac{\sqrt{2}\,L}{4}\right) = 0.18\,L$$

Example 12

A cubic box is dragged along a floor at constant velocity as shown in Fig. 11-12. The coefficient of friction is $\mu = 0.25$. Find the line of action of the normal force of the floor on the box.

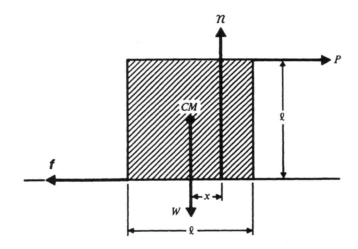

Figure 11-12

247

Solution:

From the free body diagram of Fig. 11-12, the first conditions of equilibrium are

$$0 = \Sigma F_x = P - f = P - \mu n$$

$$0 = \Sigma F_y = n - W$$

Combining these we find

$$P = \mu W$$

Taking moments about the CM, we have

$$0 = \Sigma \Gamma_{CM} = x\, n - \frac{L}{2}P - \frac{L}{2}f$$

$$0 = xW - \frac{1}{2}\mu WL - \frac{1}{2}\mu WL$$

or since W factors out;

$$x = L\mu = .25\,L$$

the point of application would need to be beyond the edge of the cube and it would tip. If the coefficient of friction $\mu \geq 0.5$, the box would tip. Why?

Example 13

A 10 kg weight is hung from a steel wire of unstretched length 1 m and diameter 2 mm. How much does the wire stretch, and what is its final diameter?

Solution:

We know the force (weight), the area and the unstretched length, so by looking up Young's modulus, Y, for steel, we can find the change in length:

$$Y = \frac{F_n/A}{\Delta L/L_0}, \quad \text{so that}$$

$$\Delta L = \frac{F_n L_0}{YA} = \frac{mgL_0}{YA} = \frac{(10\,\text{kg})(9.8\,\text{m/s}^2)(1\,\text{m})}{(2 \times 10^{11}\,\text{N/m}^2)\pi(10^{-3}\,\text{m})^2}$$

$$= 1.56 \times 10^{-4}\,\text{m} = 0.156\,\text{mm}.$$

The change in the width Δw is

$$\Delta w = -\sigma \frac{\Delta L}{L_0} w = -(0.19) \frac{(1.56 \times 10^{-4}\, m)}{(1\, m)} (2\, mm) = -5.9 \times 10^{-5}\, mm$$

$w = w + \Delta w = 2\, mm - 5.9 \times 10^{-5}\, mm$.

Example 14

What is the Young's modulus of a sample of brass wire 0.5 m long and 2mm in diameter which stretches 0.15 mm when stressed by a 10 kg weight?

Solution:

$$Y = \frac{F_n/A}{\Delta L/L_0} = \frac{F_n L_0}{A \Delta L} = \frac{mg L_0}{A \Delta L} = \frac{(10\, kg)(9.8\, m/s^2)(0.5\, m)}{\pi (10^{-3}\, m)^2 (0.15 \times 10^{-3}\, m)}$$

$$= 1.04 \times 10^{11}\, N/m^2 = 1.04 \times 10^{11}\, Pa$$

Example 15

A steel wire has a breaking stress of 7.2×10^8 Pa. Its cross section is 0.06 cm² and its length is 3 m. What is the maximum load it will support?

Solution:

$$\text{Tensile stress} = \frac{F_n}{A}$$

$$\text{Maximum stress} = 7.2 \times 10^8\, Pa = \frac{(F_n)_{max}}{(0.06\, cm^2)}$$

$(F_n)_{max} = 0.06(10^{-2}\, m)^2(7.2 \times 10^8\, N/m^2)$

$= 4,300\, N$.

Example 16

A copper wire of cross-sectional area 0.05 cm2 and length 5 m is attached end to end to a steel wire of length 3 m and cross-sectional area 0.02 cm2. The wires are stretched to a tension of 200 N. Find the stress in each wire and the length of the combination.

Solution:

(s = steel, c = copper)

$$(\text{Tensile stress})_c = \frac{F_n}{A_c} = \frac{(200\ \text{N})}{(5 \times 10^{-6}\ \text{m}^2)} = 4.0 \times 10^7\ \text{N/m}^2 = 4.0 \times 10^7\ \text{Pa}$$

$$(\text{Tensile stress})_s = \frac{F_n}{A_s} = \frac{(200\ \text{N})}{(2 \times 10^{-6}\ \text{m}^2)} = 1.0 \times 10^8\ \text{N/m}^2 = 1.0 \times 10^8\ \text{Pa}$$

$$(\Delta L)_s = \frac{(F_n/A_s)L_s}{Y_s} = \frac{(1.0 \times 10^8\ \text{Pa})(3\ \text{m})}{(2 \times 10^{11}\text{Pa})} = 1.5 \times 10^{-3}\ \text{m}$$

$$(\Delta L)_c = \frac{(F_n/A_c)L_c}{Y_c} = \frac{(4.0 \times 10^7\ \text{Pa})(5\ \text{m})}{(1.1 \times 10^{11}\text{Pa})} = 1.8 \times 10^{-3}\ \text{m}$$

$\Delta L = (\Delta L)_s + (\Delta L)_c = 3.3 \times 10^{-3}$ m.

Example 17

A steel elevator cable of breaking stress 7.2 x 108 Pa is to support a 2000 kg elevator whose maximum upward acceleration is 2 m/s2. What diameter cable should be used if the maximum stress is one-quarter the breaking stress?

Solution:

If F is the upward force of the cable on the elevator, and a is its upward acceleration,

F - Mg = Ma

where   M = 2000 kg.

Thus

$$\text{stress} = \frac{F}{A} = \frac{M(g+a)}{A} = \frac{1}{4}(\text{breaking stress})$$

$$= (1/4)(7.2 \times 10^8\ \text{Pa}) = 1.8 \times 10^8\ \text{Pa}$$

$$A = \frac{M(g + a)}{(1.8 \times 10^8 \text{ Pa})} = \frac{(2000 \text{ kg})(9.8 \text{ m/s}^2 + 2 \text{ m/s}^2)}{(1.8 \times 10^8 \text{ Pa})}$$

$$A = 1.3 \times 10^{-4} \text{ m}^2 = \pi \left(\frac{D}{2}\right)^2$$

where D is the cable diameter. Solving for D:

$$D = 2\sqrt{\frac{(1.3 \times 10^{-4} \text{ m}^2)}{\pi}} = 1.29 \times 10^{-2} \text{ m} = 1.29 \text{ cm.}$$

Example 18

What is the change in volume of a cube of lead, one inch on a side, originally at atmospheric pressure, subjected to a hydrostatic pressure of 21 atmospheres? What is its <u>new</u> edge length L?

Solution:

$$B = -V_0\left(\frac{\Delta p}{\Delta V}\right), \text{ with } \Delta p = (21 - 1) \text{ atm or } (20)(14.7 \text{ lbs/in}^2)$$

$$\Delta V = -V_0\frac{\Delta p}{B} = -\frac{(1 \text{ in})^3 (20)(14.7 \text{ lbs/in}^2)}{(1.1 \times 10^6 \text{ lbs/in}^2)} = -2.67 \times 10^{-4} \text{ in}^3$$

$L^3 = V + \Delta V = (1 - 2.67 \times 10^4) \text{ in}^3$

$L = (1 - 2.67 \times 10^{-4})^{1/3} \cong (1 - 0.89 \times 10^{-4}) \text{ in.}$

To four significant figures, L is equal to 1 in.

Example 19

Find the compressibility of a 0.1 m³ sample of oil whose decrease in volume is 2.04 x 10-4 m³ when subjected to a pressure increase of 1.02 x 10⁷ Pa. Express your result in atmospheres of pressure.

Solution:

$$B = -V_0 \left(\frac{\Delta p}{\Delta V}\right) = -(0.1 \, m^3)\left(\frac{1.02 \times 10^7 \, Pa}{-2.04 \times 10^{-4} \, m^3}\right)$$

$$B = 5.0 \times 10^9 \, Pa = \left(\frac{5.0 \times 10^9 \, Pa}{1.013 \times 10^5 \, Pa/atm}\right)$$

$$= 4.9 \times 10^4 \, atm$$

$$k = \frac{1}{B} = 2.0 \times 10^{-5} \, /atm.$$

Example 20

What is the shear stress at each of the glued wooden joints under the tensions of Fig. 11-13 if each glued area is 10 cm²?

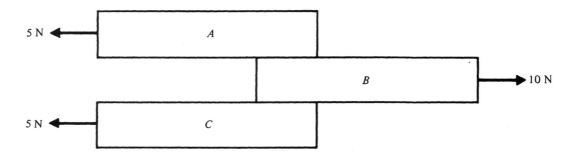

Figure 11-13

Solution:

The force of A on B is 5 N and is tangential to the joint.

$$shear \, stress = \frac{F_t}{A} = \frac{(5 \, N)}{(10 \, cm^2)} = 5.0 \times 10^3 \, N/m^2$$

252

QUIZ

1. A 18 kg child sits 1.5 m from the pivot of a weightless teeter-totter (see-saw). How far away on the opposite side should his 40 kg mother sit if she is to balance him?

Answer: 0.7 m

2. A 90 kg painter is halfway up a 8 m ladder which makes an angle of 30° with a vertical wall. What is the torque of his weight about the foot of the ladder?

Answer: 1800 N·m

3. A post weighing 500 N rests on a rough horizontal surface with a static coefficient of friction $\mu_s = 0.3$. The upper end is held by a rope fastened to the surface and making an angle of 37° with the post. A horizontal force acts at the midpoint of the post. What is the largest value this force may have if the post is not to slip?

Answer: 500 N

4. A uniform ladder 5 m long and weighing 400 N rests against a frictionless wall with its lower end 3 m from the wall. The coefficient of static friction of the ladder with the ground is 0.40. A 800 N man stands on the ladder at its midpoint. Find the frictional force.

Answer: 450 N

5. A steel rod 8 m long and 0.5 cm^2 in cross section is found to stretch 0.4 cm under a tension of 12,000 N. What is Young's modulus for this steel?

Answer: 4.8×10^{11} N/m^2

6. An aluminum wire, originally 5 m long and 0.1 cm in diameter, stretches 5.71 mm when subjected to a tension. Find the tension and its change in width.

Answer: 8×10^7 N/m^2, 1.8×10^{-5} cm

7. A cube of lead one centimeter on a side is subjected to a pressure increase of 14 atmospheres. Find the fractional change in length of one of its edges.

Answer: $\Delta L/L = 6 \times 10^{-5}$

12

GRAVITATION

OBJECTIVES

In this chapter you will develop Newton's law of gravitation and the related concepts of weight, gravitational field, and gravitational potential energy. You will also apply these concepts to satellite motion and Kepler's laws. Your objectives are to:

Find the *gravitational force* between two bodies.

Determine the *weight* of a body on the surface of the earth in terms of this gravitational force.

Find the gravitational field near a point mass or near simple mass distributions.

Analyze the *motion of satellites* in circular and other orbits and develop Kepler's laws.

Find the gravitational field near a spherically symmetric mass distribution.

Determine the effect of the rotation of the earth on an object's *apparent weight*.

REVIEW and SUPPLEMENT

Newton's Law of Gravitation and Weight

Newton's Law of Universal Gravitation states that all small bodies attract each other with a force of magnitude,

$$F_g = G\frac{m_1 m_2}{r^2}$$

where m_1 and m_2 are the masses of the bodies, r is their separation and G is a constant whose value is **G = 6.667 x 10^{-11} N·m^2/kg^2** in the SI. Strictly speaking the law is correct only for point masses but it is a good approximation for bodies whose dimensions are small compared to their separation. Additionally, if either of the bodies is a homogeneous sphere, or if it has spherical symmetry, then its gravitational force on the other particle is as if the sphere were a point particle.

The gravitational force of the earth (assumed to be spherically symmetric) on a body of mass m on its surface is its *weight*,

$$w = G\frac{mm_E}{R^2}$$

where m_E is the mass of the earth and R is the radius of the earth. Dividing each side by the mass m gives the value of g as:

$$g = G\frac{m_E}{R^2} = 9.8\,\text{m/s}^2 \left(\text{or } 32\,\text{ft/s}^2\right)$$

By Newton's Third Law, the body pulls on the earth with a force equal in magnitude and opposite in direction to its weight. The weight of a kilogram is

$$w(\text{of kilogram}) = (1\ \text{kg})(9.8\ \text{m/s}^2) = 9.8\ \text{N}.$$

The weight of a slug is

$$w(\text{of slug}) = (1\ \text{slug})(32\ \text{ft/s}^2) = 32\ \text{lbs}.$$

Another unit of mass possible in the British system is the mass whose weight is one lb. Since m = w/g this mass is

$$\text{lb–mass} = \frac{1\ \text{lb} - \text{force}}{32\ \text{ft/s}^2} = \frac{1}{32}\ \text{slug}$$

The gravitational force is a conservative force. To show this refer to Fig. 12-1 and consider the curved path connecting two points (x_1, y_1) and (x_2, y_2) in a constant gravitational field.

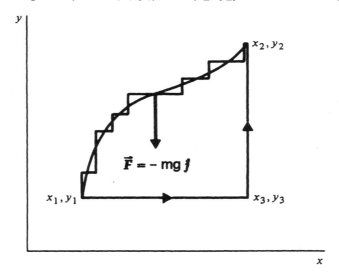

Figure 12-1

255

The path has been approximated by small horizontal and vertical staircase increments. The work of the weight is zero along the horizontal increments; no work is done in the horizontal displacements because

$$\vec{F} \cdot \vec{s} = 0$$

here. Along a vertical segment the work is $\Delta W_i = -F \Delta y_i = -mg \Delta y_i$, negative because the displacement and $\vec{F}$ are *antiparallel*. The total gravitational work is the sum of all these vertical increments

$$W_g = \Sigma_i (-mg \Delta y_i) = -mg \, \Sigma \, \Delta y_i = -mg(y_2 - y_1)$$

Any other path connecting the two end points, e.g. the 1 to 3 to 2 path of Fig. 12-1 has the same total ascent Δy and hence the same gravitational work. If the gravitational potential energy is defined as

$$U(y) = mgy$$

we find that the gravitational work is

$$W_g = mgy_1 - mgy_2 = U_1 - U_2$$

$$= -\Delta U$$

The sum $K + U$ is the *total mechanical energy* E for this system, and changes according to the rule

$$W' = E_2 - E_1 = \Delta E$$

where W' is the work of all forces other than gravity. If W' is zero, the total mechanical energy is constant and is said to be conserved,

$$W' = 0 = E_1 - E_2; \quad E_1 = E_2;$$

$$\frac{1}{2} \, mv_1^2 + mgy_1 = \frac{1}{2} \, mv_2^2 + mgy_2$$

Non-Uniform-Gravitational Field

When the displacement of a body in the earth's field is not small compared to the earth's radius, the field may not be approximated as constant; in calculating the work of the gravitational force the exact expression must be used,

$$F_g = G \frac{m_1 m_2}{r^2}$$

This force, like that in a constant gravitational field, is also conservative, that is, path independent, as shown in Fig. 12-2;

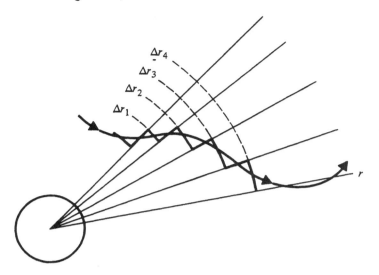

Figure 12-2

the path is divided into increments of a ray Δr_i and increments of a circular arc at constant radius. Since the force is radial the work along the arc is zero because $\vec{F}$ and $\Delta \vec{s}$ are perpendicular. Thus the total gravitational work is

$$W_g \approx \Sigma_i \, [- (F_g)_i \Delta r_i]$$

In the limit when all increments become very small

$$W_g = Lim_{\Delta r_i \to 0} \sum_i - (F_g)_i \, \Delta r_i$$

$$W_g = \int_{r_1}^{r_2} - G \frac{mm_E}{r^2} \, dr$$

$$W_g = - Gmm_E \left[\frac{-1}{r} \right]_{r_1}^{r_2} = Gmm_E \left(\frac{1}{r_2} - \frac{1}{r_1} \right)$$

$$= - U_2 + U_1$$

where

$$U(r) = - G \frac{mm_E}{r}$$

The mechanical energy for a particle of mass in the field of the earth is thus

257

$$E = K + U = \frac{1}{2}mv^2 - G\frac{mm_E}{r}$$

This energy is conserved if the work of all other forces is zero.

Satellite and Planetary Motion in Circular Orbits

Referring to Fig. 12-3, the centripetal force in this case is supplied by the gravitational attraction between the two bodies. A possible motion is uniform circular motion as indicated in the figure.

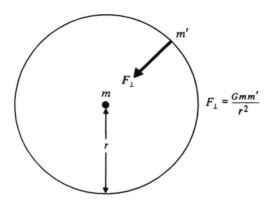

Figure 12-3

The equations of motion are

$$F_t = m'a_t = 0 = m'\frac{dv}{dt}$$

$$F_n = m'\frac{v^2}{r} = G\frac{mm'}{r^2}$$

where m' is the satellite mass and m the earth's mass. The first indicates that v is constant and the second that

$$v^2 = G\frac{m}{r}$$

The period τ of the satellite is the time for a complete revolution $2\pi r$. Thus

$$v = \frac{2\pi r}{\tau}; \text{ or } \tau = \frac{2\pi r}{v} = \frac{(2\pi r)^{3/2}}{(Gm)^{1/2}} = \sqrt{\frac{(2\pi r)^3}{(Gm)}}$$

258

If m is the earth, $m = m_E$, it is convenient to evaluate the last denominator as

$$g = G \frac{m_E}{R^2}; \quad \text{giving the identity } Gm_E = gR^2$$

where R is the radius of the earth, resulting in

$$\tau = \frac{(2\pi r)^{3/2}}{(gR^2)^{1/2}}$$

Spherical Mass Distributions

The gravitational attraction at a point P of a spherically symmetric system may be calculated as follows. If P is outside the distribution, the force at P may be calculated as if all the mass of the system were concentrated at its center. If the point P is inside the distribution, draw a shell through P concentric with the system, ignore the mass outside the shell, and proceed as if P were outside of the remaining distribution.

Effect of Earth's Rotation on g

A body at the equator when released accelerates downward (acceleration g) slightly slower than the same body released at the poles (acceleration g_0) as viewed by an observer rotating with the earth. The effect is due to the earth's rotation and is a consequence of the fact that a net force must act on the body at the equator to keep it in a circular orbit. The corresponding apparent weight (w) of an object at the equator is slightly less than the same object at the poles (weight w_0). The relation between the polar (true) weight and the equatorial (apparent) weight is:

$$w = w_0 - \frac{mv^2}{R_E}$$

and correspondingly

$$g = g_0 - \frac{v^2}{R_E}$$

where v is the velocity of the equatorial observer

$$v = \frac{2\pi R_E}{T} = \frac{2\pi R_E}{(1 \text{ day})}$$

and R_E is the radius of the earth.

Escape Speed and Black Holes

The escape speed v of a particle outside of a body of mass M may be calculated by the conservation of energy. To escape, the particle must reach a large distance $R \rightarrow \infty$ and come to rest, thus having $E_f = 0$. At the initial point then

$$E_i = \frac{1}{2} mv^2 - G \frac{mM}{R} = E_f = 0$$

$$v = \sqrt{2 \frac{GM}{R}}$$

The ingredients of this calculation include Newton's laws of motion, valid only for speeds small compared to light and Newton's law of gravitation which also fails when gravity is very strong. It is interesting that if v(the escape velocity) is put equal to c, the velocity of light, then the last expression *correctly* gives the *Schwartzschild radius* of the mass M

$$c = \sqrt{2 \frac{GM}{R_s}} \quad \text{or} \quad R_s = 2 \frac{GM}{c^2}$$

Any body less than R_s from a mass M inside R_s can never escape from it. Since nothing comes from this body and everything inside R_s falls into it, it is effectively a "black hole".

QUESTIONS AND ANSWERS

QUESTION. The moon and an artificial satellite are on opposite sides of the earth. How does the presence of the earth influence the gravitational force between the two satellites?

ANSWER. According to Newton's gravitational law, the earth's presence has no influence on the attractive force between the two satellites. That force is not altered by a third mass regardless of its location on the line between them.

QUESTION. How does the period of rotation of a satellite depend on its mass?

ANSWER. The period of rotation of a satellite does not depend on its mass but it does depend on the mass of the central body (I.e. the sun for our solar system or the earth for our moon and communication satellites)

EXAMPLES AND SOLUTIONS

Example 1

What force does a mass of 50 kg exert on the earth?

Solution:

By Newton's third law, the force of the mass on the earth is equal and opposite to the force of the earth on the mass:

$$F = mg = (50 \text{ kg})(9.8 \text{ m/s}^2) = 490 \text{ N} \quad \text{(upward)}$$

Example 2

A body weighs 20 N on the surface of the earth. What does it weigh on the surface of the moon?

Solution:

The weight of a body of mass m on the earth is equal to the force of attraction of the earth on the body. It is a vector pointing to the center of the earth. Its magnitude, w, is given by:

$$w = G\frac{m_E m}{R_E^2}$$

where m_E and R_E are the earth's mass and radius respectively. The weight, w', of the body on the moon (m_M and R_M are the moon's mass and radius) is

$$w' = G\frac{m_M m}{R_M^2}$$

In forming the ratio of the two weights, G cancels out:

$$\frac{w'}{w} = \left(\frac{m_M}{m_E}\right)\left(\frac{R_E^2}{R_M^2}\right) = \left(\frac{7.36 \times 10^{22} \text{ kg}}{5.95 \times 10^{24} \text{ kg}}\right)\left(\frac{6.37 \times 10^6 \text{ m}}{1.74 \times 10^6 \text{ m}}\right)^2 = 0.166$$

$$w' = (0.166)(20 \text{ N}) = 3.32 \text{ N}$$

Example 3

The sun is R = 1.48 x 10¹¹ m from the earth. Its mass is 1.99 x 10³⁰ kg = 330,000 m$_E$. Find the ratio of the sun's gravitational force to the earth's gravitational force *for an object on the earth's surface.*

Solution:

$$F_S = G\frac{mm_S}{R^2} = \text{force of sun on m}$$

$$F_E = G\frac{mm_E}{R_E^2} = \text{force of earth on m}$$

$$\frac{F_S}{F_E} = \left(\frac{m_S}{m_E}\right)\left(\frac{R_E}{R}\right)^2 = (330,000)\left(\frac{6.37 \times 10^6 \text{ m}}{1.49 \times 10^{11} \text{ m}}\right)^2 = 6.03 \times 10^{-4}$$

Example 4

Three 1000 kg masses are at the vertices of an equilateral triangle r = 1 m on its side. Find the force of any two of the masses on the third. (See Fig. 12-4.)

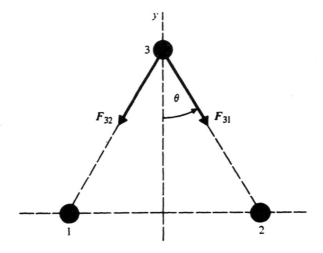

Figure 12-4

Solution:

$$\vec{F} = \vec{F}_{31} + \vec{F}_{32}; \quad \left|\vec{F}_{31}\right| = \left|\vec{F}_{32}\right| = \frac{GM^2}{R^2}$$

$$F_x = F_{31,x} + F_{32,x} = |F_{31}| \sin\theta - |F_{32}| \sin\theta = 0$$

$$F_y = F_{31,y} + F_{32,y} = |F_{31}| \cos\theta - |F_{32}| \cos\theta = 0$$

$$F_y = -\frac{2GM^2 \cos\theta}{R^2}$$

$$F_y = -2\frac{\left(6.67 \times 10^{-11} \text{ N} \cdot \text{m}^2/\text{kg}^2\right)\left(10^3 \text{ kg}\right)^2 (0.87)}{(1 \text{ m})^2}$$

$$= -1.16 \times 10^{-4} \text{ N}$$

Example 5

Communications satellites revolve in orbits over the earth's equator adjusted so that their period of rotation is the same as the period of rotation of the earth about its axis. An observer rotating with the earth thus sees the satellite fixed in the sky. Find the distance of these satellites above the surface of the earth.

Solution:

Let
  r = distance of satellite from the center of earth;
  R = radius of earth, 6.38×10^6 m; and
  T = 1 day = (24)(60)(60 s).

Then the period of the satellite is:

$$T = \frac{2\pi r^{3/2}}{\sqrt{gR^2}}$$

Solving for r,

$$r = \left[\frac{TR\sqrt{g}}{2\pi}\right]^{2/3}$$

$$r = \left[\frac{(24 \times 3600 \text{ s})(6.38 \times 10^6 \text{ m})\sqrt{9.8 \text{ m/s}^2}}{2\pi} \right]^{2/3}$$

$r = 4.23 \times 10^7$ m ≈ 7 earth radii

Distance above earth $d = r - R = 4.23 \times 10^7$ m $- 6.38 \times 10^6$ m

$d = 3.6 \times 10^7$ m

$= 36,000$ km

Example 6

A 500 lb polar bear walks from the North Pole to the equator. His mass (despite the exertion) does not change. What is the fractional loss in apparent weight?

Solution:

If w is his apparent weight at the equator and $w_0 = mg_0$ his (true) weight at the Pole,

$$\frac{w}{w_0} = 1 - \frac{mv^2}{R_E w_0} = 1 - \frac{v^2}{R_E g_0}$$

The velocity v of the bear when he rests at the equator is

$$v = \frac{2\pi R_E}{(1 \text{ day})}$$

The fractional weight loss is

$$\frac{w - w_0}{w_0} = -\frac{v^2}{R_E g_0} = -\frac{4\pi^2 R_E}{g_0 (1 \text{ day})^2}$$

$$\frac{w - w_0}{w_0} = -\frac{4\pi^2 (6.38 \times 10^6 \text{ m})}{(9.8 \text{ m/s}^2)(24 \times 60 \times 60 \text{ s})^2} = -3.4 \times 10^{-3}$$

Example 7

What velocity must a rocket have at the surface of the earth if it is to rise to a height equal to an earth radius before it begins to descend? Neglect air resistance.

Solution:

We have

$$W' = 0 = E_2 - E_1; \quad E = K + U; \quad U = -G\frac{mm_E}{r}$$

$$E_1 = \frac{1}{2}mv_1^2 - G\frac{mm_E}{R}$$

$$E_2 = 0 - G\frac{mm_E}{2R}$$

where R is the radius of the earth. Since $Gm_E = R^2 g$, $E_1 = E_2$ implies that

$$\frac{1}{2}mv_1^2 - mgR = -\frac{mgR}{2}; \quad \text{or} \quad \frac{1}{2}mv_1^2 = \frac{mgR}{2}$$

$$v_1 = (gR)^{1/2} = [(9.8 \text{ m/s}^2)(6.37 \times 10^6 \text{ m})]^{1/2}$$

$$= 7.9 \times 10^3 \text{ m/s}$$

Example 8

What is the escape velocity of a particle at the surface of the sun?

Solution:

The expression for the escape velocity, modified for the sun's gravity is:

$$v_e = \sqrt{\frac{2GM_s}{R_s}}$$

where M_s and R_s are the mass and radius of the sun respectively. Thus:

$$v_e = \sqrt{\frac{2\left(6.67 \times 10^{-11} \text{ Nm}^2/\text{kg}^2\right)\left(1.99 \times 10^{30} \text{ kg}\right)}{\left(6.95 \times 10^8 \text{ m}\right)}}$$

$$v_e = 6.18 \times 10^5 \text{ m/s}.$$

Example 9

By what fraction would the sun's radius have to be compressed for the surface escape velocity of to be c (the speed of light)?

Solution:

At a compressed radius R, the escape velocity would be c. Using Example 8,

$$c = \sqrt{\frac{2GM_S}{R}}$$

$$v = \sqrt{\frac{2GM_S}{R_S}} = 6.18 \text{ x } 10^5 \text{ m/s}$$

Thus

$$\frac{c}{v} = \sqrt{\frac{R_S}{R}}$$

or

$$\frac{R}{R_S} = \left(\frac{v}{c}\right)^2 = \left(\frac{6.18 \text{ x } 10^5 \text{ m/s}}{3.00 \text{ x } 10^8 \text{ m/s}}\right)^2 = 4.24 \text{ x } 10^{-6}$$

QUIZ

1. What would be the period of a satellite in circular orbit just above the earth's surface if air resistance could be neglected?

Answer: 1.4 hours.

2. If you drill a hole through the earth from North Pole to South Pole and drop a bowling ball down at the North Pole, how long does it take to reach the South Pole?

Answer: 42 minutes.

3. Find the escape velocity of a grain of dust on the surface of the moon.

Answer: 2.4 km/s.

13
PERIODIC MOTION

OBJECTIVES

In this chapter you will study an oscillatory motion along a straight line called *simple harmonic motion* (SHM). It occurs when a body is attracted to an equilibrium position by a force proportional to its displacement from the equilibrium position. An example is a body on the end of a spring. Since the acceleration in this motion is *not* constant, the kinematical equations of Chapter 2 are *not* applicable. Your objectives are to:

Obtain, solve, and apply the *equations of motion* of SHM.

Define the *amplitude, frequency*, and *period* for SHM.

Find the position and velocity of a body in SHM, given the initial position and velocity.

Solve problems involving SHM, using the *conservation of energy*.

Apply the equations of SHM to a body attached to a spring and to a *simple* and *physical* pendulum.

Treat *damped* and *forced oscillations*.

REVIEW

A body of mass m attached to the origin by a spring of force constant k has the equation of motion

$$F = -kx = ma, \qquad a = -\frac{k}{x}m$$

where x is the displacement from the origin. Since the spring force is conservative, the total energy is conserved (constant),

$$E = \frac{mv^2}{2} + \frac{kx^2}{2} = \text{constant} = K + U_e$$

where the elastic potential energy is

$$U_e = \frac{kx^2}{2}$$

The maximum displacement, or *amplitude* $A = x_{max}$, occurs when $v = 0$, yielding

$$E = \frac{mv^2}{2} + \frac{kx^2}{2} = \frac{kA^2}{2}$$

Solving this equation for v, we have

$$v = \pm \sqrt{\left[\frac{k}{m}(A^2 - x^2)\right]}$$

$$v_{max} = \pm \sqrt{\left[\frac{k}{m}\right]} A$$

$$\frac{mv_{max}^2}{2} = \frac{kA^2}{2} = E$$

The general solution to the equation of motion is

$$x = A \cos \left(\sqrt{\frac{k}{m}}\, t + \theta_0\right)$$

$$= A \cos (\omega t + \theta_0) = A \cos (2\pi ft + \theta_0)$$

$$x = A \cos \left(\frac{2\pi t}{\tau} + \theta_0\right)$$

where A is the amplitude and the other constants are related according to

$$\omega = \sqrt{\frac{k}{m}} = 2\pi f = \frac{2\pi}{\tau}$$

The motion repeats itself in the period

$$\tau = 2\pi \sqrt{\frac{m}{k}}$$

which is the time for one complete *cycle*. The frequency, or number of cycles per unit time, is

$$f = \frac{1}{\tau} = \frac{\omega}{2\pi}$$

with units

one cycle/s = one hertz = 1 hz.

The velocity of the body is the time rate of change of the displacement,

$$v = \frac{dx}{dt} = -\sqrt{\frac{k}{m}}\, A \sin \left(\sqrt{\frac{k}{m}}\, t + \theta_0\right) = -\omega A \sin \left(\omega t + \theta_0\right)$$

The constants A (= amplitude) and θ_0 (= phase) in $x(t)$ and $v(t)$ must be determined by the initial conditions of the motion.

If $x = 0$ at $t = 0$, we have

$$x = \pm A \sin \omega t$$

$$v = \pm A\omega \cos \omega t$$

where the sign is chosen according to the initial direction of the velocity.

If $v = 0$ at $t = 0$, we have

$$x = \pm A \cos \omega t$$

$$v = -(\pm) \omega A \sin \omega t$$

where, again, the sign is chosen according to the initial conditions.

If neither x nor v is zero at $t = 0$, then, from the general form for x and v evaluated at $t = 0$, we have

$$x_0 = A \cos \theta_0$$

$$v_0 = -\omega A \sin \theta_0,$$

with the solution

$$\theta_0 = \arctan\left(-\frac{v_0}{\omega}\right)$$

$$A^2 = x_0^2 + \frac{v_0^2}{\omega^2}$$

A *simple pendulum* (point mass suspended by a light string) undergoes SHM if its displacement is small, with

$$\omega = \sqrt{\frac{g}{L}} = 2\pi f = \frac{2\pi}{\tau}$$

where L is the length of the string. In a simple pendulum all the mass is concentrated at a single point.

An extended object pivoted to swing about a fixed axis is a *physical pendulum*. If its moment of inertia is I and the angular displacement from equilibrium is small, the body undergoes SHM with

$$\omega = \sqrt{\frac{mgh}{I}} = 2\pi f = \frac{2\pi}{\tau}$$

where h is the distance from the pivot to the center-of-gravity and I is the moment of inertia.

HINTS and PROBLEM-SOLVING STRATEGIES

Review Simple Harmonic Motion Problem-Solving Strategies I and II of the main text. Choose the form $x = \pm A \sin \omega t$ when the SHM motion starts at the origin. Choose the form $x = \pm A \cos \omega t$ when the motion starts at maximum or minimum displacement. Otherwise the general form $x = A \sin(\omega t + \theta)$ or $A \cos(\omega t + \theta_0)$ must be used.

Commit to memory: $\omega = 2\pi f = 2\pi/\tau$.

When the restoring force is provided by a spring of constant k, $\omega = (k/m)^{1/2}$. In a simple pendulum $\omega = (g/L)^{1/2}$. In a physical pendulum $\omega = (mgh/I)^{1/2}$.

QUESTIONS AND ANSWERS

QUESTION. A real ball bounces up and down, losing 1/10 of its energy each time it hits the floor. Does the motion have a uniform frequency?

ANSWER. Since there is an energy loss of 10% each time it strikes the floor, the ball rises to only 90% of its previous height. Since the time to go up (or down) depends on the square-root of the height, the times get shorter so the motion does not have a uniform frequency.

QUESTION. Two springs, each of constant k are attached end to end (in "series.") Compare the oscillation frequency of a weight suspended from one end of the two spring system to the frequency of the same weight suspended from one of the springs.

ANSWER. For two springs in series, the effective force constant $K_{eff} = k/2$, where k is the force constant of each spring. This can be seen by an energy analysis where the stored energy in the combination is $(1/2)K_{eff}(Y)^2 = (1/2)k(Y/2)^2 + (1/2)k(Y/2)^2$. This gives the above result. Thus the oscillation frequency is reduced by a factor $\sqrt{2}$.

EXAMPLES AND SOLUTIONS

Example 1

A body of mass 0.1 kg is attached to a wall by a spring of constant k = 15 N/m (see Fig. 13-1a). It is initially pulled a distance x = 0.2 m from its equilibrium position and released from rest (see Fig. 13-1b).
    (a) What is its initial potential energy?
    (b) What is its initial kinetic energy?
    (c) What is its initial total energy?
    (d) What is its velocity when it passes through the equilibrium position?

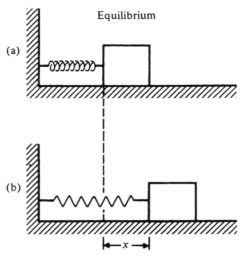

Equilibrium

(a)

(b)

$\leftarrow x \rightarrow$

Figure 13-1

Solution:

Note again, in Fig. 13-1 that we start with a sketch and a choice of coordinate system with displacement x from the equilibrium position. This solution illustrates the usefulness of the energy equation to solve for velocity at a given position.

(a) The initial elastic potential energy is

$$U_e = \frac{kx_i^2}{2} = \frac{(15\,\text{N/m})(0.2\,\text{m})^2}{2} = 0.30\,\text{J}.$$

(b) The initial kinetic energy is

$$K = \frac{mv_i^2}{2} = 0$$

(c) The total energy is

$$E = K + U = 0.30\,\text{J},$$

and is conserved.

(d) When the body passes through its equilibrium position (x = 0 where U_e = 0) its energy is

$$E = \frac{mv^2}{2} + 0 = 0.30\,\text{J}$$

$$v = \sqrt{\frac{2E}{m}} = \sqrt{\frac{2(0.30\,\text{J})}{(0.1\,\text{kg})}} = 2.5\,\text{m/s}.$$

Example 2

In the last example, what are the amplitude, frequency and period of the motion?

Solution:

The amplitude is the maximum displacement. Since the body is released from rest its amplitude is its initial displacement,

A = 0.2 m.

The frequency of the motion is

$$f = \frac{1}{2\pi}\sqrt{\frac{k}{m}} = \frac{1}{2\pi}\sqrt{\frac{(15\,\text{N/m})}{(0.1\,\text{kg})}} = 1.95\,\text{Hz} = 1.95\,\text{s}^{-1}$$

The period of the motion is

$\tau = f^{-1} = (1.95\,/\text{s})^{-1} = 0.51$ s.

The angular frequency ω is

$\omega = 2\pi f = 2\pi(1.95\,/\text{s}) = 12.25\,/\text{s}$.

Example 3

In the last example, find the displacement x and the velocity v for time t = 0.1 s after release.

Solution:

Since the displacement x is maximum at t = 0 the appropriate choice among the forms

x = A sin ωt

x = A cos ωt

x = A cos (ωt + θ_0)

is

x = A cos ωt [x_{max} = x(o) = A]

with

ω = 12.25 /s, and A = 0.2 m.

(If the spring had been compressed in Fig. 13-1b then, with x defined in Fig. 13-1b, the correct form would have been x = -A cos ωt. Choose the sign according to the initial conditions and your sign convention for positive x.)

At time t = 0.1 s,

$$x(0.1 \text{ s}) = (0.2 \text{ m}) \cos [(12.25 \text{ /s})(0.1 \text{ s})]$$

$$= 0.07 \text{ m}$$

$$v(t) = \frac{dx}{dt} = -A\omega \sin \omega t$$

$$v(0.1 \text{ s}) = -(0.2 \text{ m})(12.25 \text{ /s}) \sin [(12.25 \text{ /s})(0.1 \text{ s})]$$

$$= -2.3 \text{ m/s}.$$

Note that in the above formula, ωt, the argument of the trigonometric functions, is measured in radians. Your calculator must be informed accordingly when evaluating sin ωt or cos ωt.

Example 4

In the examples 1 through 3, find x and v at t = nτ/8, when n = 1,2,3,...,8 and plot the results.

Solution:

To calculate the displacement, we will use

$$x = A \cos \omega t = (0.2 \text{ m})\cos [(12.25 \text{ /s})t] \quad \text{where} \quad \tau = 0.51 \text{ s}$$

and for the velocity we use

$$v(t) = \frac{dx}{dt} = -\omega A \sin \omega t = -(12.25 \text{ /s})(0.2 \text{ m}) \sin [(12.25 \text{ /s})t]$$

The calculated values are presented in the Table below.

n	t(sec)	x(meters)	v(meters/s)
0	0	0.2	0
1	$\tau/8$	0.14	− 1.72
2	$\tau/4$	0	− 2.45
3	$3\tau/8$	− 0.14	− 1.72
4	$\tau/2$	− 0.20	0
5	$5\tau/8$	− 0.14	1.72
6	$3\tau/4$	0	2.45
7	$7\tau/8$	0.14	1.72
8	τ	0.20	0

The results are plotted in Fig. 13-2.

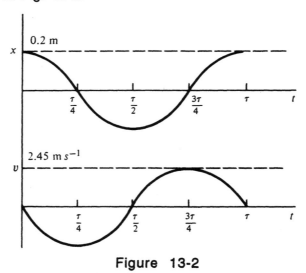

Figure 13-2

Example 5

A body of mass 0.5 kg is attached, as in Fig. 13-1a, to a wall by a spring of constant k = 100 N/m. It is given an initial velocity, at x = 0, of 5 m/s.
      (a) Find the total energy of the body.
      (b) Find the amplitude of oscillation.
      (c) Find the velocity when the displacement is half the amplitude.
      (d) Find the displacement when the velocity is half the initial velocity.
      (e) Find the displacement when the kinetic and potential energy are the same.
      (θ) Find the frequency and the period of the motion.
      (g) Plot the displacement, velocity, and acceleration against time for one period.

Solution:

Here again it is useful to use the energy relation. The total conserved energy is

$$E = \frac{mv^2}{2} + \frac{kx^2}{2}.$$

(a) Evaluating the energy at the initial time, when x = 0,

$$E = \frac{mv^2}{2} = \frac{(0.5\,\text{kg})(5\,\text{m/s})^2}{2} = 6.25\,\text{J}.$$

(b) The maximum displacement, or amplitude is achieved when v = 0, when

$$E = \frac{kx^2_{max}}{2} = \frac{kA^2}{2} = 6.25\,\text{J}.$$

$$A = \sqrt{\frac{2E}{k}} = \sqrt{\frac{2(6.25\,\text{J})}{(100\,\text{N/m})}} = 0.35\,\text{m}.$$

(c) When x = A/2,

$$E = 6.25\,\text{J} = \frac{mv^2}{2} + \frac{kx^2}{2} = \frac{mv^2}{2} + \frac{1}{2}k\left(\frac{A}{2}\right)^2$$

$$v = \sqrt{\frac{2E - k(A/2)^2}{m}} = \sqrt{\frac{2(6.25\,\text{J}) - (100\,\text{N/m})(0.35\,\text{m}/2)^2}{(0.5\,\text{kg})}}$$

$$= \pm\,4.34\,\text{m/s}.$$

(d) When

$$v = \frac{v_{max}}{2} = \frac{(5 \text{ m/s})}{2} = 2.5 \text{ m/s},$$

$$E = 6.25 \text{ J} = \frac{mv^2}{2} + \frac{kx^2}{2}.$$

$$x = \sqrt{\frac{2E - mv^2}{k}} = \sqrt{\frac{2(6.25 \text{ J}) - (0.5 \text{ kg})(2.5 \text{ m/s})^2}{(100 \text{ N/m})}}$$

$$= \pm 0.31 \text{ m}$$

(e) When the kinetic and potential energies are the same,

$$\frac{mv^2}{2} = \frac{kx^2}{2}$$

$$E = \frac{mv^2}{2} + \frac{kx^2}{2} = 2\frac{kx^2}{2} = kx^2$$

$$x = \sqrt{\frac{E}{k}} = \sqrt{\frac{(6.25 \text{ J})}{(100 \text{ N/m})}} = 0.25 \text{ m}$$

(f) The frequency is

$$f = \frac{1}{2\pi}\sqrt{\frac{k}{m}} = \frac{1}{2\pi}\sqrt{\frac{(100 \text{ N/m})}{(0.5 \text{ kg})}} = 2.25 \text{ Hz.}$$

The period is

$$\tau = f^{-1} = (2.25 \text{ Hz})^{-1} = 0.44 \text{ s}$$

$$\omega = 2\pi f = 14.14 \text{ /s.}$$

(g) Since the minimum displacement and maximum velocity are at $t = 0$, the appropriate forms for x and v are

$$x = A \sin\omega t \qquad x(0) = 0$$

$$v = \frac{dx}{dt} = \omega A \cos \omega t \qquad v(0) = \omega A.$$

The acceleration a is

$$a = \frac{dv}{dt} = -\omega^2 A \sin \omega t = -\omega^2 x = -\frac{k}{m}x.$$

276

The constants are (see (b) and (f) above)

$$A = 0.35 \text{ m} \qquad \omega = 14.14 \text{ /s}$$

$$a_{max} = \omega^2 A = (14.14 \text{ /s})^2 (0.35 \text{ m}) = 70 \text{ m/s}^2.$$

For a plot of x, v and a, see Fig. 13-3.

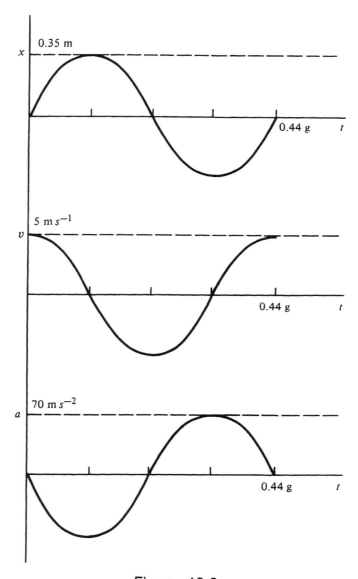

Figure 13-3

Example 6

A 200 g mass vibrates in SHM with amplitude 0.5 m and frequency 5 Hz. Find
- (a) the energy of the motion.
- (b) the maximum values of the velocity and acceleration.
- (c) the time it takes to move from 25 cm below to 25 cm above the equilibrium position.

Solution:

(a) The total energy is

$$E = \frac{mv^2}{2} + \frac{kx^2}{2}.$$

At maximum displacement, $x = A = 0.5$ m and $v = 0$, so

$$E = \frac{kA^2}{2}.$$

The force constant may be evaluated from the known frequency, since

$$f = \frac{1}{2\pi} \sqrt{\frac{k}{m}} \qquad \text{or} \quad (2\pi f)^2 = \frac{k}{m},$$

$$k = m(2\pi f)^2 = (0.2 \text{ kg})(2\pi \cdot 5 \text{ /s})^2 = 197 \text{ N/m}.$$

Thus

$$E = \frac{kA^2}{2} = \frac{(197 \text{ N/m})(0.5 \text{ m})^2}{2} = 24.7 \text{ J}.$$

(b) The maximum value of the velocity may be evaluated in two ways. First, from the energy expression,

$$E = \frac{mv^2}{2} + \frac{kx^2}{2} = \frac{mv_{max}^2}{2},$$

because v is maximum at the origin where $x = 0$:

$$v_{max} = \sqrt{\frac{2E}{m}} = \sqrt{\frac{2(24.7 \text{ J})}{(0.2 \text{ kg})}} = 15.7 \text{ m/s}$$

Another way is to use the general form $x = A \cos(\omega t + \theta_0)$ so that the velocity is:

$$v = \frac{dx}{dt} = -\omega A \sin\left(\omega t + \theta_0\right)$$

$$v_{max} = \omega A = 2\pi f A = 2\pi(5 \text{ /s})(0.5 \text{ m}) = 15.7 \text{ m/s}.$$

The acceleration is

$$a = \frac{dv}{dt} = -\omega^2 A \cos\left(\omega t + \theta_0\right) = -\omega^2 x$$

$$a_{max} = \omega 2A = (2\pi f)2A$$

$$= [2\pi(5 \ /s)]^2(0.5 \ m) = 493 \ m/s^2.$$

(c) Here we must choose a convenient form for x(t) among those listed in Example 3. If we start the clock (take t = 0) as the body passes through x = 0 with positive velocity,

$$x = A \sin \omega t.$$

Referring to Fig. 13-4, we wish to find the times t_+ and t_- when the body has displacement of ± 25 cm. Since x = A sin ωt

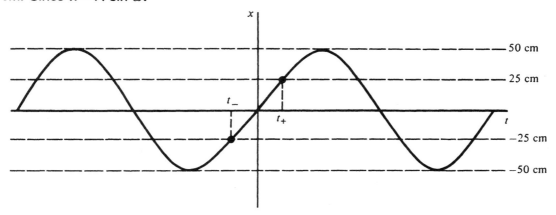

Figure 13-4

$$t = \frac{1}{\omega} \text{ arc sin } \frac{x}{A} = \frac{1}{2\pi f} \text{ arc sin } \frac{x}{A}$$

$$t_+ = \frac{1}{2\pi\left(5 \ s^{-1}\right)} \text{ arc sin } \frac{25}{50} = 0.017 \ s$$

$$t_- = \frac{1}{2\pi\left(5 \ s^{-1}\right)} \text{ arc sin } \frac{-25}{50} = -0.017 \ s$$

The total time is

$$t_+ - t_- = 0.033 \ s.$$

Example 7

A body of mass 200 g stretches a spring 10 cm when suspended from it in a gravitational field. What is the period of oscillation if a 500 g mass is suspended from the spring?

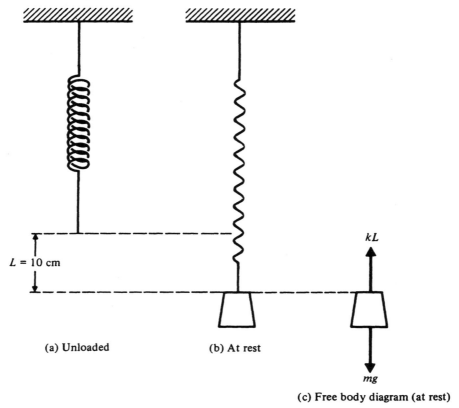

(a) Unloaded   (b) At rest

(c) Free body diagram (at rest)

Figure 13-5

Solution:

Referring to Fig. 13-5, we see that when the mass is at rest it is in equilibrium:

kL = mg

$$k = \frac{mg}{L} = \frac{(0.2\,\text{kg})(9.8\,\text{m/s}^2)}{(0.1\,\text{m})} = 19.6\,\text{N/m}.$$

If now a mass m' = 500 g is suspended from the spring of force constant k,

$$f = \frac{1}{2\pi}\sqrt{\frac{k}{m'}} = \frac{1}{2\pi}\sqrt{\frac{(19.6\,\text{N/m})}{(0.5\,\text{kg})}} = 0.99\,/\text{s}.$$

$$\tau = f^{-1} = 1.01\ s$$

Example 8

A spring stretches 10 cm when its tension is 50 N. A body of mass 5 kg is hung from the spring. When at rest the body is given an initial upward velocity of 1 m/s.

(a) Find the amplitude and the frequency of the motion.
(b) Find the acceleration of the mass when it is 5 cm above its equilibrium position.
(c) Find the force of tension in the spring at this point.
(d) What is the displacement and velocity 0.4 s after the initial time?

Solution:

(a) First we find the spring constant,

$$k = \frac{F}{x} = \frac{(50\ N)}{(0.1\ m)} = 500\ N/m$$

from which we determine the frequencies

$$\omega = 2\pi f = \sqrt{\frac{k}{m}} = \sqrt{\frac{(500\ N/m)}{(5\ kg)}} = 10\ /s$$

$$f = \frac{10\ /s}{2\pi} = 1.6\ Hz.$$

To find the amplitude, use the energy method:

$$E = \frac{mv^2}{2} + \frac{kx^2}{2} = \frac{kA^2}{2}$$

In this case the initial energy is $E_i = (mv_i{}^2/2)$; $(x_i = 0)$ so that

$$\frac{mv_i^2}{2} = \frac{kA^2}{2}$$

$$A = \sqrt{\frac{m}{k}}\ v_i = \frac{v_i}{\omega} = \frac{(1\ m/s)}{(10\ /s)} = 0.1\ m = 10\ cm.$$

(b)

$$a = -\frac{k}{m}x = -\omega^2 x = -(10\ /s)^2\ (0.05\ m) = -5\ m/s^2.$$

281

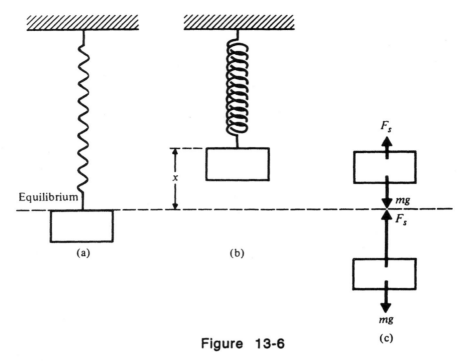

Figure 13-6

(c) Referring to Fig. 13-6, which includes free body diagrams for the cases in which the mass is above and below its equilibrium position, we have, for the *total force* F in terms of the spring force F_s,

$$F = - kx = F_s - mg$$

$$F_s = mg - kx$$

$$= (5 \text{ kg})(9.8 \text{ m/s}^2) - (500 \text{ N/m})(0.05 \text{ m}) = 24 \text{ N}$$

(d) Since the initial displacement is zero, the forms

$$x = A \sin \omega t; \quad x(o) = 0; \quad A = 0.1 \text{ m (from part a)}$$

$$v = \omega A \cos \omega t; \quad v(o) = \omega A = (10 \text{ /s})(0.1 \text{ m}) = 1 \text{ m/s}$$

obey the initial conditions. Thus we have

$$x = (0.10 \text{ m}) \sin [(10 \text{ /s})t]$$

$$x(0.4 \text{ s}) = (0.10 \text{ m}) \sin [(10 \text{ /s})(0.4 \text{ s})] = - 0.076 \text{ m} = - 7.6 \text{ cm}$$

$$v = (1 \text{ m/s}) \cos [(10 \text{ /s})t]$$

$$v(0.4 \text{ s}) = (1 \text{ m/s}) \cos [(10 \text{ /s})(0.4 \text{ s})]$$

v(0.4 s) = -0.65 m/s = -65 cm/s.

The block is 7.6 cm below its equilibrium position moving downward with speed 65 cm/s.

Example 9

A body in SHM with angular frequency ω = 0.5 /s is initially 10 cm from its equilibrium position and moving back toward its equilibrium position with a velocity 5 cm/s, as shown in Fig 13-7.
  (a) Find the period of the motion.
  (b) Find the coordinate and velocity of the body as a function of time.
  (c) How long does it take the body to return to its equilibrium position?

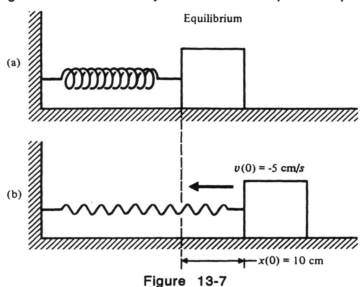

Figure 13-7

Solution:

(a) The period is

$$\tau = \frac{1}{f} = \frac{2\pi}{\omega} = \frac{2\pi}{(0.5 \text{ /s})} = 12.6 \text{ s.}$$

(b)   Since both initial displacement and initial velocity are non-zero, we must use the general forms

$$x = A \cos (\omega t + \theta_0); \quad \omega = 0.5 \text{ /s}$$

$$v = \frac{dx}{dt} = -\omega A \sin \left(\omega t + \theta_0\right)$$

and evaluate the unknown constants A and θ_0 from the initial conditions .

283

x(0) = 10 cm and v(0) = - 5 cm/s.

(The initial displacement has been taken as positive. This amounts to a choice of coordinate system. See Fig. 13-7. The initial velocity is then negative because the body is moving back toward the origin from a positive coordinate.)

Thus we have, at t = 0

$$10 \text{ cm} = A \cos \theta_0 \qquad \text{(displacement condition)}$$

$$- 5 \text{ cm/s} = - \omega A \sin \theta_0 \qquad \text{(velocity condition)}$$

yielding

$$\tan \theta_0 = \frac{1}{\omega} \frac{(5 \text{ cm/s})}{(10 \text{ cm})} = \frac{1}{(0.5 \text{ /s})} \frac{(5 \text{ cm/s})}{(10 \text{ cm})} = 1$$

$$\theta_0 = \text{arc tan}(1) = 0.79 \text{ rad}$$

$$A = 10 \text{ cm}(\cos \theta_0)^{-1} = 14.1 \text{ cm}.$$

Thus

$$x = (14.1 \text{ cm}) \cos [(0.5 \text{ /s})t + 0.79]$$

$$v = - (0.5 \text{ /s})(14.1 \text{ cm}) \sin [(0.5 \text{ /s})t + 0.79].$$

(c) Solving the first equation for t when x = 0,

$$0 = (14.1 \text{ cm}) \cos [(0.5 \text{ /s})t + 0.79]$$

For positive t starting at t = 0, the cosine function vanishes when its argument passes through $\pi/2$:

$$(0.5 \text{ /s})t + 0.79 = \frac{\pi}{2}$$

$$t = \frac{(\pi/2 - 0.79)}{(0.5 \text{ /s})} = 1.56 \text{ s}.$$

Example 10

A clock pendulum is approximately a simple pendulum. How long should the suspension be if its period is to be 2 s and its mass is 5 kg?

Solution

For a simple pendulum

$$\omega = 2\pi f = \frac{2\pi}{\tau} = \sqrt{\frac{g}{L}}$$

$$\left(\frac{2\pi}{\tau}\right)^2 = \frac{g}{L}$$

Solving for the length, L,

$$L = g\left(\frac{\tau}{2\pi}\right)^2 = (9.8 \text{ m/s}^2)\left(\frac{2 \text{ s}}{2\pi}\right)^2 = 0.99 \text{ m}$$

Note the mass is irrelevant in the small oscillation approximation.

Example 11

A simple pendulum consists of a 5.0000 kg mass suspended by a 2.0000 m wire. It is pulled 10.0000 cm from the vertical and released. Find the maximum velocity to 5 significant figures, using g = 9.8000 m/s^2:
  (a) exactly, and
  (b) making the small oscillation approximation.

Solution

Refer to Fig. 13-8:

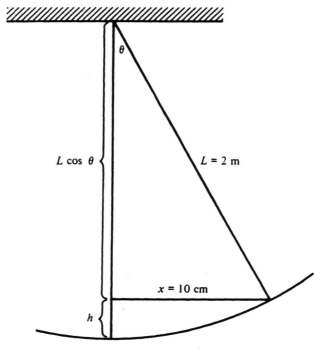

Figure 13-8

(a) The initial energy is

$$E = mgh = mgL(1 - \cos \theta)$$

and the energy at the bottom of the swing with v at its maximum value is

$$E = \frac{1}{2} mv_{max}^2$$

By energy conservation

$$\frac{1}{2} mv_{max}^2 = mgL \left(1 - \cos \theta \right)$$

$$v_{max} = [2gL(1 - \cos \theta)]^{1/2}$$

$$\cos \theta = .99875$$

$$v_{max} = [2(9.8000 \text{ m/s}^2)(2.0000 \text{ m})(1.2508 \times 10^{-3})]^{1/2}$$

$$= 0.22142 \text{ m/s}$$

(b) Using the small oscillation approximation

$$x = A \sin \omega t$$

$$v = \omega A \cos \omega t$$

$$v_{max} = \omega A; \quad \text{with } \omega = \sqrt{\frac{g}{L}}$$

$$v_{max} = \sqrt{\frac{g}{L}} A = \sqrt{\frac{(9.8 \, \text{m/s}^2)}{(2.000 \, \text{m})}} (0.100000 \, \text{m}) = 0.22136 \, \text{m/s}$$

Example 12

A thin uniform rod is pivoted at a point one quarter of its length from one end. The period is measured. Then it is pivoted at its end and the period is again measured. Find the ratio of the initial to the final period.

Solution

The moment of inertia of a rod about its center of mass is

$$I_{cm} = \frac{1}{12} mL^2.$$

By the parallel axis theorem

$$I_x = I_{cm} + mx^2$$

where x is the distance from the 'cm' axis to the parallel axis.

Initially we have

$$I_i = I_{cm} + m\left(\frac{L}{4}\right)^2 = \frac{1}{12} mL^2 + \frac{1}{16} mL^2 = \frac{7}{48} mL^2$$

and finally

$$I_f = I_{cm} + m\left(\frac{L}{2}\right)^2 = \frac{1}{12} mL^2 + \frac{1}{4} mL^2 = \frac{1}{3} mL^2$$

The period of a physical pendulum is

$$\tau = 2\pi \sqrt{\frac{I}{mgh}} \ ,$$

where h is the distance from the CM to the pivot point.

Thus

$$\frac{\tau_i}{\tau_f} = \frac{2\pi \sqrt{\dfrac{I_i}{mgh_i}}}{2\pi \sqrt{\dfrac{I_f}{mgh_f}}}$$

$$\frac{\tau_i}{\tau_f} = \sqrt{\frac{I_i h_f}{I_f h_i}} = \sqrt{\frac{(7/48)(mL^2)(L/2)}{(1/3)(mL^2)(L/4)}} = \sqrt{\frac{7 \cdot 3 \cdot 4}{48 \cdot 2}} = 0.94$$

Example 13

A body of mass 2 kg is suspended at a point 3 cm from its center-of-mass and observed to oscillate with a period of 2 s. Find its moment of inertia.

Solution

$$\tau = 2\pi \sqrt{\frac{I}{mgh}} \ ,$$

$$\left(\frac{\tau}{2\pi}\right)^2 = \frac{I}{mgh} \quad \text{or } I = mgh\left(\frac{\tau}{2\pi}\right)^2$$

$$I = (2 \text{ kg})(9.8 \text{ m/s}^2)(0.03 \text{ m})\left(\frac{2 \text{ s}}{2\pi}\right)^2$$

$$= 5.96 \times 10^{-2} \text{ kg·m}^2$$

Example 14

A body of mass 0.3 kg hangs by a spring of force constant 50 N/m. By what factor is the frequency of oscillation reduced if the oscillation is damped and reaches $e^{-1} = 0.37$ of its original amplitude in 100 oscillations?

Solution:

The fractional frequency shift is

$$\frac{\Delta\omega}{\omega} = \frac{\omega' - \omega}{\omega} = \frac{\sqrt{[(k/m) - (b^2/4m^2)]} - \sqrt{(k/m)}}{\sqrt{(k/m)}}$$

$$\frac{\Delta\omega}{\omega} = \sqrt{\left[1 - \left(\frac{m}{k}\right)\left(\frac{b^2}{4m^2}\right)\right]} - 1$$

For small damping, the second term under the square-root is smaller than the first and we make the approximation

$$\sqrt{1-x} - 1 = (1-x)^{1/2} - 1 \cong 1 - \frac{1}{2}x - 1 \cong -\frac{1}{2}x$$

This gives for the fractional frequency shift the approximate expression

$$\frac{\Delta\omega}{\omega} = -\frac{1}{2}\left(\frac{m}{k}\right)\left(\frac{b^2}{4m^2}\right)$$

The amplitude is diminished by the factor exp(-bt/2m). This factor is e^{-1} when $t = 2m/b = $ 100 periods $= 100(2\pi)[m/k]^{1/2}$. Thus we can evaluate the factor $b/2m = [k/m]^{1/2}[1/(100)(2\pi)]$ and determine the frequency shift

$$\frac{\Delta\omega}{\omega} = -\frac{1}{2}\left(\frac{m}{k}\right)\left(\frac{k}{m}\right)\left(\frac{1}{100 \cdot 2\pi}\right)^2$$

$$= -1.3 \times 10^{-6}$$

QUIZ

1. A body of mass 1 kg rests on a frictionless table and is attached to the wall with a spring of spring constant 10 N/m. It is pulled 10 cm from its equilibrium position and released from rest.
    (a) Find the period of the motion.
    (b) Find the velocity when the displacement is 5 cm.

Answer: 1.99 s, $\pm$ 0.27 m/s

2. A simple pendulum consists of a 10 kg mass suspended by an 8 m long wire. It is pulled 40 cm from the vertical and released. Find the maximum velocity of the pendulum bob.

Answer: 0.44 m/s

3. A thin solid uniform disk of mass 0.8 kg and radius 4 cm is pivoted at a point on its edge so that it can swing freely in its plane. Find the period of oscillation.

Answer: 0.5 s

4. An automobile spring has an undamped period of 1 s and supports one quarter of the weight of a 1000 kg car. What is the damping constant b of the damping force $F = - bv$ supplied by the car's shock absorber if the system is critically damped?

Answer: 3.1 x 10^3 kg/s

5. A mass oscillates on the end of a spring. Its maximum speed is 0.5 m/s. Its maximum displacement is 2 cm. What is the frequency of oscillation?

Answer: 4 Hz.

14
FLUID MECHANICS

OBJECTIVES

In this chapter you will learn about density, pressure, buoyancy and Archimedes' Principle, surface tension, streamline flow of an ideal fluid, Bernoulli's equation, viscosity, and turbulence. Your objectives are to:

Define *density* and *pressure*.

Convert the *pressure units* of atmospheres, mm of mercury and N/m^2, to each other.

Solve problems involving the *variation* of *pressure* with *depth*.

Apply *Archimedes' Principle* to floating bodies and determine their apparent weight.

Calculate the *forces* on the *walls* of a vessel containing a fluid.

Derive and apply the *equation of continuity* and *Bernoulli's equation* to the flow of fluids out of tanks and through tubes of varying diameter.

Define the coefficient of viscosity; develop *Poiseuille's law* for the flow of a viscous fluid through a circular pipe. Apply *Stokes' law* to a sphere falling in a viscous fluid.

REVIEW

Pressure and Density

The density of a substance is its mass per unit volume

$\rho = m/V$

The SI units of density are $kg/m^3 = 10^3 \ g/cm^3$. In the British system, since weight $w = mg$, mass has units of w/g or

British units of mass = $lb/g = lb/(32 \ ft/s^2)$ = slug

Thus in the British system of units, $\rho g = mg/V$ = weight per unit volume has units lb/ft^3.

The hydrostatic pressure in a fluid (liquid or gas) is, as we learned in the last chapter, the normal force per unit area against a surface ΔA within the fluid,

$$p = \frac{\Delta F}{\Delta A}$$

It is independent of the direction of the area but increases with depth. If the density ρ is a constant,

$$p = p_a + \rho gh,$$

where p is the pressure a depth h below a level where the pressure is atmospheric pressure. More generally the pressure difference between two levels at elevation y above a reference level is

$$p_2 - p_1 = -\rho g(y_2 - y_1).$$

A pressure gauge usually measures the difference between absolute pressure p and atmospheric pressure p_a,

$$\text{gauge pressure} = p - p_a = \rho gh.$$

Given a fluid of density ρ, the gauge pressure is *proportional* to the height h and may be quoted in units of h referred to the specific substance, such as "millimeters of mercury". Common pressure units are

$$1 \text{ N/m}^2 = 1 \text{ Pa} = 10^{-5} \text{ bar}$$

$$\text{one Torr} = 1 \text{ mm mercury.}$$

A pressure of one Torr corresponds to

$$p = \rho gh = (13.6 \times 10^3 \text{ kg/m}^3)(9.8 \text{ m/s}^2)(10^{-3} \text{ m})$$

$$= 133 \text{ N/m}^2 = 133 \text{ Pa}$$

where ρ is the density of mercury. A pressure of 30 in = 0.76 m mercury is an average atmospheric pressure, p_a,

$$p_a = 760 \text{ Torr} = 1.013 \times 10^5 \text{ Pa} = 14.7 \text{ lb/in}^2.$$

Archimedes' Principle

When a body is immersed in a fluid, the fluid exerts an upward force on the body equal to the weight of the fluid that is displaced by the body. If the *fluid* has density ρ and the *body* a volume V, the upward buoyant force is

$$F_B = mg = \rho Vg.$$

Surface Tension

A surface tension is a force on the surface of a liquid tending to minimize its area. It may be measured by the thin film experiment of the text, Fig. 14-8, with

$$\gamma = \frac{F}{2L}$$

in units N/m = 1000 dyne/cm.

For a soap bubble of radius R the difference between the interior air pressure p and the exterior pressure p_a is

$$p - p_a = \frac{4\gamma}{R}.$$

For a liquid drop

$$p - p_a = \frac{2\gamma}{R}$$

where p is the pressure within the fluid.

An *ideal* fluid is *incompressible* and has no *viscosity*. In *steady* or *stationary* flow the velocity at each point in space is constant. A *streamline* is a curve along which a particle would flow if the flow were stationary. The tangent to a streamline at any point is the direction of the fluid velocity at that point. The streamlines passing through the boundary of an area enclose a *tube of flow*. In a *streamline* or *laminar* flow adjacent layers of fluid slide smoothly past each other. A pattern of streamlines is steady or changes smoothly with time. In *turbulent flow* the streamline pattern is always changing in a seemingly random way.

An incompressible fluid obeys the *equation of continuity*,

$$A_1 v_1 = A_2 v_2,$$

where A_1 and A_2 are two cross-section areas of a flow tube where the corresponding velocities of flow are v_1 and v_2. The volume V of fluid passing through the surfaces per unit time, or discharge rate is:

$$\text{Discharge rate} = AV = \frac{dV}{dt}.$$

Bernoulli's equation is a relation among the pressure, elevation and velocity at two points in a tube of streamline flow of an ideal fluid,

$$p_1 + \rho g y_1 + \frac{\rho v_1^2}{2} = p_2 + \rho g y_2 + \frac{\rho v_2^2}{2}$$

where ρ is the density, y is the elevation and v the velocity.

In a viscous fluid, adjacent layers of flowing fluid exert shear stress on each other. If a layer a perpendicular distance L from a stationary layer has velocity v, the coefficient of viscosity is

$$\eta = \frac{F/A}{v/L}$$

where F is the shear force and A the area of the layer. (See the text, Fig. 14-24).

In steady viscous flow a pressure difference must be maintained to balance the viscous forces on a tube of flow. The volume rate of flow in a pipe of radius R and length L is given by *Poiseuille's law*

$$\frac{dV}{dt} = \frac{\pi}{8} \frac{R^4}{\eta} \left(\frac{p_1 - p_2}{L} \right)$$

A sphere falling in a viscous medium where r is the radius and v the velocity (Stokes' Law) experiences a retarding force $F = 6\pi\eta rv$. A raindrop in the atmosphere eventually reaches a terminal velocity v where the downward weight is balanced by the upward viscous force and buoyant force. If r is its radius, ρ the density of the sphere, and ρ' the density of the fluid, then

$$v = \frac{2}{9} \frac{r^2 g}{\eta} \left(\rho - \rho' \right).$$

HINTS and PROBLEM-SOLVING STRATEGIES

The density of water is $\rho_w = 10^3$ kg/m^3 = 1.9 slug/ft^3.

Add 14.7 lb/in^2 = 1.013 x 10^5 Pa to gauge pressure to get absolute pressure.

QUESTIONS AND ANSWERS

QUESTION. A diver sits inside a rigid diving bell that it is connected to the surface of the ocean 10 m above with a tube open to the atmosphere. Under what pressure is the diver? Under what pressure is the diving bell?

ANSWER. The diver is under pressure $p_a + \rho_{air}g(10$ m$)$ which is approximately p_a (or 1 atm.) while the diving bell is under pressure $p_a + \rho_{water}g(10$ m$)$ which is about 2 atmospheres.

QUESTION. A high speed jet of air is blown between the two dangling sheets of a page of newspaper folded over a curtain rod. Do the sheets "attract" or "repel" each other? Why?

ANSWER. The sheets "attract" each other because the region (in between) where the high speed air jet exists has a pressure lower than atmospheric so the atmospheric pressure on the other side of the sheet pushes each sheet in the direction of lower pressure.

EXAMPLES AND SOLUTIONS

Example 1

32 grams of a gas occupy a volume of 22 liters. What is the density of the gas in kg/m3?

Solution:

$$1 \text{ liter} = 1000 \text{ cc} = 1000(10^{-2} \text{ m})^3$$

$$\rho = \frac{m}{V} = \frac{32 \text{ g}}{22 \text{ L}} = \frac{\left(32 \times 10^{-3} \text{ kg}\right)}{(22 \text{ L})\left(10^3 \text{ cm}^3/\text{L}\right)\left(10^{-2} \text{ m}/\text{cm}\right)^3}$$

$$= 1.45 \text{ kg/m3}$$

Example 2

Referring to Table 14-1 in the text, what is the density of mercury in kg/m3? Of water in kg/m3?

Solution:

$$\rho_{Hg} = 13.6 \text{ g/cm3} = (13.6 \times 10^{-3} \text{ kg})(10^{-2} \text{ m})^{-3}$$

$$= 13.6 \times 10^3 \text{ kg/m3}$$

$$\rho_w = 1 \text{ g/cm3} = (10^{-3} \text{ kg})(10^{-2} \text{m})^{-3}$$

$$\rho_w = 10^3 \text{ kg/m3}.$$

Example 3

A bicycle pump has a piston of diameter one inch. What force on the piston is necessary to add air to a tire at gauge pressure 60 lb/in²?

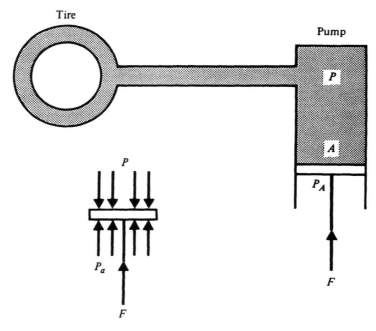

Figure 14-1

Solution:

Referring to Fig. 14-1, we see that the net upward force on the piston is

$$F + p_aA - pA.$$

When this force is positive, air enters the tire,

$$F + p_aA - pA > 0$$

$$F > (p - p_a)A = \text{gauge pressure} \times A$$

$$(p - p_a)A = (60 \text{ lb/in}^2)\pi(0.5 \text{ in})^2$$

$$= 47 \text{ lb.}$$

296

Example 4

Water pressure in a town is maintained by a water tower 100 ft high, open to the atmosphere at the top.

        (a) What is the gauge pressure at ground level, in atmospheres?

        (b) A garden hose in this town, of diameter 5/8 inch, is open at an end and spilling water. What force at the end of the hose is necessary to close the leak?

Solution:

(a) $p - p_a = \rho g h$ = gauge pressure

at ground level, where the density of water is

$$\rho = 1 \text{ g/cm}^3 = 10^3 \text{ kg/m}^3.$$

In the British system the weight per unit volume is

$$\frac{w}{V} = \frac{mg}{V} = 62 \text{ lb/ft}^3$$

so that the mass density is

$$\rho = \frac{m}{V} = \frac{62 \text{ lb/ft}^3}{g}$$

Thus

$$p - p_a = \rho g h = \left(\frac{62 \text{ lb/ft}^3}{g} \right) g \, (100 \text{ ft})$$

$$= 6200 \text{ lb/ft}^2$$

$$= (6200 \text{ lb})/(12 \text{ in})^2 = 43 \text{ lb/in}^2.$$

$$p - p_a = \frac{\left(43 \text{ lb/in}^2 \right)}{\left(14.7 \text{ lb/in}^2 \cdot \text{atm} \right)} = 2.93 \text{ atm}$$

(b) The force needed to contain the water, referring to Fig. 14-2, is

$$pA = p_aA + F$$

$$F = (p - p_a)A = (43 \text{ lb/in}^2)\rho(5/16 \text{ in})^2$$

$$= 13.2 \text{ lb.}$$

Figure 14-2

Example 5

(a) What is the pressure 100 ft below the ocean surface if sea water weighs 64 lb/ft³?
(b) A Dutch submarine descends to this depth and springs a leak with a hole of diameter one inch. A sea scout plugs the hole with his thumb. What force must he apply?

Solution:

(a) $p = p_a + \rho gh$

$$\rho = \frac{m}{V} = \frac{64 \text{ lb/ft}^3}{g}$$

$$p - p_a = \rho gh = \left(\frac{64 \text{ lb/ft}^3}{g}\right) g\,(100 \text{ ft})$$

$$= 6400 \text{ lb/ft}^2 \quad \text{(about 3 atm)}$$

(b) Assuming the interior of the submarine is at atmospheric pressure, a force pA of water pushes in and p_aA of air pushes out. Thus the force F necessary to keep the sea out is

$$F = (p - p_a)A \quad \text{with A} = \pi\left(\frac{D}{2}\right)^2 = \pi\left(\frac{1 \text{ inch}}{2}\right)^2$$

$$F = (6400 \text{ lb/ft}^2)\rho(1/24 \text{ ft})^2$$

$$F = 35 \text{ lb.}$$

Example 6

A water barometer consists of a column of water below a vacuum as shown in Fig. 14-3.
   (a) What is the height in feet of the column?
   (b) What is the maximum height that a suction pump can raise water?

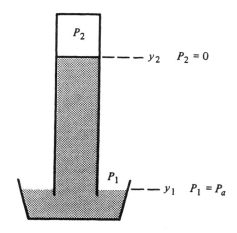

Figure 14-3

Solution:

The pressure difference between the two points is ($p_2 = 0$ and $p_1 = p_a$)

$$p_2 - p_1 = -\rho g(y_2 - y_1) = -\rho gh = -p_a.$$

Thus

$$h = \frac{p_a}{\rho g} = \frac{\left(14.7 \text{ lb/in}^2\right)}{\left(\dfrac{62 \text{ lb/ft}^3}{g}\right)g}$$

$$h = \frac{\left(14.7 \text{ lb/in}^2\right)\left(12 \text{ in/ft}\right)^2}{\left(62 \text{ lb/ft}^3\right)} = 34 \text{ ft.}$$

(b) The answer is 34 ft. The best the suction pump can do is create a perfect vacuum, as in case (a).

299

Example 7

What fraction of an ice-cube is submerged when floating in glycerin?

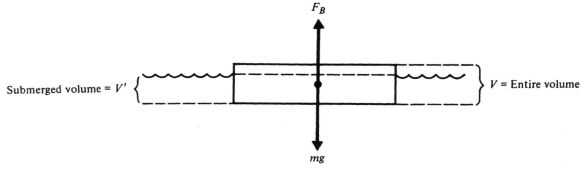

Figure 14-4

Solution:

Referring to Fig. 14-4, the buoyant force is

$$F_B = \rho_G V'g$$

where ρ_G is the density of the glycerin and V' is the submerged volume, equal to the volume of displaced glycerin. The weight mg of the cube is

$$mg = \rho_I V g$$

where ρ_I is the density of ice and V the volume of the ice-cube. Thus

$$\rho_I V g = \rho_G V'g$$

Using the known densities of ice (0.92 g/cm³) and glycerin (1.26 g/cm³),

$$\frac{V'}{V} = \frac{\rho_I}{\rho_G} = \frac{\left(0.92 \text{ g}/\text{cm}^3\right)}{\left(1.26 \text{ g}/\text{cm}^3\right)} = 0.73$$

Example 8

A dumpling floats two-thirds submerged in water. What is its density?

Solution:

The buoyant upward force is

$$F_B = \rho_w V'g$$

where ρ_w = density of water and V' is the submerged volume of the dumpling. The weight of the dumpling is

$$w = \rho_0 Vg$$

where ρ_0 is the dumpling density and V its volume. Thus

$$\rho_w V'g = \rho_0 Vg$$

$$\rho_0 = \rho_w \frac{V'}{V} = \frac{2}{3}\rho_w = \frac{2}{3}\left(1 \text{ gm/cm}^3\right)$$

$$= 0.66 \text{ g/cm}^3.$$

Example 9

A sphere of volume 10 cm³ "weighs" 50 g in water. (The spring balance reads 50 g.)
   (a) What does it "weigh" in air?
   (b) What is its density?

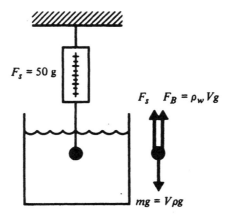

Figure 14-5

Solution:

Referring to Fig. 14-5, the equilibrium condition is

$$F_s + F_B = mg$$

$$F_s + \rho_w Vg = \rho Vg$$

$$\rho = \frac{F_s + \rho_w Vg}{Vg} = \frac{F_s}{Vg} + \rho_w$$

$$= 6 \text{ g/cm}^3.$$

The "weight" in air is ρV or 60 g as measured by the scale. Its weight is of course

$$(60 \text{ g})(980 \text{ cm/s}^2) = 58,800 \text{ dynes.}$$

Example 10

A block of mass m = 15 kg and volume 0.01 m³ hangs by a cord from a spring balance and is submerged in an unknown liquid as shown in Fig. 14-6. The spring scale reads 50 N.
    (a) Draw a diagram indicating all forces acting on the mass m.
    (b) Find the density of the liquid.
    (c) If the combined weight of the liquid and its container is W = 500 N, find the force of the table against the container.

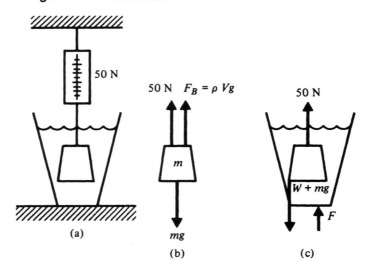

Figure 14-6

Solution:

(a) See Fig. 14-6b.

(b) Since the block is in equilibrium,

$$50 \text{ N} + \rho Vg = mg$$

$$\rho = \frac{mg - 50 \text{ N}}{Vg} = \frac{\left[(15 \text{ kg})(9.8 \text{ m/s}^2) - 50 \text{ N} \right]}{(0.01 \text{ m}^3)(9.8 \text{ m/s}^2)} = 990 \text{ kg/m}^3.$$

(c) We consider all forces acting on the system composed of the liquid, its container, and the mass m. These are indicated in Fig. 14-6c. Thus we have

$$F + 50 \text{ N} = W + mg$$

$$F = W + mg - 50 \text{ N} = 500 \text{ N} + (15 \text{ kg})(9.8 \text{ m/s}^2) - 50 \text{ N} = 597 \text{ N}.$$

Example 11

An aluminum block floats at the interface between water and mercury. What fraction of the block is submerged in the mercury?

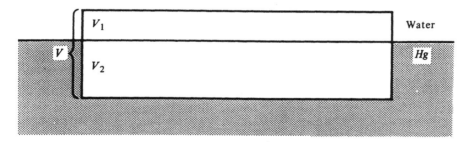

Figure 14-7

Solution:

Referring to Fig. 14-7, we see that the weight of displaced water is $\rho_w V_1 g$ and the weight of displaced mercury is $\rho_{Hg} V_2 g$. Their sum is the net buoyant force and is equal to the weight, $mg = \rho_{Al} Vg$:

$$\rho_{Al} Vg = \rho_w V_1 g + \rho_{Hg} V_2 g$$

$$= \rho_w (V - V_2)g + \rho_{Hg}(V_2)g$$

$$(\rho_{Al} - \rho_w)V = (\rho_{Hg} - \rho_w)V_2$$

$$\frac{V_2}{V} = \frac{\rho_{Al} - \rho_w}{\rho_{Hg} - \rho_w} = \frac{2.7 - 1.0}{13.6 - 1.0} = 0.13$$

Example 12

The vessel shown in Fig. 14-8 is filled with water. Find the forces of the walls against the water contained in the vessel. Neglect atmospheric pressure.

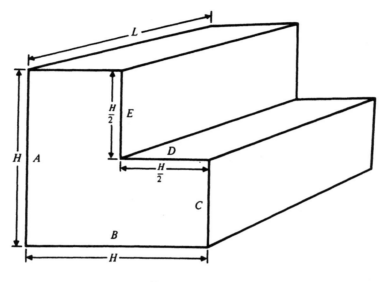

Figure 14-8

Solution:

The forces on the walls A, C and E are found by integrating $dF = pdA = \rho L dy$:

$$F_A = \int_0^H \rho g L \left(H - y\right) dy = \rho g L \frac{H^2}{2}$$

$$F_C = \int_0^{H/2} \rho g L \left(H - y\right) dy = \rho g L \frac{3H^2}{8}$$

$$F_E = \int_{H/2}^H \rho g L \left(H - y\right) dy = \rho g L \frac{H^2}{8}$$

Note $F_A = F_C + F_E$ so that the vessel does not spontaneously move sideways.

The vertical forces are

$$F_B = p_B HL \quad \text{(up)}$$

$$F_D = p_D HL/2 \quad \text{(down)}$$

The net upward force is

$$F = F_B - F_D = p_B HL - p_D \frac{HL}{2} = \left(p_B - \frac{p_D}{2} \right) HL$$

Neglecting the atmospheric pressure,

$$p_B = \rho g H$$

$$p_D = \rho g \frac{H}{2}$$

Thus

$$F = \rho g H (HL) - \rho g \frac{H}{2} \frac{HL}{2}$$

$$F = \rho g \left[H^2 L - \left(\frac{H}{2} \right)^2 L \right] = \rho g V$$

where V is the volume of water. The net upward force is equal to the weight of the water.

Example 13

Find the gauge pressure inside a raindrop of radius 1 mm at 0° C.

Solution:

$$p - p_a = \frac{2\gamma}{R} \quad \left(2\gamma \text{ because the raindrop is solid} \right)$$

$$p - p_a = \frac{2 \left(72.8 \text{ dyne}/\text{cm} \right)}{\left(0.1 \text{ cm} \right)}$$

$$= 1456 \text{ dyne/cm}^2.$$

Example 14

What is the gauge pressure, in atmospheres, necessary to blow a soap bubble of radius 5 cm?

Solution:

$$p - p_a = \frac{4\gamma}{R} = \frac{4\,(25\ \text{dyne}/\text{cm})}{(5\ \text{cm})} \qquad \left(4\gamma\ \text{because the bubble is hollow}\right)$$

$$= 20\ \text{dyne/cm}^2 = (20 \times 10^{-5}\ \text{N})/(10^{-4}\ \text{m}^2)$$

$$p - p_a = 2\ \text{N}/\text{m}^2 = \left(2\ \text{N}/\text{m}^2\right)\frac{(1\ \text{atm})}{\left(10^5\ \text{N}/\text{m}^2\right)}$$

$$= 2 \times 10^{-5}\ \text{atm.}$$

Example 15

A 30 ft high cylindrical tank of area 1 ft² is filled with water.
     (a) Find the velocity of discharge as a function of the height of water remaining in the tank when a hole of area 0.5 ft² is opened in the bottom of the tank.
     (b) Find the initial discharge velocity.
     (c) Find the initial volume rate of discharge.

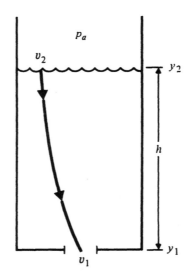

Figure 14-9

Solution:

(a) Referring to Fig. 14-9, we apply Bernoulli's equation to a flow tube which starts at the upper surface and ends just outside the hole:

$$p_1 + \rho g y_1 + \frac{\rho v_1^2}{2} = p_2 + \rho g y_2 + \frac{\rho v_2^2}{2}$$

$$p_1 = p_2 = p_a$$

$$\rho g (y_2 - y_1) = \frac{\rho \left(v_1^2 - v_2^2\right)}{2}$$

$$v_1^2 - v_2^2 = 2gh.$$

We now use the equation of continuity to determine a relation between v_1 and v_2

$$v_2 A_2 = v_1 A_1 \quad \text{or } v_2 = \frac{A_1}{A_2} v_1$$

yielding

$$(v_1)^2 - \left(\frac{A_1}{A_2}\right)^2 (v_1)^2 = 2gh$$

$$v_1 = \sqrt{(2gh)} \left\{ 1 - \left(\frac{A_1}{A_2}\right)^2 \right\}^{-1/2}$$

Often in problems of this type, $A_1 << A_2$, and the last factor may be neglected. This is <u>not</u> now the case:

$$[1 - (A_1/A_2)^2]^{-1/2} = [1 - (0.5/1)^2]^{-1/2} = 1.15,$$
and thus

$$v_1 = 1.15 \, [2(32 \text{ ft/s}^2)h]^{1/2} = 9.24 \, [h \text{ ft/s}^2]^{1/2}.$$

The velocity is greatest when the "head" h is the greatest. At the beginning

(b) $v_1 = 9.24 \, [30 \text{ ft}^2/\text{s}^2]^{1/2} = 51 \text{ ft/s}.$

(c) The discharge rate at the beginning is:

$$A_1 v_1 = (0.5 \text{ ft}^2)(51 \text{ ft/s}) = 25 \text{ ft}^3/\text{s}.$$

Example 16

Water inside an enclosed tank is subjected to a pressure of two atmospheres at the top of the tank. What is the velocity of discharge from a small hole 3 m below the top surface of the water?

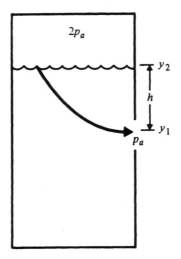

Figure 14-10

Solution: -

Referring to Fig. 14-10, we apply Bernoulli's equation to the flow tube from the top of the water to the hole,

$$p_1 + \rho g y_1 + \frac{\rho v_1^2}{2} = p_2 + \rho g y_2 + \frac{\rho v_2^2}{2}$$

$$p_a + \rho g y_1 + \frac{\rho v_1^2}{2} = 2p_a + \rho g y_2$$

where we neglect the velocity v_2,

$$v_2 = \frac{A_1}{A_2} v_1 \cong 0$$

because the hole is small relative to the cross-section of the tank,

$$\frac{A_1}{A_2} << 1.$$

Thus we find

$$\frac{\rho v_1^2}{2} = p_a + \rho g (y_2 - y_1) = p_a + \rho g h$$

$$v_1^2 = \frac{2}{\rho} \left[p_a + \rho g h \right] = \frac{2p_a}{\rho} + 2gh$$

The density of water is $\rho = 1$ g/cm^3 = 10^3 kg/m^3 and atmospheric pressure is $p_a = 1.013 \times 10^5$ pa. Thus we find

$$v_1 = \sqrt{\frac{2\left(1.01 \times 10^5 \text{ Pa}\right)}{\left(10^3 \text{ kg/m}^3\right)} + 2\left(9.8 \text{ m/s}^2\right)\left(3 \text{ m}\right)}$$

$$v_1 = [203 \text{ m}^2/\text{s}^2 + 59 \text{ m}^2/\text{s}^2]^{1/2} = 16 \text{ m/s}$$

The pressure term is approximately three times the gravity term. This is reasonable because an atmosphere of pressure corresponds to a water height of 34 ft (about 10 m), approximately three times the gravity head of 3 m in this problem.

Example 17

Water discharged from a hose reaches a maximum height of 10 m. What is the gauge pressure in the water system at the hose nozzle level?

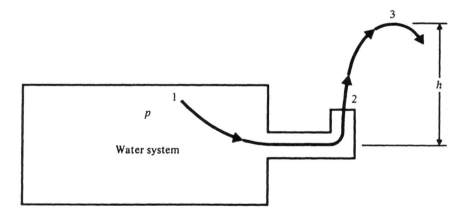

Figure 14-11

Solution:

First apply Bernoulli's equation to a flow line from somewhere inside the water system at the same height as the nozzle to the point of discharge. Referring to Fig. 14-11, we find

$$p_1 + \rho g y_1 + \frac{\rho v_1^2}{2} = p_2 + \rho g y_2 + \frac{\rho v_2^2}{2}$$

$$p_1 = p, \quad p_2 = p_a, \quad y_1 = y_2, \quad \text{and } v_1 = 0$$

$$p - p_a = \frac{\rho v_2^2}{2} = \text{gauge pressure}$$

To rise to a height h above the nozzle the water must have an initial velocity

$$v_2{}^2 = 2gh.$$

Thus

$$p - p_a = \frac{\rho(2gh)}{2} = \rho gh.$$

Alternatively apply Bernoulli's equation to the tube of flow 1 to 2 to 3,

$$p_1 + \rho g y_1 + \frac{\rho v_1^2}{2} = p_2 + \rho g y_2 + \frac{\rho v_2^2}{2}$$

$$p_1 = p; \quad p_3 = p_a; \quad y_3 - y_1 = h; \quad v_1 - v_3 = 0$$

$$p - p_a = \rho g(y_3 - y_1) = \rho gh$$

$$= (10^3 \text{ kg/m}^3)(9.8 \text{ m/s}^2)(10 \text{ m}) = 9.8 \times 10^4 \text{ Pa.}$$

Example 18

A toy rocket of diameter 2 in. consists of water under the pressure of compressed air created by pumping up a nose chamber. When the gauge air pressure is 60 lbs/in² the water is ejected through a hole of diameter 0.2 in. Find the propelling force or "thrust" of the rocket.

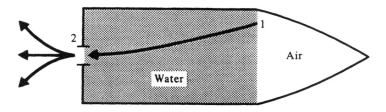

Figure 14-12

Solution:

Applying Bernoulli's equation between points 1 and 2 of the flow line in Fig. 14-12,

$$p_1 + \rho g y_1 + \frac{\rho v_1^2}{2} = p_2 + \rho g y_2 + \frac{\rho v_2^2}{2}$$

$$p_2 = p_a; \quad y_1 = y_2; \quad \text{and } v_1 = \frac{A_2}{A_1} v_2 = 0.$$

$$\frac{\rho v_2^2}{2} = p_1 - p_2 = \text{gauge pressure inside rocket} = p - p_a$$

The thrust (F) is

$$F = v_2 \frac{dm_2}{dt} = v_2 \rho \frac{dV_2}{dt} = v_2 \rho A_2 v_2$$

$$= \rho A_2 v_2^2,$$

where dV_2/dt = volume discharge rate = $v_2 A_2$.

$$F = \rho A_2 (v_2)^2 = 2A_2 \frac{\rho (v_2)^2}{2} = 2A_2 (p - p_a)$$

$$= 2\pi (0.1 \text{ in})^2 (60 \text{ lbs/in}^2)$$

$$= 3.8 \text{ lbs.}$$

311

Example 19

Water at gauge pressure of an atmosphere is flowing in a pipe of 2 cm diameter at a velocity of 6 cm/s. Find the pressure drop when it meets an obstruction of diameter 1 cm.

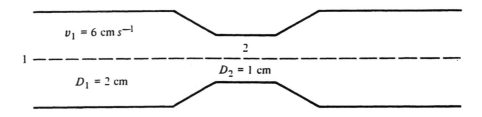

$v_1 = 6 \text{ cm s}^{-1}$

2

$D_2 = 1 \text{ cm}$

1

$D_1 = 2 \text{ cm}$

Figure 14-13

Solution:

Referring to Fig. 14-13, we apply Bernoulli's equation along the flow line from point 1 to 2,

$$p_1 + \rho g y_1 + \frac{\rho v_1^2}{2} = p_2 + \rho g y_2 + \frac{\rho v_2^2}{2}$$

$y_1 = y_2$

$$p_1 + \frac{\rho v_1^2}{2} = p_2 + \frac{\rho v_2^2}{2}$$

The velocities v_1 and v_2 are related by the continuity equation

$$v_1 A_1 = v_2 A_2 \quad \text{so } v_2 = \frac{A_1}{A_2} v_1$$

so that

$$p_1 + \frac{\rho v_1^2}{2} = p_2 + \frac{\rho (A_1/A_2)^2}{2} v_1^2$$

$$p_2 = p_1 + \frac{\rho v_1^2}{2}\left[1 - \left(\frac{A_1}{A_2}\right)^2\right].$$

The pressure *drop* is

$$p_1 - p_2 = (\rho v_1^2/2)[(A_1/A_2)^2 - 1]$$

$$= (1/2)(10^3 \text{ kg/m}^3)(0.06 \text{ m/s})^2[(2/1)^2 - 1]$$

$$p_1 - p_2 = 5.4 \text{ Pa}.$$

Example 20

Referring to Fig. 14-14, find the manometer height difference $h_1 - h_2$ when water of velocity $v_1 = 15$ cm/s enters a tube of area $A_1 = 2 \times 10^{-4}$ m² and then meets a constriction of area $A_2 = 1 \times 10^{-4}$ m².

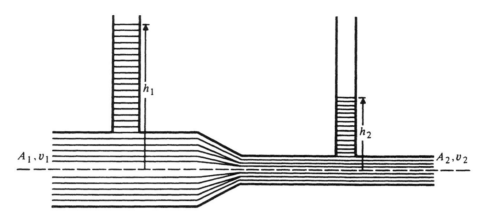

Figure 14-14

Solution:

First we check that the flow is laminar so that we may apply Bernoulli's equation. The test is that the Reynolds number $N_R < 2000$,

$$N_R = \frac{\rho vD}{\eta}$$

where in this case the parameters are:

$$\rho = 10^3 \text{ kg/m}^3; \quad v = 15 \text{ cm/s}; \quad A = \pi r^2 = \pi(D/2)^2$$

$$D = 2[A/\pi]^{1/2} \text{ and } \eta = 1.0005 \times 10^{-3} \text{ N·s/m}^2.$$

Thus the product ρvD is

$$\rho vD = 2(10^3 \text{ kg/m}^3)(0.15 \text{ m/s})(2 \times 10^{-4} \text{ m}^2/\pi)^{1/2}.$$

Dividing by the viscosity η gives for N_R

$$N_R = 2390.$$

The flow is in the transition region where it may be laminar or turbulent. To proceed we assume laminar flow and apply Bernoulli's equation on a streamline:

313

$$p_1 + \rho g y_1 + \frac{\rho v_1^2}{2} = p_2 + \rho g y_2 + \frac{\rho v_2^2}{2}$$

Since $y_1 = y_2$,

$$p_1 - p_2 = \frac{\rho\left(v_2^2 - v_1^2\right)}{2}.$$

By the equation of continuity,

$$v_2 A_2 = v_1 A_1 \text{ so } v_2/v_1 = A_1/A_2$$

and

$$p_1 - p_2 = \frac{\rho v_1^2}{2}\left[\left(\frac{A_1}{A_2}\right)^2 - 1\right].$$

However, in the hydrostatic columns

$$p_1 - p_a = \rho g h_1$$

$$p_2 - p_a = \rho g h_2,$$

so that

$$p_1 - p_2 = \rho g\left(h_1 - h_2\right) = \frac{\rho v_1^2}{2}\left[\left(\frac{A_1}{A_2}\right)^2 - 1\right].$$

or

$$h_1 - h_2 = (v_1^2/2g)[(A_1/A_2)^2 - 1]$$

$$h_1 - h_2 = \frac{(0.15 \text{ m/s})^2}{2(9.8 \text{ m/s}^2)}\left[2^2 - 1\right]$$

$$= 3.4 \times 10^{-3} \text{ m} = 0.34 \text{ cm}.$$

Example 21

Castor oil at 20°C flows through a pipe of radius 3 cm and length 1 m. The flow velocity at the center of the pipe is 5 cm/s. Find the pressure drop along the pipe.

Solution:

The velocity at a point r from the center of a pipe of radius R is

$$v = \frac{p_1 - p_2}{4\eta L}\left(R^2 - r^2\right)$$

where in this case

$$R = 0.03 \text{ m}; \quad L = 1 \text{ m}; \quad r = 0; \quad v = 0.05 \text{ m/s}$$

and

$$\eta = 9.86 \text{ poise} = 0.986 \text{ N·s/m}^2$$

Thus

$$p_1 - p_2 = \frac{4\eta L v}{R^2}$$

$$p_1 - p_2 = \frac{4\left(0.986 \text{ N·s/m}^2\right)(1 \text{ m})(0.05 \text{ m/s})}{(0.03 \text{ m})^2}$$

$$= 219 \text{ N/m}^2.$$

Example 22

A steel sphere of radius 0.5 cm falls through castor oil at 20° C. Find its terminal velocity.

Solution:

The terminal velocity is given by

$$v_T = \frac{2}{9} \frac{r^2 g}{\eta}\left(\rho - \rho'\right)$$

where in this case

$$r = 0.005 \text{ m}; \quad \eta = 0.986 \text{ N·s/m}^2,$$

315

$$\rho = \rho_s = 7.8 \text{ g/cm}^3; \quad \text{and } \rho' = \rho_{air} \ll \rho_s.$$

Neglecting the density of air compared to steel,

$$v_T = \frac{2}{9} \frac{(0.005 \text{ m})^2 (9.8 \text{ m/s}^2)}{(0.986 \text{ N} \cdot \text{s/m}^2)} (7.8 \text{ x } 10^3 \text{ kg/m}^3)$$

$$= 0.43 \text{ m/s}.$$

QUIZ

1. A block of wood floats in water, one third submerged. What is its density?

Answer: $(1/3)$ g/cm^3.

2. A body whose density is three times that of water is released at rest at the surface of a pond. Find its downward acceleration.

Answer: 6.5 m/s

3. A 10 m high standpipe, open to the atmosphere, is filled with water. Find the pressure at the bottom of the standpipe.

Answer: 1.99 x 10^5 N/m^2

4. A 10 m high standpipe, filled with water and open to the atmosphere at its top, has a small hole at its base. Find the velocity of discharge.

Answer: 14 m/s

15

MECHANICAL WAVES

OBJECTIVES

In this chapter the transverse vibrations of a wave or a pulse on a string are considered in detail. The mathematical results for this simple physical system can be used to develop intuition that will be of great use in analyzing more complicated and more interesting physical systems. Your objectives here are to:

Classify the various types of wave motions.

Define frequency, period, and wavelength.

Describe mathematically simple harmonic traveling waves, and find the displacement, given position and time.

Recognize the form of the general wave equation and apply it to problems.

Calculate the velocity of propagation for transverse waves.

Apply the rules for reflection of a wave or pulse on a string (from a free end or a fixed end) to a variety of problems.

Apply the boundary conditions responsible for reflection to produce standing waves when the initial wave train is a harmonic wave and obtain the normal modes of vibration.

Superimpose two or more waves with the same frequency and a constant phase relationship to produce interference phenomena.

REVIEW

In contrast to the random, chaotic motion studied in the last chapter, this chapter is concerned with the organized, cooperative motion of a large group of particles in a wave that propagates in space. This motion results from a non-equilibrium situation and the departure from equilibrium can be perpendicular to the direction of propagation: transverse waves (e.g. waves on a string and electromagnetic waves such as light, radio, T.V.); or the departure from equilibrium can be in the same direction as the wave propagation: longitudinal waves (e.g. sound waves).

The waves considered in this chapter are periodic waves (to be contrasted with pulses) such that the departure from equilibrium y repeats after a definite time interval τ, the period. Here y can represent a coordinate that gives the displacement of a particular segment of a string or it can represent a pressure variation in a gas. If we choose the x direction as the direction of propagation, then y(x,t) can be called the wave function as it gives the departure from equilibrium, y, at each point x as a function of the time. If, for fixed x, the departure from equilibrium is a simple harmonic oscillation, then the wave function for a traveling wave in the x direction is of the form

$$y(x,t) = A \sin[2\pi(ft - x/\lambda) + \phi]$$

where A is the amplitude (maximum excursion from equilibrium) and must carry the same units as y; f is the frequency with $2\pi f = \omega$ the angular frequency; and λ is the wavelength with $2\pi/\lambda = k$ the wave number. The entire quantity contained inside the [ ] bracket is called the phase of the wave. The angle ϕ is called the phase constant. If $\phi = 0$ the oscillation follows the sine function, if $\phi = \pi/2$ we get a cosine. In general it allows us to adjust the form of the wave to fit the starting conditions of the problem. In terms of the wave vector and angular frequency, the wave function becomes:

$$y(x,t) = A \sin(\omega t - kx + \phi)$$

If we pick some particular value of the phase (e.g. if the phase is $\pi/2$ we have a crest of the wave but $3\pi/2$ produces a trough) then the speed at which that point of constant phase advances in the x direction is called the phase velocity, c. To find it, we simply take the phase in the above expression (ϕ is a constant), equate it to a constant, and calculate the first time derivative.

$$\frac{d}{dt}\left(\omega t - kx + \phi\right) = \frac{d}{dt}(constant) = 0$$

Then we have

$$c = \frac{dx}{dt} = \frac{\omega}{k} = \frac{2\pi f}{2\pi/\lambda} = f\lambda$$

Had we written the phase as $\omega t + kx + \phi$, the above procedure would have given $c = -\omega/k$ which corresponds to a wave propagating in the - x direction. See Examples 1 and 2.

Thus at a given point in space the disturbance from equilibrium is a simple harmonic oscillation whereas (if ϕ is set equal to zero) at a given time the disturbance is a sine wave in space. This separation, which fixes x but allows time to vary in the first case and then fixes time while x is allowed to vary, is accomplished mathematically by means of the *partial derivative*.

For a wave propagating on a string in the x direction, the *particle velocity*, v, is actually in the y direction and calculated by taking the partial derivative of the wave function y(x,t) with respect to time. For instance:

$$v = \frac{\partial y}{\partial t} = \frac{\partial}{\partial t}\left[A \sin\left(\omega t - kx\right)\right] = \omega A \cos\left(\omega t - kx\right)$$

The maximum particle velocity is equal to ωA and is not the same as the propagation velocity of the wave, c.

By calculating the second partial derivatives with respect to x and t of the wave function in the harmonic form given, it can be shown that $y(x,t)$ obeys the wave equation:

$$\frac{\partial^2 y}{\partial x^2} = \frac{1}{c^2}\left(\frac{\partial^2 y}{\partial t^2}\right)$$

This form is very easy to remember if you note that the speed c must be paired with time t to give a quantity with the dimensions of x (i.e. the product ct has dimensions of a length). Various solutions to this very important equation are explored in Examples 3 and 4.

The velocity of propagation c depends on specific parameters that characterize a particular problem. For each situation, this velocity must be calculated. Several such examples are given in the text, such as the speed of transverse waves on a string (see Example 5). Although it is a challenging problem to calculate the velocity of propagation of a wave in a given medium, a thoughtful analysis based on the dimensions (units) of the pertinent physical parameters usually gives at least an approximate value for the propagation velocity.

Dimensionally we have seen (recall the formula for kinetic energy) that an energy divided by a mass is a velocity squared. For a string of mass M and length L, the only other physical quantity that is used to characterize the string is its tension, S (a force). The tension divided by the mass per unit length ($\mu = M/L$) has the same units as energy (force times distance) divided by mass, that is a velocity squared. As shown in the text, the velocity of propagation of a wave on a string is equal to the square root of the ratio of the tension to the mass per unit length.

$$c = \sqrt{\frac{SL}{M}} = \sqrt{\frac{S}{\mu}}$$

As seen in this chapter, there are many possible solutions to the wave equation. Even sums and differences of these solutions are themselves solutions. Restrictions placed on these solutions by the physical problem at hand pick out the one appropriate solution from the numerous possible ones. These restrictions are called boundary conditions and are somewhat like the "initial conditions" we used in studying the simple harmonic oscillator. The transverse waves on a string are used to obtain the most important results. These results are:

(1) A wave (or pulse) reflected from a "fixed end" of a string suffers a 180° phase reversal. If the fixed end is at $x = 0$, the boundary condition is that $y(0,t) = 0$.

(2) A wave (or pulse) reflected from an "open end" or unattached end suffers no phase change. If the open end is at $x = 0$, the boundary condition is that $\partial y/\partial x = 0$ at $x = 0$.

319

In both cases the principle of superposition was used to get the most general solution before applying the boundary condition. A function of the form f(x - ct) or g(x + ct) by itself satisfies the wave equation. Furthermore, addition (or superposition) of these solutions also gives a solution. The most general solution is the one that covers all the possibilities while the boundary conditions pick out the specific choice for the problem at hand.

A node in a wave form is a place where the displacement vanishes. For harmonic waves when the string is fixed at one end (say at x = 0), there is a node in the wave function at that point. The next node occurs one half wavelength away and in between these two nodes is an antinode. An antinode is a place where the displacement is a maximum. If we specify a boundary condition at the other end of the string as well, this restriction, coupled with the one from the fixed end, is so stringent that only certain wave vectors k are possible, leading to *standing waves*. We examine three cases.

(1) The string is fixed at x = 0 and x = L so that it has at least two displacement nodes. The mathematical condition for this case is that the allowed wavelengths, λ_n, are given by:

$$\lambda_n = \frac{2L}{n} \qquad \text{where } n = 1, 2, ..., \text{ etc.}$$

It is instructive to sketch these solutions.

(2) The string has unattached or open ends at x = 0 and x = L. The string then has at least two displacement antinodes (with one or more nodes in between). The allowed wavelengths in this case are precisely the ones given above although at first glance the sketches appear different. While this case is not an especially interesting one for a string, it is quite useful in discussing pressure-displacement longitudinal waves in a gas in the next chapter.

(3) The string has one fixed end at x = 0 and one open end at x = L. The boundary condition at the open end is that $\partial y/\partial x = 0$ at x = L. The allowed wavelengths, λ_n, for standing waves are:

$$\lambda_n = \frac{4L}{2n - 1} \qquad \text{where } n = 1, 2, ..., \text{ etc.}$$

For the longest wavelength possible, only a quarter wavelength fits into L, with a displacement node at the fixed end and a displacement antinode at the open end.

If two travelling waves with the same frequency and a constant difference of phase constants are superimposed, the waves may reinforce or cancel each other. When this occurs, it is termed *interference*. Constructive interference occurs (reinforcement of waves) when the two waves have a path difference of 0, λ, 2λ,... etc., implying a phase difference of 0, 2π, 4π, ...etc. Destructive interference occurs whenever the path difference of the two waves is $\lambda/2$, $3\lambda/2$, $5\lambda/2$, etc., implying a phase difference of π, 3π, 5π, etc. A useful result is the following:

$$\frac{(\text{path difference})}{\lambda} = \frac{(\text{phase difference})}{2\pi}$$

PROBLEM-SOLVING STRATEGIES

A review of the problem solving strategies of the main text is made here. The standard form of the waves (traveling in the x direction) considered here is:

$$y = A \sin [2\pi ft \pm kx + \phi]$$

where A is the wave amplitude; f is the frequency, k is the wavevector, and ϕ is the starting phase. The starting phase is used to "adjust" the standard form to fit the specific conditions of a given problem. The wave traveling in the positive x direction has the sign ($\pm$) replaced by ($-$) but the wave traveling in the negative x direction has the sign ($\pm$) replaced by ($+$), and this is sometimes confusing.

Since wave speed (c), wavelength (λ), and period (T) are related by $\lambda = cT$ or distance traveled in one period equals (constant) speed multiplied by period, knowing any two gives the third quantity.

The interference of waves that produces standing waves is further simplified by imposing boundary conditions as done in the chapter. The two major classes of problems follow from the results derived for (1) a string fixed at both ends (similar to a musical instrument); or (2) a string fixed at one end but free at the other (like a bull-whip). Identify which set of boundary conditions apply to the problem at hand and then use the results derived for the transverse waves on a string. The pattern of nodes and anti-nodes depends on the choice of boundary conditions.

QUESTIONS AND ANSWERS

QUESTION. A pulse is described by the function $y = f(x - v_0 t)$. What is the function that describes a pulse of half the amplitude? What is the function that describes a pulse traveling twice as fast? What is the function that describes a pulse that leads the first by the distance x_0?

ANSWER. The function describing a pulse with half the amplitude is the same as that given but multiplied by the factor (0.5). For a pulse traveling twice as fast, change the velocity v_0 to $2v_0$. For a pulse that leads the first by the distance x_0, change the variable x to $x - x_0$.

QUESTION. A wave generator send waves along a stretched elastic cord. The frequency (f) and amplitude (A) of the generator and the tension (S) in the cord are adjustable. How would you increase the wave length of the wave? How would you increase the particle speed on the cord? How would you increase the propagation speed of the wave?

ANSWER. To increase the wavelength, one can decrease f, for fixed S, increase S for fixed f or change both. To increase the particle speed on the cord, one can increase the amplitude or increase the frequency. To increase the propagation speed, increase the tension.

EXAMPLES AND SOLUTIONS

Example 1

Given that a transverse wave disturbance is of the form

$$y = (0.015 \text{ m}) \sin 2\pi \left(\frac{t}{[10^{-3} \text{ s}]} - \frac{x}{[1.2 \text{ m}]} \right)$$

where t is expressed in seconds and x in meters, identify the following quantities:
- (a) amplitude
- (b) wavelength
- (c) frequency
- (d) speed of propagation
- (e) direction of propagation

Solution:

We compare the above form with the standard form,

$$y = A \sin \left(2\pi ft - \frac{2\pi x}{\lambda} \right)$$

and see that

(a) the amplitude is A = 0.015 m

(b) the wavelength is λ = 1.2 m

(c) the frequency is f = $(10^{-3} \text{ s})^{-1}$ = 1000 s^{-1} = 1000 Hz.

(d) the velocity of propagation is

$$v = \lambda f = (1.2 \text{ m})(1000 \text{ s}^{-1}) = 1.2 \times 10^3 \text{ m/s}$$

(e) the direction of propagation is the + x direction

Example 2

Write the equation y(x,t) describing a traveling transverse wave that propagates in the + x direction and satisfies the following conditions:
- (a) The maximum disturbance from equilibrium at any point is 1 cm.
- (b) The wavelength is 2 m.
- (c) The period is 0.02 s.
- (d) At t = 0 and x = 0.5 m, the instantaneous particle velocity is $\pi/2$ m/s down (or negative).

322

Solution:

The form to be used is

$$y = A \sin [2\pi ft - kx + \phi]$$

where the minus sign was chosen to make the wave go in the + x direction. Rewritten in terms of the period and wavelength, the displacement is

$$y = A \sin \left(2\pi \left[\frac{t}{\tau} - \frac{x}{\lambda} \right] + \phi \right)$$

Since τ and λ are given in the problem, only ϕ must be found. The amplitude A is equal to the maximum disturbance from equilibrium or 0.01 m. To determine ϕ, we must calculate the instantaneous particle velocity.

$$v_y = \frac{\partial y(x,t)}{\partial t} = \left(\frac{2\pi}{\tau} \right) A \cos \left(2\pi \left[\frac{t}{\tau} - \frac{x}{\lambda} \right] + \phi \right)$$

This expression must be evaluated at t = 0 and x = 0.5 m.

$$v_y(0.5 \text{ m}, 0 \text{ s}) = \left(\frac{2\pi A}{\tau} \right) \cos \left(-\frac{\pi}{2} + \phi \right) = \left(\frac{2\pi A}{\tau} \right) \sin \phi$$

$$- \pi/2 = \pi \sin \phi$$

$$\sin \phi = - (1/2) \text{ so that } \phi = - 30° = - \pi/6 \text{ radians}$$

The final result is:

$$y(x,t) = (0.01 \text{ m}) \sin 2\pi \left[\frac{t}{(0.02 \text{ s})} - \frac{x}{(2 \text{ m})} - \frac{1}{12} \right]$$

Example 3

Calculate $\dfrac{\partial^2 y}{\partial x^2}$ and $\dfrac{\partial^2 y}{\partial t^2}$ for the functions y(x,t) given below.

   (a) $y(x,t) = A(x - vt)^n$ where A is a constant and n > 1
   (b) $y(x,t) = A \exp(-kx)\sin \omega t$ where A, k, and ω are constants

Solution:

(a) To calculate the partial derivative of y with respect to x, treat A, n, v, and t as constants. Thus

$$\frac{\partial y}{\partial x} = nA(x - vt)^{n-1}$$

and

$$\frac{\partial}{\partial x}\frac{\partial y}{\partial x} = \frac{\partial^2 y}{\partial x^2} = n(n-1)A(x-vt)^{n-2}$$

To calculate the partial derivative of y with respect to t, treat everything except t as a constant.

$$\frac{\partial y}{\partial t} = nA(x - vt)^{n-1}(-v)$$

$$\frac{\partial}{\partial t}\frac{\partial y}{\partial t} = \frac{\partial^2 y}{\partial t^2} = n(n-1)A(x-vt)^{n-2}(-v)^2$$

In this case note that

$$\frac{1}{v^2}\frac{\partial^2 y}{\partial t^2} = \frac{\partial^2 y}{\partial x^2}$$

so that this function *is* an acceptable solution of the wave equation.

(b) For the function y = A exp(-kx) sin ωt, the partial derivative with respect to x is taken by treating all other variables (such as t) as constants.

$$\frac{\partial y}{\partial x} = -kA\exp(-kx)\sin(\omega t) = -kAe^{-kx}\sin(\omega t)$$

$$\frac{\partial^2 y}{\partial x^2} = k^2 A\exp(-kx)\sin(\omega t)$$

For the partial derivative with respect to t, treat x as constant.

$$\frac{\partial y}{\partial t} = \omega A\exp(-kx)\cos(\omega t)$$

$$\frac{\partial^2 y}{\partial t^2} = -\omega^2 A \exp{(-kx)} \sin{(\omega t)}$$

$$-\frac{1}{\omega^2}\frac{\partial^2 y}{\partial t^2} = \frac{1}{k^2}\frac{\partial^2 y}{\partial x^2}$$

The minus sign is critical here and we *do not* have a function that obeys the wave equation.

Example 4

A rope under tension provided by a hanging mass of 20 kg has a velocity of wave propagation of 30 m/s. If the velocity is to be the same, what mass should be hung from a rope of the same material with
  (a) half the diameter of the original rope?
  (b) twice the diameter of the original rope?

Solution:

The velocity of propagation v is given by

$$v^2 = \frac{S}{\mu}$$

The tension S is provided by the hanging mass M, where we have

$$S = Mg$$

The mass per unit length, μ, is equal to

$$\mu = \frac{M}{L} = \frac{\rho AL}{L} = \rho A$$

where ρ is the density and is the same for all the ropes in this problem. To make v a constant, we must have M/A equal to a constant.

(a) If the diameter is halved, the area decreases by a factor 4 so M = 5 kg.

(b) If the diameter is doubled, the area increases by a factor 4 so M = 80 kg.

Example 5

Two strings of mass per unit length μ_1 and μ_2 are joined together at point A. The tension in the two strings is the same. If the wavelength in the string with $\mu_1 = 0.005$ kg/m is 0.03 m, what is the mass per unit length, μ_2, if the wavelength in that string is .05 m? Assume the wave frequencies are the same in the two strings.

Solution:

The velocity of propagation, c, is given by

$$c = \sqrt{\frac{S}{\mu}}$$

where S is the tension and μ is the mass per unit length. Since the tension is the same in the two strings, squaring the above equation gives:

$$c_1{}^2\mu_1 = c_2{}^2\mu_2.$$

The speeds are not given but the frequency is the same in each string and the wavelengths are known $(c_1 = \lambda_1 f, c_2 = \lambda_2 f)$ so we have

$$\lambda_1{}^2 f^2 \mu_1 = \lambda_2{}^2 f^2 \mu_2.$$

leading to

$$\frac{\mu_2}{\mu_1} = \left(\frac{\lambda_1}{\lambda_2}\right)^2 = \left(\frac{0.03\text{ m}}{0.05\text{ m}}\right)^2 = 0.36$$

and yielding

$$\mu_2 = 1.8 \times 10^{-3} \text{ kg/m}.$$

Example 6

A uniform string of length L = 0.5 m is fixed on both ends. Calculate the wavelength of the fundamental mode of vibration. If the wave speed is 300 m/s, calculate the frequency of the fundamental mode and the next possible mode.

326

Solution:

For the fundamental mode we have $L = \lambda/2$, so since $L = 0.5$ m, the wavelength of the fundamental mode is:

$\lambda_1 = 2L = 1$ m.

For a wave speed of 300 m/s = c, the frequency of the fundamental mode, f_1, will be:

$$f_1 = \frac{c}{\lambda_1} = \frac{(300 \text{ m/s})}{(1 \text{ m})} = 300 \text{ Hz}$$

The next possible mode will have $L = \lambda_2$.

$\lambda_2 = L = 0.5$ m

The associated frequency is

$$f_2 = \frac{c}{\lambda_2} = \frac{(300 \text{ m/s})}{(0.5 \text{ m})} = 600 \text{ Hz}$$

Example 7

The string in the previous example, of length $L = 0.5$ m, is fixed at one end but free at the other end. Calculate the wavelength of the fundamental mode of vibration. For a wave speed of 300 m/s calculate the frequency of the fundamental mode and the next possible mode.

Solution:

For the string fixed at one end but free at the other, the fundamental mode is characterized by:

$$\frac{\lambda_1}{4} = L$$

$\lambda_1 = 4(0.5 \text{ m}) = 2$ m.

The corresponding frequency is:

$$f_1 = \frac{c}{\lambda_1} = \frac{(300 \text{ m/s})}{(2 \text{ m})} = 150 \text{ Hz}$$

For the next possible mode,

$$\frac{3}{4} \lambda_2 = L$$

$$\lambda_2 = \frac{4L}{3} = \frac{4(0.5\,\text{m})}{3} = \frac{2}{3}\,\text{m.}$$

The frequency of this first overtone is:

$$f_2 = \frac{c}{\lambda_2} = \frac{(300\,\text{m/s})}{(2/3\,\text{m})} = 450\,\text{Hz}$$

The second harmonic is missing and the third harmonic is the first overtone.

QUIZ

1. Given that the transverse wave disturbance is of the form

$$y(x,t) = (0.025 \text{ m}) \cos 2\pi \left[\frac{t}{(10^{-2} \text{ s})} + \frac{x}{(1.5 \text{ m})} \right]$$

where t is in seconds and x is in meters, identify or calculate the following quantities:
(a) the amplitude; (b) the wavelength; (c) the frequency; (d) the velocity of propagation; and
(e) the direction of propagation.

Answer: (a) 0.025 m, (b) 1.5 m, (c) 100 s^{-1}, (d) 150 m/s, (e) - x direction.

2. Write an equation that correctly describes a traveling wave disturbance y(x,t) propagating
in the + x direction that satisfies the following conditions:
(a) The maximum disturbance from equilibrium is 0.02 m; (b) The wave velocity is 50 m/s;
(c) The frequency is 80 Hz; and   (d) At t = 0 and x = 0, y(0,0) = - 0.02 m.

Answer: $y(x,t)$ = (0.02 m) sin 2π[(80 s^{-1})t - (1.6 m^{-1})x + 0.75]
$\qquad\qquad$ = - (0.02 m) cos 2π[(80 s^{-1})t - (1.6 m^{-1})x]

3. A mass of 10 kg hangs vertically, supported by a string of length 1.5 m and mass of 45
grams. (a) Calculate the velocity of propagation, c, for this string; and (b) Find the mass of a
second string of the same length that would give a wave speed double that of the first string (for
the same tension).

Answer: (a) c = 57.2 m/s, (b) m_2 = 11.25 grams

4. A wave traveling in the -x direction is described by the wave function

$$y(x,t) = (0.05 \text{ m}) \sin 2\pi[\frac{t}{(0.01 \text{ s})} + \frac{x}{(0.5 \text{ m})}].$$

$\qquad$ (a) Calculate the maximum particle speed, v.
$\qquad$ (b) Calculate the particle speed at x = 0.3 m and t = 1.5 x 10^{-3} s.

Answer: (a) v_{max} = 10π m/s, (b) v(0.3 m, 1.5 x 10^{-3} s) = 0.

5. Write the equation of a wave $y_2(x,t)$ that when superimposed with the wave

$\qquad y_1(x,t)$ = A sin(ωt - kx)

will produce a standing wave, $y_t = y_1 + y_2$, with nodes at x = 0 and x = λ/2.

Answer: y_2 = -A sin(ωt +kx)

16
SOUND AND HEARING

OBJECTIVES

In this chapter concepts of waves in gases developed in the last chapter are applied to sound waves or acoustics. Your objectives are to:

Relate the pressure amplitude of a sound wave to the displacement amplitude.

Calculate the intensity in a wave from either the pressure or displacement amplitude (and the bulk modulus).

Calculate the speed of sound waves and that of sound waves in gases.

Apply the boundary conditions used for transverse waves on a string to the longitudinal standing waves produced by reflections at open and closed ends in organ pipes and obtain the normal modes of vibration.

Calculate the beat frequency resulting from the superposition of two harmonic waves with different frequencies.

Calculate the apparent frequency change due to source and/or listener motion arising from the Doppler effect.

REVIEW

Sound waves in gases can be characterized by the displacement y of a given mass from an equilibrium position or by a pressure difference Δp from the equilibrium or average pressure. The amplitude of the pressure wave is proportional to the local change in the displacement wave function, $\partial y/\partial x$, the proportionality constant being the adiabatic bulk modulus B of the gas:

$$\Delta p = -B\left(\frac{\partial y}{\partial x}\right)$$

For an ideal gas where $pV^\gamma = K$ for an adiabatic process we have $B = \Delta p$, where p is the average pressure. If we express the wave function $y(x,t)$ as a harmonic wave, then the maximum value of the partial derivative will be kA where k is the wave vector and A is the amplitude of the displacement wave. This means that the maximum pressure amplitude can be related to the maximum displacement amplitude by $\Delta p_m = \Delta pkA$, so that a calculation of either one completely characterizes the problem. The smallness of A is illustrated in Example 1. The intensity I of travelling waves of the harmonic type considered here is

$$I = \frac{1}{2} \omega k B A^2$$

The intensity in general is equal to the time average power per unit area (units are watts per square meter). The factor 1/2 in the above expression comes from the fact that the time average of a sine (or cosine) squared over a cycle is equal to 1/2. An alternative form for the intensity, written in terms of the pressure amplitude p_{max} is:

$$I = \frac{1}{2} c \frac{(p_{max})^2}{B}$$

The range of sound intensities that the human ear can tolerate is so large that a logarithmic scale (base 10) is used to describe the intensity level. The intensity level ß is defined by:

$$\text{ß} = (10 \text{ db}) \log_{10} (I/I_0)$$

where $I_0 = 10^{-12}$ W/m² is the hearing threshold. See Example 2. The intensity levels that the human ear responds to, from the hearing threshold to the level where pain begins, span twelve orders of magnitude or 120 db.

Longitudinal waves propagate in fluids with a velocity v given by:

$$v = \sqrt{\frac{B}{\rho}}$$

where B is the bulk modulus and ρ is the mass density. In a solid, the expression for the speed of a longitudinal wave is similar to that in fluids except the bulk modulus is replaced by Young's modulus, Y, defined in Chapter 11.

$$v = \sqrt{\frac{Y}{\rho}}$$

For a gas, the parameters that characterize its state, T, p, V, and the density ρ are not all independent but are connected by an equation of state. Since RT is dimensionally an energy, the ratio RT/M where M is the mass per mole is a velocity squared. Also the ratio of pressure to density has the dimensions of a velocity squared. For an ideal gas, this ratio is identical to RT/M. Because the sound oscillations are fast compared to the times necessary to reach local thermal equilibrium in a gas, the compressions and rarefactions are adiabatic, giving for v:

$$v = \sqrt{\gamma \left(\frac{p}{\rho}\right)} = \sqrt{\gamma \left(\frac{RT}{M}\right)}$$

where $\gamma = C_p/C_v$.

A vibrating string, set into motion by plucking or bowing, can support many frequencies of vibration at the same time. Our perception of the sound that originates from a stringed instrument depends on the various frequencies present and on their relative intensities. The 'quality' of a sound depends, in a subjective way, on this intensity distribution. The 'pitch' of a sound is subjectively connected with the frequency but also depends somewhat on the intensity especially at low frequencies.

Superposition of waves of the same frequency traveling in opposite directions can lead to standing waves. For a string with an open end, $\partial y/\partial x = 0$, so a wave function using Δp as the variable would have a node at the open end. When Δp has an antinode, the displacement wave, $y(x,t)$ has a node. This makes it really quite simple to analyze the vibrations of organ pipes which are totally analogous to the three cases considered for the vibrating string. The three possibilities are:

(1) The organ pipe is closed at both ends (small openings are needed to let the sound escape). This is like the string fixed at both ends. There must be a displacement node (and hence a pressure antinode) at each end. Since $\lambda_n = 2L/n$, the allowed frequencies, f_n, are:

$$f_n = \frac{c}{\lambda_n} = n\left(\frac{c}{2L}\right).$$

The fundamental frequency is c/2L and all multiples (or harmonics) are present.

(2) The organ pipe is open at both ends. This is like the string with both ends unattached. Since Δp is the departure of the pressure from the average value and the bulk air at both ends of the pipe is at the average pressure inside the pipe, then there must be a pressure node (and hence a displacement antinode) at each end. Again, the allowed frequencies are:

$$f_n = n\left(\frac{c}{2L}\right).$$

Just like the pipe closed on both ends, the fundamental frequency is c/2L and all harmonics are allowed.

(3) The organ pipe has one closed end and one open end. This is analogous to the string with one end free and one end fixed. At the closed end there will be a displacement node (and a pressure antinode) whereas at the open end there will be a pressure node (and a displacement antinode). Since the allowed wavelengths are

$$\lambda_n = \left(\frac{4L}{2n-1}\right) \qquad \text{where } n = 1, 2, ..., \text{etc.}$$

the allowed frequencies are:

$$f_n = \left(\frac{2n-1}{4L}\right)c.$$

The fundamental frequency is $c/4L$ (lower by a factor of two than the previous cases) and only the odd harmonics are allowed.

The systems treated here, the stretched string and the organ pipe, can exhibit the phenomenon of resonance if the driving "force" happens to oscillate at or near one of the special frequencies (called normal modes) of the system. If you blow air across the top of a bottle, several possible sound frequencies are present in the stream of air but the bottle will pick out one (usually) that corresponds to one of its normal modes (dependent on the level of liquid in the bottle) and make a relatively loud sound at that frequency. This is one example of a resonance.

Superposition of waves of the same frequency traveling in the same direction leads to constructive and destructive interference. Superposition of waves traveling in the same direction with nearly equal frequencies leads to a phenomenon known as "beats".

The mathematical expression that predicts the beat frequency for two frequencies f_1 and f_2 contains a sum term $(f_1 + f_2)/2$, which represents the average frequency, and a difference term that gives the beats $(f_1 - f_2)/2$. The ear is a power detector--it senses the square of the amplitude--so we hear two maxima and two zeroes per cycle. Thus the beat frequency we hear is $\Delta f = f_1 - f_2$. See Example 10. Beats result from a superposition in the time domain.

The Doppler effect for sound waves is very complicated as it involves three reference frames and three velocities: the speed of sound c, the velocity of the sound source, v_s, and the velocity of the observer or listener, v_L. The sign convention employed in the text can be visualized in the following way: designate the position of the observer to be the <u>origin</u> and draw the axis from the observer to the source. That direction is <u>positive</u> for the velocities v_L and v_s. If v_L and v_s point in the positive direction (we only consider motion on this line in the text), they are positive, otherwise negative.

One way of obtaining the formula for the Doppler effect is to imagine that the source emits sharp sound pulses with a period T so the source frequency $f_s = T^{-1}$. If the origins of the coordinate systems on the listener and source coincide at $t = 0$ when the first pulse is emitted, there is no delay between sending and receiving the first pulse. When the second pulse is emitted by the source (at time T), the distance between source and observer is equal to $(v_s - v_L)T$. The velocity of this sound pulse relative to the listener is $c + v_L$ so if $\Delta t = t - T$ is the time interval necessary for the pulse to reach the listener, we have:

$$\Delta t = \frac{\left(v_s - v_L\right)T}{\left(c - v_L\right)}$$

Substituting $\Delta t = t - T$ and rearranging the above equation gives:

$$t(c + v_L) = T(c + v_s)$$

Since the observer received the first pulse from the source at t = 0 and the second at time t, then t is the period of the source according to the listener. Accordingly $f_L = t^{-1}$ and since $f_s = T^{-1}$ the frequencies are related by:

$$\frac{f_L}{\left(c + v_L\right)} = \frac{f_s}{\left(c + v_s\right)}$$

This expression is correct for sound waves but would be incorrect for light waves (or electromagnetic waves). The problem with light occurs because the method used to find the relative velocity here breaks down at high speeds. Light travels with a velocity c in any inertial frame whereas sound velocity is frame dependent. The Doppler effect for light depends only on the relative velocity between source and observer, which is not the case for sound waves. The sign convention used in the Doppler effect is illustrated in Examples 11 and 12.

PROBLEM-SOLVING STRATEGY

The main text gives an excellent problem-solving strategy for Doppler Effect problems that should be reviewed now. Other difficulties may be encountered in comparing intensities in decibels. Carefully review the defining equation and practice with intensities that are 10 times greater (100 times greater) and 10 times smaller that a chosen reference intensity.

QUESTIONS AND ANSWERS

QUESTION. Explain how to measure the bulk modulus of air using a tuning fork of known frequency and a tube of variable length.

ANSWER. Using the tuning fork, which produces a wave of constant frequency (f) and the tube of variable length (L), measure the distance between two successive intensity resonances ΔL. The sound velocity (c) is then obtained from c = 2f ΔL. Since c^2 = B/r, where B is the bulk modulus and r is the density of air, the value of B is obtained if the density of air is known.

QUESTION. By what factor must one amplify the intensity of a whisper to make it reach the threshold of pain?

ANSWER. For a whisper with intensity $I/I_0 = 10^3$ (or 30 dB), the intensity must be amplified to $I/I_0 = 10^{12}$ (or 120 dB) to reach the pain threshold. Thus the amplification needed is 10^9 (or 90 dB).

QUESTION. Is a velocity node also a pressure node? Why?

ANSWER. Yes. The wave functions that describe the velocity and the pressure are both 90° out of phase with the displacement wave function.

334

QUESTION. Estimate the lowest two frequencies that you can achieve by blowing across the top of a coke can.

ANSWER. Treating the can (length L) as a pipe open at one end and closed at the other, the first resonance occurs when $L = l/4$ so that $f_0 = c/4L$ and the second when $L = 3\, l/4$ so that $f_1 = 3c/4L = 3\, f_0$. Estimating $L = 11.4$ cm (4.5 inches) and taking $c = 343$ m/s gives $f_0 = 750$ Hz.

EXAMPLES AND SOLUTIONS

Example 1

The "sound intensity level" of ordinary conversation at an average frequency of 500 Hz is 65 db above the hearing threshold (where the intensity is 10^{-12} W/m^2). Calculate the displacement amplitude A and compare it with the mean spacing between air molecules at S.T.P.

Solution:

(a) To find the intensity, I, write

$$65 \text{ dB} = (10 \text{ dB}) \log \left(\frac{I}{I_0} \right)$$

where $I_0 = 10^{-12}$ W/m^2. Solving this gives $I = 3.16 \times 10^{-6}$ W/m^2. The intensity is related to the displacement amplitude by:

$$I = \frac{1}{2} \omega k B A^2 = \frac{1}{2} (2\pi f) \left(\frac{2\pi}{\lambda} \right) B A^2$$

If we take $v = 340$ m/s, $\lambda = 0.68$m, and $B = 1.42 \times 10^5$ Pa, then A is equal to:

$$A = \frac{1}{2\pi} \sqrt{\frac{2 I \lambda}{fB}} = \frac{1}{2\pi} \sqrt{\frac{2 \left(3.16 \times 10^{-6} \text{ W}/\text{m}^2 \right) (0.68 \text{ m})}{(500 \text{ Hz}) \left(1.42 \times 10^5 \text{ N}/\text{m}^2 \right)}}$$

$$A = 3.92 \times 10^{-8} \text{ m}$$

(b) To find the average spacing (a) between air molecules at S.T.P., since 1 mole occupies $22.4 = 22.4 \times 10^{-3}$ m^3 and $N_A = 6.02 \times 10^{23}$ mol^{-1}, use the result: $N_A a^3 = 22.4 \times 10^{-3}$ to obtain the value for (a), the average spacing between molecules $a = 3.34 \times 10^{-9}$ m.

The displacement amplitude for ordinary conversation is more than 10 times larger than the average spacing between gas molecules at S.T.P.

Example 2

Two sound waves are characterized by pressure amplitudes of $P_1 = 20$ Pa and $P_2 = 0.20$ Pa. Calculate the ratio of their intensities and the difference in their sound intensity levels.

Solution:

Since the intensities are proportional to the squares of the respective pressure amplitudes,

$$\frac{I_1}{I_2} = \left(\frac{P_1}{P_2}\right)^2 = \left(\frac{20}{0.2}\right)^2 = 10^4$$

To obtain the difference in sound intensity levels, calculate:

$$\beta = \beta_1 - \beta_2 = (10 \text{ dB})\left[\log\left(\frac{I_1}{I_0}\right) - \log\left(\frac{I_2}{I_0}\right)\right]$$

$$\beta = (10 \text{ dB})\log\left(\frac{I_1}{I_2}\right)$$

Since we have

$$\log\frac{I_1}{I_2} = 4$$

$\Delta\beta = 40$ dB. The absolute intensities can be calculated if desired from the known pressure amplitude, the speed of sound, and the bulk modulus.

Example 3

It is desired to make an organ pipe that will produce middle C on the "even tempered scale", or f = 261.6 Hz. Take c = 345 m/s.

    (a) If the tube is open at both ends, how long should it be?

    (b) If the tube is open at one end but closed at the other, how long should it be?

Solution:

(a) A tube open at both ends has the same relationship between λ and L as a tube closed at both ends or a string fixed at both ends.

$$\frac{\lambda_1}{2} = L$$

The wavelength corresponding to middle C is:

$$\lambda_1 = \frac{c}{f_1} = \frac{(345 \text{ m/s})}{(261.6 \text{ /s})} = 1.319 \text{ m}$$

Solving for the length, we have

$$L = \frac{\lambda_1}{2} = \frac{1.319 \text{ m}}{2} = 0.659 \text{ m}$$

(b) For a tube open at one end but closed at the other, the wavelength is related to the length by

$$\frac{\lambda_1}{4} = L$$

The wavelength is unchanged from part (a) so

$$L = \frac{\lambda_1}{4} = \frac{(1.319 \text{ m})}{4} = 0.330 \text{ m}.$$

Example 4

Obtain the allowed frequencies (normal modes) of the two organ pipes in the previous problem where $f_1 = 261.6$ Hz and $c = 345$ m/s.

Solution:

(a) For the pipe open on both ends, of length $L = 0.659$ m, the possible wavelengths satisfy:

$$\frac{\lambda_1}{2} = L; \lambda_2 = L; \text{ and } \frac{3\lambda_3}{2} = L$$

giving the general relationship between wavelength and pipe length:

$$\lambda_n = \frac{2L}{n} \qquad \text{where } n = 1, 2, ..., \text{ etc.}$$

The frequencies of the harmonics are:

$$f_n = \frac{c}{\lambda_n} = n\left(\frac{c}{2L}\right) \quad \text{where } n = 1, 2, 3, \text{ etc.}$$

$$f_1 = \left(\frac{c}{2L}\right) = \left(\frac{345 \text{ m/s}}{2(0.659 \text{ m})}\right) = 261.6 \text{ Hz}$$

$$f_2 = \left(\frac{c}{L}\right) = \left(\frac{345 \text{ m/s}}{0.659 \text{ m}}\right) = 523.2 \text{ Hz}$$

$$f_3 = \left(\frac{3c}{2L}\right) = 3f_1 = 784.8 \text{ Hz}$$

All harmonics are present.

(b) For the pipe open at one end,

$$\frac{\lambda_1}{4} = L \; ; \; \frac{3\lambda_3}{4} = L \; ; \text{ and } \; \frac{5\lambda_3}{4} = L$$

The general term is:

$$\lambda_n = \frac{4L}{2n - 1} \qquad \text{where } n = 1, 2, \dots, \text{ etc.}$$

This gives for the normal mode frequencies

$$f_n = \frac{c}{\lambda_n} = \frac{c}{4L}(2n - 1) \qquad \text{where } n = 1, 2, \dots, \text{ etc.}$$

$$f_1 = 261.6 \text{ Hz}$$

$$f_2 = 784.8 \text{ Hz}$$

$$f_3 = 1308 \text{ Hz.}$$

All odd harmonics are present but the even ones are missing.

Example 5

In a Kundt's tube (see figure in the text), the spacing between mounds of powder is measured to be 4.4 cm when the source frequency is 4000 Hz.

   (a) Calculate the sound velocity in the tube.
   (b) Assuming the gas in the tube to be air ($M = 29 \times 10^{-3}$ kg·mol^{-1}, $\gamma = 1.4$) calculate the Kelvin temperature.

Solution:

(a) The distance d between pressure antinodes where the powder mounds accumulate is equal to one-half wavelength ($\lambda/2$).

$$d = \frac{\lambda}{2}$$

Thus we have

$$v = \lambda f$$

$$= 2df$$

$$= 2(0.044 \text{ m})(4000 \text{ s}^{-1})$$

$$= 352 \text{ m/s}$$

(b) The speed of sound in a gas is related to the temperature by:

$$v^2 = \gamma \frac{RT}{M}$$

Solving for T, we have

$$T = \frac{Mv^2}{\gamma R}$$

$$T = \frac{(29 \times 10^{-3} \text{ kg/mol})(352 \text{ m/s})^2}{(1.40)(8.314 \text{ J/molK})}$$

$$= 308.7 \text{ K} \quad (\text{or } 35.7 \text{ C})$$

Example 6

The gas in a Kundt's tube is a mixture of H_2 and O_2 ($\gamma = 1.4$ for each). The temperature of the gas mixture is 20 C. At a source frequency of 5000 Hz, the distance between mounds of powder is 5.57 cm. Calculate the fraction of H_2 present in the gas.

Solution:

First the speed of sound must be calculated from the given data. In the previous example it was shown that:

$$v = 2df$$

Substituting numerically we have

$$v = 2(0.0557 \text{ m})(5000 \text{ s}^{-1}) = 557 \text{ m/s}.$$

The molecular mass, M, can be calculated from the equation

339

$$v^2 = \gamma \frac{RT}{M}$$

As v, T, and γ are known, we can solve for M:

$$M = \gamma \frac{RT}{v^2}$$

$$M = (1.40)\frac{(8.314\ J/molK)(293\ K)}{(557\ m/s)^2}$$

$M = 10.99 \times 10^{-3}$ kg/mol.

$\quad = 10.99$ g.

The molecular mass M can be written in terms of the known molecular masses of H_2 and O_2 and the fraction x of H_2 present.

$$M = x(2\ g/mol) + (1 - x)(32\ g/mol)$$

For 1 mole we have

$$10.99\ g = 32\ g - x(30\ g)$$

Solving for x, we have

$$x = \frac{(21.0\ g)}{(30.0\ g)} = 0.700$$

or 70% of the mixture is H_2.

Example 7

An acoustic interferometer consists of two "U-shaped tubes" (containing air) driven by a common source with the superimposed sound waves incident on a common detector. One tube is of fixed length while the second one can "slide" (like a trombone slide), varying its length. A horizontal motion of this tube through a distance d produces a path change of 2d because of the shape of the tube. For an acoustic interferometer filled with air calculate the frequency of vibration of the source that will produce successive maxima and minima for a horizontal motion of the slide of 2 cm.

Solution:

The path change corresponding to a horizontal motion of the slide through a distance d is

$$\Delta x = 2d$$

To go from a minimum to a maximum, this path change must correspond to one half wavelength:

$$\Delta x = 2d = \frac{\lambda}{2}$$

or

$$\lambda = 4d$$

$$\lambda = 4(0.02 \text{ m}) = 0.08 \text{ m}$$

The frequency will be

$$f = \frac{v}{\lambda} = \frac{(345 \text{ m/s})}{(0.08 \text{ m})} = 4125 \text{ Hz}.$$

Example 8

In an attempt to find where a plug is in a tube containing air, a plumber blows air across the opening to the tube and hears a resonance at a frequency of 80 Hz. If this is the fundamental mode, how far away is the plug? (Take v = 345 m/s.)

Solution:

Treating this as an "organ pipe" with one end open and one end closed, let L be the distance from the open end to the plug. This distance is related to the fundamental wavelength, λ_1, by

$$L = \frac{\lambda_1}{4}$$

The fundamental frequency, f_1, is given by:

$$f_1 = \frac{v}{\lambda_1} = \frac{(345 \text{ m/s})}{4L}$$

Taking $f_1 = 80$ Hz, we have

$$L = \frac{(345 \text{ m/s})}{4(80 \text{ Hz})} = 1.08 \text{ m}.$$

Example 9

A one meter long tube open at one end and closed at the other contains water to a depth d. Assuming the sound waves have a displacement node at the water surface and an antinode at the open end, calculate the liquid depth that makes the tube resonate at C' above middle C, middle C, and a note an octave below middle C. Take the speed of sound to be 340 m/s. See Fig. 16-1.

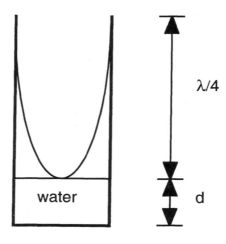

Figure 16-1

Solution:

The frequencies can be obtained from the main text. For ease they are given here. C' has a frequency of 528 Hz, middle C has a frequency of 264 Hz, and a note one octave below middle C would have a frequency of 1/2(264)Hz = 132 Hz.

For resonance, we must be able to set up a standing wave in the tube. The lowest mode that can be excited has one quarter wavelength just fitting in the available space.

$$\frac{\lambda}{4} = 0.161 \text{ m} \quad \text{for C'}$$

$$= 0.322 \text{ m} \quad \text{for C}$$

$$= 0.644 \text{ m} \quad \text{for the note at 132 Hz.}$$

When using the note at 132 Hz, only one resonance will occur and that is when

$$d = 1 \text{ m} - \frac{\lambda}{4} = (1 - 0.644) \text{ m} = 0.356 \text{ m}.$$

When using C, we can obtain two resonances, one occurring for

$$d_1 = 1 \text{ m} - \frac{\lambda}{4} = (1 - 0.322) \text{ m} = 0.678 \text{ m}.$$

and a second one occurring for

$$d_2 = 1 \text{ m} - \frac{3\lambda}{4} = (1 - 0.966) \text{ m} = 0.034 \text{ m}.$$

since the next mode will also fit into one meter.

For the note at 528 Hz, we hear resonances at:

$$d_1 = 1 \text{ m} - \frac{\lambda}{4} = 0.839 \text{ m}.$$

$$d_2 = 1 \text{ m} - \frac{3\lambda}{4} = 0.517 \text{ m}.$$

$$d_3 = 1 \text{ m} - \frac{5\lambda}{4} = 0.195 \text{ m}.$$

Since $(7\lambda/4)$ is greater than 1 m, we only obtain three resonances.

Example 10

A detector at $x = 0$ responds to two traveling waves

$$y_1 = A \sin(k_1 x - \omega_1 t)$$

$$y_2 = A \sin(k_2 x - \omega_2 t)$$

where $f_1 = 400$ Hz and $f_2 = 396$ Hz. How many beats per second are obtained if the detector is a power detector?

Solution:

Superimposing these two solutions to give the total displacement y_T, we have

$$y_T = y_1 + y_2 = A[\sin(k_1 x - \omega_1 t) + \sin(k_2 x - \omega_2 t)]$$

Using the identity

$$\sin C + \sin D = 2 \sin \left(\frac{C + D}{2} \right) \cos \left(\frac{C - D}{2} \right)$$

the quantity $[\sin(k_1 x - \omega_1 t) + \sin(k_2 x - \omega_2 t]$ can be rewritten as

$$y_T = 2A \cos\left[\frac{1}{2}\left(\Delta k x - \Delta \omega t\right)\right] \sin\left(\bar{k}x - \bar{\varpi}t\right)$$

where

$$\bar{k} = \frac{k_1 + k_2}{2}; \quad \bar{\varpi} = \frac{\omega_1 + \omega_2}{2}; \text{ and } \Delta k = k_1 - k_2 \text{ with } \Delta\omega = \omega_1 - \omega_2$$

For the detector at $x = 0$, the displacement is:

$$y_T = -2A \cos\left[\frac{1}{2}\left(\Delta\omega t\right)\right] \sin\left(\bar{\varpi}t\right)$$

Since the detector is sensitive to power, it measures y_T^2. For instance, zeroes in the intensity occur whenever $\cos(\Delta\omega t/2) = 0$. The term $\sin^2 \bar{\varpi}t$ can be considered to supply a constant time averaged background signal that is modulated by the term $\cos(\Delta\omega t/2)$. The times, t_n, where we get nulls, obey the equation:

$$\frac{\Delta\omega t_n}{2} = \frac{(2n-1)}{2}\pi \quad \text{where } n = 1,2,3, \text{ etc.}$$

Explicitly we have

$$t_1 = \frac{\pi}{\Delta\omega} \text{ and } t_2 = \frac{3\pi}{\Delta\omega}$$

The time difference between these nulls is

$$\Delta t = \frac{2\pi}{\Delta\omega} = \frac{1}{\Delta f}$$

The number (n) of beats in a time T is

$$n = (T/\Delta t) = T(\Delta f)$$

So if T is one second, we hear Δf beats per second. Numerically, for this example, we have $\Delta f = (400 - 396)$ Hz = 4 Hz so there are four beats per second.

Example 11

Consider a source that produces a sound with frequency of 500 Hz. If the speed of sound in air is denoted by c, and if c = 340 m/s, and the source and listener both move along the line joining them with speeds of 25 m/s (about 56 mph), what are the frequencies heard by the listener for all possible signs of the velocities? The directions are specified below.

Solution:

The formula for the Doppler shift given in the text is:

$$\frac{f_L}{\left(c + v_L\right)} = \frac{f_S}{\left(c + v_S\right)}$$

This example is an exercise in using the sign conventions on v_L and v_S.

(a) Let both v_L and v_S point in the positive x direction. They are both then positive as the positive direction is from the listener to the source. Since both speeds are numerically equal, f_L = f_S = 500 Hz. Here there is no relative motion of the source and listener and there is no Doppler shift.

(b) Let v_L point in the negative x direction and v_S point in the positive x direction. Then we have

$$\frac{f_L}{\left(340 \text{ m/s} - 25 \text{ m/s}\right)} = \frac{f_S}{\left(340 \text{ m/s} + 25 \text{ m/s}\right)}$$

$$f_L = \left(500 \text{ Hz}\right)\left(\frac{315}{365}\right) = 432 \text{ Hz}$$

(c) Let v_L point in the positive x direction and v_S point in the negative x direction. Then we have

$$\frac{f_L}{\left(340 \text{ m/s} + 25 \text{ m/s}\right)} = \frac{f_S}{\left(340 \text{ m/s} - 25 \text{ m/s}\right)}$$

$$f_L = \left(500 \text{ Hz}\right)\left(\frac{365}{315}\right) = 579 \text{ Hz}$$

(d) If both v_L and v_S point in the negative x direction, then again there is no relative motion of source and listener so f_L = f_S = 500 Hz.

Example 12

A "bullet" car traveling at a speed with respect to the ground of 20 m/s signals a "bullet" train traveling at a speed with respect to the ground of 30 m/s with a sound wave of frequency 600 Hz. Take the speed of sound in air to be 340 m/s. The sound wave bounces off the bullet train and is then received back at the car at a new frequency. Calculate that frequency for (a) the train going in the same direction as the car as shown in Fig. 16-2a; and (b) the train going in the opposite direction as the car as shown in Fig. 16-2b.

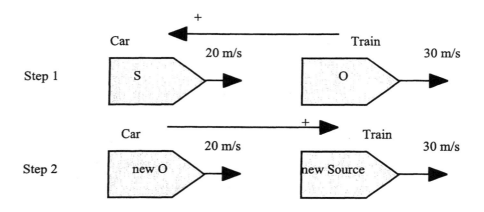

Figure 16-2a

Solution:

The sign convention is illustrated in the figure. The axis from observer (listener) to the source defines the positive direction for the source and listener velocities used to find the unknown frequency. We will work this problem in two steps as suggested by the problem-solving strategy of the main text.

(a) Step 1. Using the signs for the velocities based on Fig. 16-2a; v_L and v_S are both negative.

$$\frac{f_L}{(c - 30 \text{ m/s})} = \frac{f_S}{(c - 20 \text{ m/s})}$$

Solving for the "listener" frequency--which we will label "intermediate frequency" because there is another step,

$$f_L = f_S \frac{(c - 30 \text{ m/s})}{(c - 20 \text{ m/s})} = f_I; \text{ the intermediate frequency}$$

Step 2. The bullet train is now the new source of sound, sending its signal back to the car at the intermediate frequency just calculated. Referring to Fig. 16-2a, "Step 2", we see that both velocities are now positive.

$$\frac{f_I}{(c + 30 \text{ m/s})} = \frac{f_{car}}{(c + 20 \text{ m/s})}$$

Solving for the new listener frequency, the frequency detected at the car,

$$f_{car} = f_I \frac{(c + 20 \text{ m/s})}{(c + 30 \text{ m/s})} = f_s \frac{(c - 30 \text{ m/s})}{(c - 20 \text{ m/s})} \frac{(c + 20 \text{ m/s})}{(c + 30 \text{ m/s})}$$

Substituting $c = 340$ m/s and $f_S = 600$ Hz, we find $f_{car} = 565$ Hz.

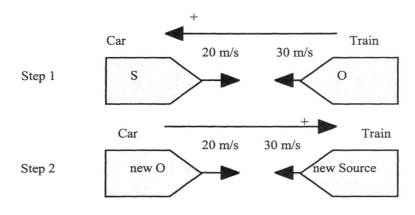

Figure 16-2b

(b) Step 1. Refer to Fig. 16-2b where now v_S is negative but v_L is positive giving,

$$f_L = f_s \frac{(c + 30 \text{ m/s})}{(c - 20 \text{ m/s})} = f_I; \text{ the intermediate frequency}$$

Step 2. Here, v_S is negative but v_L is positive because the roles of source and listener have been reversed.

$$f_{car} = f_I \frac{(c + 20 \text{ m/s})}{(c - 30 \text{ m/s})} = f_s \frac{(c + 30 \text{ m/s})}{(c - 20 \text{ m/s})} \frac{(c + 20 \text{ m/s})}{(c - 30 \text{ m/s})}$$

Evaluating this numerically, $f_{car} = 806$ Hz. Thus the frequency detected at the car by the sound reflected from the train is very sensitive to whether the train is approaching the car or receding.

QUIZ

1. The pressure amplitude (P) is related to the velocity amplitude (V) by $P = ZV$ where $Z = (B\rho)^{1/2}$ is the mechanical impedance. If Z, for air, is 429 kg/m$^2 \cdot$s, and the pressure amplitude for a 100 Hz sound is 3×10^{-5} Pa, (a) find the velocity amplitude (V) and (b) the displacement amplitude (A).

Answer: (a) $V = 7.0 \times 10^{-8}$ m/s, (b) $A = 1.11 \times 10^{-10}$ m.

2. A sound source radiates isotropically in all directions. The acoustic power of the source is 200 Watts. (a) Find the intensity at a distance of 1 meter from the source. (b) At what distance from the source will the intensity level be 20 dB lower than at R = 1 m?

Answer: (a) $I = 15.9$ W/m^2, (b) R = 10 m.

3. Two identical loudspeakers that radiate isotropically in all directions are oscillating in phase at 300 Hz. Take c = 330 m/s. Their acoustic power output is 10^{-3} W. (a) What is their minimum separation if a point 1.1 m from one of the speakers, on the line joining the speakers, is a point of *minimum sound intensity*? (b) Determine the sound intensity at that point if the power to the nearest speaker is shut off. (c) Determine the sound intensity at that point if the power to the most distant speaker is shut off.

Answer: (a) $d_{min} = 1.65$ m, (b) $I = 6.58 \times 10^{-5}$ W/m^2, (c) $I = 2.63 \times 10^{-4}$ W/m^2.

4. Calculate the speed of propagation at 300 K of a wave
   - (a) in a mixture of diatomic gases ($\gamma = 7/5$) containing 80% N_2 and 20% O_2 (nominally "air");
   - (b) in neon, a monatomic gas ($\gamma = 5/3$); and
   - (c) in helium (^{4}He) gas ($\gamma = 5/3$).

Answer: (a) 347 m/s; (b) 456 m/s; and (c) 1019 m/s.

5. In an experiment to test Doppler sound equipment to detect speeding motor vehicles, all test cars are equipped with a 500 Hz (f_s) transmitter and the police cars have a detector that measures the frequency shift (Doppler shift), $\Delta f = f_L - f_s$. Take $c_{air} = 345$ m/s. Calculate the shift Δf detected by a *stationary* police car for a test vehicle moving toward the police car at (a) 120 km/hr; (b) 60 km/hr; and (c) 240 km/hr. (d) Is the frequency shift *linear* in the vehicle velocity?

Answer: (a) 53.5 Hz; (b) 25.4 Hz; (c) 120 Hz; (d) No.

6. Assume each automobile below carries a sound source with frequency of 20,000 Hz. Take c = 330 m/s and assume the listener is at rest. (a) What frequency would be "heard" by a listener if the car moved toward the listener at a speed of 112 km/hr (about 70 mph)? (b) What

frequency would be "heard" by a listener if the car moved away from the listener at the same speed? (c) If the listener also had a 20,000 Hz source, what would be the beat frequency in the two cases?

Answer: (a) 22,081 Hz, (b) 18,277 Hz, (c) 2081 Hz and 1722 Hz.

7. Calculate the distance between pressure antinodes (or displacement nodes) in a Kundt's tube filled with a mixture containing 60% Helium (mass 4) and 40% Neon (mass 20) at 20 C for a source frequency of 5000 Hz.

Answer: d = 6.25 cm.

8. A long thin pipe connects the water in a well with the surface. Sound resonates in this pipe at a lowest frequency of 24 Hz. If c = 345 m/s, calculate how deep the well is by treating this system as an "organ pipe" open on one end but closed on the other.

Answer: depth = 3.59 m.

9. A pipe open on both ends is 2m long. Taking the speed of sound in air to be 345 m/s, calculate:
      (a) the wavelength of the lowest frequency mode;
      (b) the fundamental frequency; and
      (c) the frequency of the second harmonic.

Answer: (a) 4 m; (b) 86.25 Hz; and (c) 172.5 Hz.

10. An organ pipe open at one end but closed at the other produces a frequency f = 230 Hz. This frequency is the second overtone (or fifth harmonic) of the fundamental. Take the speed of sound in air to be 345 m/s.
      (a) Calculate the length of the pipe.
      (b) Calculate the fundamental frequency.

Answer: (a) L = 1.875 m, (b) f_1 = 46 Hz.

17
TEMPERATURE AND HEAT

OBJECTIVES

In this chapter you begin the study of the equilibrium properties of large numbers of particles that constitute macroscopic systems. The system's temperature is of central importance to this discussion. You will calculate the quantity of heat transferred in several specific processes and relate it to temperature changes and phase changes in the system under consideration. Also, you will be introduced to the processes by which heat energy can be transferred from one body to another; conduction, convection, and radiation. Your objectives are to:

Apply the different numerical values of the fixed temperature points to establish the most common temperature scales.

Familiarize yourself with a few common thermometers.

Calculate thermal expansion of common materials.

Calculate stresses that can be set up by temperature changes.

Relate the temperature change of an object to the quantity of heat given to or taken from the object.

Calculate the quantity of heat required to convert a pure substance from one phase (e.g. solid) to another (e.g. liquid).

Calculate the temperature changes and/or phase changes of pure substances resulting from the conversion of some other form of energy (mechanical, electrical, etc.) into thermal energy.

Calculate the heat energy per unit time (the heat current) transferred by conduction due to a temperature difference in a given material.

Calculate the heat current radiated from a given surface at temperature T.

REVIEW

Three undefined physical quantities (mass, length, and time) were introduced when we studied the mechanics of a particle. We were assumed to be familiar with these quantities (i.e. fast, slow, long, short, heavy, light being frequently used comparisons) and were given only operational definitions (told how to measure these quantities in terms of an agreed upon standard). A fourth such undefined basic quantity, temperature, is introduced in this chapter.

Two macroscopic bodies A and B placed in contact with each other in such a way that they can exchange energy (diathermal wall) eventually reach a state where the variables describing the two bodies stop changing with time. The two bodies are said to be in *thermal equilibrium*. The temperature of body A is now equal to that of body B. If A and B are now isolated from each other and a third body C is put in contact with A, A and C will be in thermal equilibrium if the temperature of body C is equal to that of body A. Thus there exists a scalar quantity, temperature, that is a property of all macroscopic systems in equilibrium states such that temperature equality is a necessary and sufficient condition for thermal equilibrium. Alternatively: two systems each in thermal equilibrium with a third system are in thermal equilibrium with each other. This is a statement of the zeroth law of thermodynamics.

A thermometer can be any instrument or device capable of assigning a numerical value to the temperature of a body which it is in contact with. Various physical properties can be measured in order to infer temperature: length of a column of liquid, pressure of a gas, resistivity of a metal or semi-conductor, magnetic properties of certain materials, etc., so long as they give adequate sensitivity, accuracy, reproducibility, speed of measurement, and ease of interpolation. Concerning this last point, thermometers are usually calibrated at a few fixed temperature points and other values of the temperature are either interpolated or extrapolated. It is desirable that the material property to be measured, the state coordinate, change linearly with temperature if the interpolation is to be reliable. This is not always the case.

Various temperature scales are in use: the Fahrenheit scale, the Celsius scale, the Rankine scale, and the Kelvin scale. Thus a particular temperature could have four different numerical values on the four different scales. These temperature scales are established by assuming the temperature (t) is linearly proportional to the physical parameter (X) being measured by the thermometer,

$$t(X) = AX + B$$

Measurement of X at two known temperatures (X_1 at t_1 and X_2 at t_2) determines both A and B and fixes the temperature scale:

$$t_1 = AX_1 + B$$

$$t_2 = AX_2 + B$$

so that A and B are now known functions of t_1, t_2, X_1, and X_2.

For the Celsius scale, the temperature of the melting point of ice is assigned the value 0°C and the temperature of the boiling point of water is assigned the value 100°C. For the Fahrenheit scale, these same two fixed points are assigned the values 32°F and 212°F respectively. Thus both the size of the degree and the zero of the scale are different for the two scales. From just the information given here, it is possible to derive the conversion equation (see Example 1 for one method):

$$t_c = \frac{5}{9}\left(t_f - 32°\,F\right)$$

The Kelvin and Rankine scales are both absolute scales established such that the zero of the temperature scale coincides with zero value of the thermometric property, X, being measured (think of X as the gas pressure of a constant volume thermometer rather than the length of a mercury column). This is equivalent to setting B equal to zero in our previous linear relationship between t and X. Now only one other fixed point is necessary to establish the temperature scale. That fixed point is chosen by international agreement to be the triple point of water (t_c = 0.01°C).

The Kelvin scale uses the same size of temperature unit (degree) as the Celsius scale. The Rankine scale uses the same size degree as the Fahrenheit scale. Thus $A_K = A_c$ and $A_R = A_f$. Capital T's are used to denote these absolute temperatures.

$$T_K(X) = A_K X = A_c X$$

$$T_R(X) = A_R X = A_f X$$

Extrapolation of these linear relationships (T verses X) using gas-law thermometry yields the result that the triple point of water (0.01°C) corresponds to a temperature of 273.16 K on the Kelvin scale. This establishes the conversion formula:

$$t_c + 273.15\ K = T_K.$$

With very few exceptions (water below 4°C happens to be one), liquids and solids expand when heated. If the temperature change ΔT is not too large, the increase in a typical linear dimension, ΔL, of a homogeneous material is found to be proportional to the mean length and the temperature change. The proportionality constant (coefficient of linear expansion α) is material dependent and tabulated in reference books over wide ranges of temperature for most materials:

$$\Delta L = \alpha L (\Delta T)$$

For most practical problems, it does not matter significantly whether the value of L in the above equation is taken to be the initial value, the final value, or the mean value (see Example 4) since α is extremely small (10^{-5} per C°) and not known to many significant figures. When considering expansions (or contractions) of areas or volumes of complicated objects, it is useful to imagine the expansions (or contractions) as photographic enlargements (or reductions). Thus in a donut shaped object, when heated, both the inner and outer radius have the same fractional increase ($\Delta R/ R = \alpha \Delta T$).

Since different materials have different thermal expansion coefficients, stresses can be set up as a result of temperature changes in heterogeneous bodies. These stresses arise from either real elongations or compressions of the materials or length changes that would occur if not otherwise prevented. To estimate the magnitude of such stresses, one usually finds the fractional change in length, $\Delta L/L = \alpha \Delta T$, from the temperature change and then uses the definition of Young's modulus to find the force per unit area (stress). See Examples 6 and 7.

Heat is introduced in this chapter as another form of energy. A heat flow is an energy transfer that takes place exclusively due to a temperature difference. When two objects at different temperatures are placed in contact, the final temperature is intermediate between the two initial temperatures. The conclusion is that heat energy flowed from the hotter body to the colder body.

Not all energy transfers involve a heat flow. Heat flow and performance of work are equivalent in the sense that an energy increase (decrease) occurs and no later experiment can tell how this change took place.

The ratio of the temperature change (dT) of a given object to the input of a specified infinitesimal quantity of heat energy, dQ, depends on the mass of the object and a material dependent property, the specific heat capacity, c. An essential definition is:

$$c = \frac{1}{m}\left(\frac{dQ}{dT}\right)$$

Another quantity defined in the text is the molar heat capacity (C) or the heat capacity per mole of matter. Since the number of particles per mole is constant, comparison of the molar heat capacities of different substances highlights differences that arise from physical effects not differences in number of particles. The relationship between c and C is: $C = Mc$ where M is the mass per mole. As pointed out in the text, c is generally a function of temperature and this equation must be integrated to find the quantity of heat Q involved in a given finite temperature change. For simplicity we usually assume that c is independent of T over the temperature range of interest.

If c is independent of T, then we can write:

$$\Delta Q = mc(\Delta T)$$

where ΔQ is the heat energy added to (or taken from) the system and ΔT is the temperature change.

Since heat is a form of energy and the joule (J) is the S.I. unit of energy, the proper S.I. units for c are J/kg C° as seen from either of the above equations. Two other units of heat energy are still used frequently. Both units were originally introduced to make the specific heat capacity of water numerically equal to 1.00. The calorie (cal) is the amount of heat energy required to raise the temperature of 1 g of water 1 C°. Thus c for water is 1 cal/g C°. One calorie is found experimentally to be the equivalent of 4.186 J so for water c = 4186 J/kg C°. The kilocalorie (kcal) is 10^3 calories, so for water c = 1 kcal/kg C°.

The Btu is defined such that the specific heat capacity of unit weight (1 pound) of water is equal to unity. Thus 1 Btu heat energy input will raise the temperature of 1 pound of water by 1 F°. Conversions between the Btu and cal units are illustrated in Example 8 but joules are used in all other examples.

Change of phase is still a topic of considerable interest in current research. Melting of a pure substance is a complicated physical phenomenon and not completely understood (at least in detail) on the microscopic level. Here we concentrate on the gross aspects of this problem which are well understood.

Melting of ice and the boiling of water are two common examples of a change of phase. In this context, we identify three phases of matter: solid, liquid and gaseous. There are other phases at higher temperatures (energies), such as plasmas and nuclear matter, but we will focus on these three.

The remarkable fact is that the phase changes, melting, freezing, vaporization, or liquefaction take place at *fixed temperature*. For instance, in melting as more heat is supplied, more of the solid is converted into the liquid phase but no temperature change occurs until all the solid disappears. This provides an excellent fixed temperature point for calibration of thermometers. The quantity of heat energy required for a given phase change, such as fusion (or melting), depends on the mass of material present and a material dependent property of the substance called the heat of fusion (L_F). If the phase change being made involves vaporization (or boiling), the heat of vaporization (L_V) is used. In either case, the heat required is:

$$Q = mL$$

where L_F or L_V is used, depending on the process. "Phase change salts" are now of great practical importance as they may be useful in storing energy conveniently for solar houses.

Conduction of heat energy is a transport of energy (without a net motion of matter) from a higher temperature to a lower temperature. This energy transport requires matter (some solid, liquid, gas or a combination of these) to be present. For a single substance, the heat energy transported per unit time (H) by conduction depends on the cross-sectional area, A, the temperature gradient (dT/dx), and a material dependent property, the thermal conductivity, κ:

$$H = -\kappa A \left(\frac{dT}{dx} \right)$$

Since heat is transported from "hot" to "cold", the origin of coordinates can be placed at the high temperature end of the material so dT/dx is negative. If H, κ, and A in this equation are constants, then T changes linearly with x as shown for the steady-state temperature in Fig. 17-15 in the text. If the area is not constant, as would be the case for radial heat flow in a cylinder or sphere, the temperature distribution will depend on the particular geometry. This point is illustrated in Example 18.

Conduction of heat through various combinations of objects is treated by extending the basic equation of conduction. If two objects are joined at a common interface (a series connection), the steady-state heat current is constant and the same through both objects. Example 15 treats this case. If two objects are connected side-by-side (a parallel combination) between two temperature reservoirs, the total heat current is the sum of the individual heat currents.

Heat transfer by convection involves the actual motion of matter and hence is of no practical importance in solids but is important in both liquids and gases. One important example of heat transport by convection is the warming of the Scandinavian Peninsula by the Gulf Stream. Convection is much more complicated than conduction (and hence less well understood) because the heat transport depends on details of the matter transport such as whether the flow is laminar or turbulent.

Heat transfer by radiation (meaning electromagnetic radiation or waves) is different from conduction or convection in that no intervening material is required--it works perfectly well in a vacuum. Thus in outer space where conduction and convection are negligible (since the necessary matter is not there), radiation becomes the dominant heat transfer mechanism. The power radiated by an object has been found to depend on the fourth power of its absolute temperature (T^4), its surface area, A, and specific properties of its surface through a factor e (called the emissivity). The proportionality constant here is denoted σ and called the Stefan-Boltzmann constant. Summarizing this:

$$H = Ae\sigma T^4 \quad \text{(Stefan Law)}$$

If a body is a perfect radiator (e = 1) it is called a "blackbody". A perfect reflector would have e = 0. All real surfaces lie somewhere in between these two idealizations. The power absorbed by a body depends on its particular environment. In general a good radiator is a good absorber also. Conversely a body that reflects most of the radiation incident on it is a poor absorber and a poor radiator.

When the heat current is radiated isotropically (uniformly in all directions), the energy per unit area per unit time (power per unit area) falls off as r^{-2} where r is the distance from the source. This is necessary to keep the power crossing an imaginary surface of area $4\pi r^2$ independent of r and equal to the power being radiated.

PROBLEM-SOLVING STRATEGIES

The main text includes Problem-Solving Strategies for (1) Thermal Expansion; (2) Calorimetry Problems; and (3) Heat Conduction. Now is an appropriate time to review those strategies. Values of the necessary elastic constants such as Young's modulus are given in the text. To calculate the thermal stress in a given piece of material, it is frequently useful to calculate the length change that would occur if the motion was unrestricted and then compute the force per unit area needed to compress or stretch the material from that hypothetical length to the real length.

When only thermal energy transfer occurs, the sum of the heat energy increases for all the bodies is zero. Be careful with signs; Q > 0 means in our sign convention that heat is added to a body, Q < 0 that heat is taken away from it. For many of the problems the specific heat must be expressed in J/kg C°. More so than most sets of problems met in this course, these problems require early substitution of numerical values since the number of physical processes that must be taken into account depend on how much heat energy is available. Examples 12 and 13 illustrate this point.

The quantity (H) used here to symbolize a "heat current" is related to the quantity of heat (ΔQ) per unit time (Δt) by $H = (\Delta Q/\Delta t)$.

The problems in this chapter are further examples of energy conservation and conversion. When the temperature at each point is no longer changing with time, the heat current into an element is equal to the heat current out of the same element.

The problems involving radiation are based on the Stefan law. Typical errors that arise involve not using an absolute (Kelvin) temperature or in not using the correct area in the formula. Frequently the effective area for absorption is *different* from that used for radiation.

QUESTIONS AND ANSWERS

QUESTION. Since the properties of water were so important (historically) in the study of heat and thermal processes, why wasn't water (rather than mercury) used for a thermometer?

ANSWER. Because a water thermometer, cooled to the ice point and placed on a stove, would have the "water based temperature" first fall and then rise since the volume of water (for constant mass) first decreases and then increases, having a minimum at 4 C.

QUESTION. Water has about five times the heat capacity of soil. How does this fact account for the differences between maritime and continental climates?

ANSWER. Water warms more slowly than soil and also cools more slowly than soil. In the summer, when the soil is warm, the water can be cooler while in winter, when the soil is cold, the water can give up heat to the air. Therefore, in coastal regions, the temperature differentials are smaller than inland in the continental regions.

EXAMPLES AND SOLUTIONS

Example 1

Given the melting point of ice, $t_C = 0$ °C and $t_F = 32$ °F and the boiling point of water, $t_C = 100$°C and $t_F = 212$°F, derive a formula that converts temperature on the Celsius scale into temperature on the Fahrenheit scale.

Solution:

Writing generally that $t_C = A_C X + B_C$ and $t_F = A_F X + B_F$, we can eliminate the common state coordinate to obtain a linear relationship between t_F and t_C.

$$X = \frac{t_0 - B_0}{A_0}$$

so that

$$t_F = A_F \left(\frac{t_0 - B_0}{A_0} \right) + B_F = C_1 t_0 + C_2$$

where C_1 and C_2 are constants. Using the given numerical values, we have

$$212° \text{ F} = C_1(100 \text{ °C}) + C_2$$

$$32° \text{ F} = C_1(0 \text{ °C}) + C_2$$

The second equation yields

$$C_2 = 32° \text{ F}.$$

Subtracting the second equation from the first gives $180 \text{ F}° = C_1(100 \text{ C}°)$, or

$$C_1 = 1.8 \text{ F}°/(\text{C}°).$$

Rewriting the linear relationship, we have

$$t_F - 32°\text{F} = (1.8 \text{ F}°)/(\text{C}°)t_C$$

or

$$t_F = \frac{9}{5} t_C + 32 ° \text{F}$$

Example 2

At what temperature, t, do the two scales of the last example give an identical reading?

Solution:

In the previous formula, set $t_C = t_F = t$ and rearrange to give:

$$t\left(1 - \frac{9}{5}\right) = 32°$$

$$t\left(-\frac{4}{5}\right) = 32°$$

$$\frac{4}{9}t = -\frac{5}{9}(32)$$

$$t = -40.$$

Thus when $t_C = -40°C$, $t_F = -40°F$, and the temperatures are identical. Above this temperature, a given number of degrees Celsius (say 20) is warmer than the same number of degrees Fahrenheit whereas $-50°C$ is colder than $-50°F$.

Example 3

What temperature does 0 K correspond to on the Celsius scale and on the Fahrenheit scale ?

Solution:

Since $T_K = t_C + 273.15$ K, then if $T_K = 0$ K, one has $t_C = -273.15$ K. Using

$$t_C = \frac{5°C}{9°F}\left(t_F - 32°F\right)$$

and then substituting $t_C = -273.15$ K, we can solve for t_F:

$$t_F = \left(\frac{9°F}{5°C}\right)t_C + 32°F = \left(\frac{9°F}{5°C}\right)(-273.15°C) + 32°F$$

$$= -459.67°F.$$

Example 4

Compute the length change for a brass bar ($\alpha = 2 \times 10^{-5}$ per C°) originally 1 meter long for a 100° C temperature rise from the expression $\Delta L = \alpha L \Delta T$ using for L on the right hand side (a) the initial length, (b) the final length, and (c) the mean length.

Solution:

(a) $L - L_0 = \Delta L = \alpha L_0 \Delta T$

$L = L_0(1 + \alpha \Delta T)$, this is the form nearly always used.

We have

$L = (1 \text{ m})[1 + (2 \times 10^{-5})(10^2)]$

or

$L - 1 \text{ m} = 2 \times 10^{-3} \text{ m}.$

(b) $L - L_0 = \Delta L = \alpha L \Delta T$

$L = L_0/(1 - \alpha \Delta T)$

Numerically we now have

$L = 1 \text{ m}/(1 - 2 \times 10^{-3}) = 1.002004 \text{ m}$

but the last 3 digits are not significant, so $L - 1 \text{ m} = 2 \times 10^{-3} \text{ m}$ as before.

(c) $L - L_0 = \Delta L = \alpha \left[\dfrac{L + L_0}{2}\right] \Delta T$

$L = L_0 \left[\dfrac{\left(1 + \alpha \Delta T/2\right)}{\left(1 - \alpha \Delta T/2\right)}\right]$

Numerically we have

$L = 1 \text{ m} (1 + 10^{-3})/(1 - 10^{-3}) = 1.002002 \text{ m}$ or

$\Delta L = L - 1 \text{ m} = 2 \times 10^{-3} \text{ m}.$

Thus in the formula $\Delta L = \alpha L \Delta T$, for most problems, it won't matter which value of L is used. The form seen in part (a) where L is taken to be L_0 is most frequently encountered.

Example 5

A flat sheet of material with coefficient of linear expansion $\alpha = 3 \times 10^{-5}$ per C° of original dimensions L_0, w_0, at temperature T_0 is heated to temperature T. Calculate the new area if $L_0 = 0.80$ m, $w_0 = 0.60$ m, $T_0 = 20°$ C, and $T = 120°$ C.

Solution:

Let A(T) be the area at temperature T. Then $A(T) = L(T)w(T)$.

$$A(T) = L_0[1 + \alpha(T - T_0)]w_0[1 + \alpha(T - T_0)] = L_0 w_0[1 + \alpha(T - T_0)]^2$$

$$= L_0 w_0[1 + 2\alpha(T - T_0) + \alpha^2(T - T_0)^2].$$

We will evaluate this square bracket numerically. In order, the three factors are: 1, 6×10^{-3}, and 9×10^{-6}. The sum of these terms is: 1.006009. Thus to the desired accuracy of this problem, $(1 + \alpha\Delta T)^2 \cong 1 + 2\alpha\Delta T$.

$$A(120°C) = 0.48 \text{ m}^2 (1.006) = 0.483 \text{ m}^2.$$

This order of approximation leads to a simple interpretation if we write:

$$A(T) - A(T_0) = A(T_0)(2\alpha)\Delta T.$$

Thus 2α (twice the coefficient of linear expansion) plays the role of the coefficient of area expansion. In the text, these considerations are carried over into three dimensions where it is shown, with the same approximations, that the coefficient of volume expansion is equal to three times the coefficient of linear expansion.

Example 6

In Fig. 17-1, consider an aluminum wire stretched across a steel yoke. Assume there are no stresses in the wire at 20°C and that the whole system is now cooled by 50°C ($\Delta T = -50°C$). If the area of contact of the wire to the yoke is 9×10^{-6} m², what force is exerted on the wire?

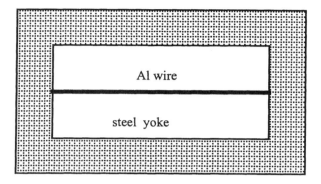

Figure 17-1

Solution:

For the wire,

$$\Delta L/L = \alpha_{Al}\Delta T = -2.4 \times 10^{-5} \times 0.5 \times 10^2 = -1.2 \times 10^{-3}.$$

For the yoke,

$$\Delta L/L = \alpha_s \Delta T = -1.2 \times 10^{-5} \times 0.5 \times 10^2 = -0.6 \times 10^{-3}.$$

The thin wire will contract by the amount dictated by the massive yoke. Thus $\Delta L/L$ will be -0.60×10^{-3} so tensile stresses are set up to take the length from the value it would like to have ($\Delta L/L = -1.2 \times 10^{-3}$) to the length determined by the yoke ($\Delta L/L = -0.60 \times 10^{-3}$).

Since
$$Y = \frac{F/A}{\Delta L/L}$$

we have

$$\frac{F}{A} = Y_{Al}\left(\frac{\Delta L}{L}\right) = \left(0.7 \times 10^{11}\ \text{N/m}^2\right)\left(-1.2 \times 10^{-3} + 0.6 \times 10^{-3}\right)$$

After multiplying by the given area, the force is found to be:

$$F = 3.8 \times 10^2\ \text{N, tending to stretch the wire.}$$

Example 7

For the piece shown in Fig. 17-2 where the aluminum bar is so large that its effect on the steel member cannot be ignored, calculate the forces on the Al for a temperature change of -50 C° and then of +50 C°. Take the area of contact (of each end) to be 9 x 10^{-4} m^2 and assume the pieces are welded at the ends and don't break loose. Take Y_{Al} = 0.7 x 10^{11} N/m^2 and Y_s = 2 x 10^{11} N/m^2.

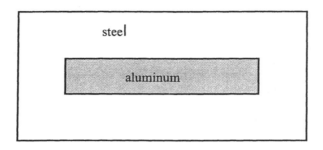

Figure 17-2

Solution:

Let the actual fractional change in length be = $\Delta L/L$. Let the fractional change in length of an isolated Al bar = $\Delta L_{Al}/L$. Let the fractional change in length of an isolated steel bar = $\Delta L_s/L$.

Recalling that Y = stress/ strain, we have

$$\text{strain in Al} = \frac{\Delta L_{Al} - \Delta L}{L} = \frac{1}{Y_{Al}}\left(\frac{F}{A}\right).$$

$$\text{strain in steel} = \frac{\Delta L - \Delta L_s}{L} = \frac{1}{Y_s}\left(\frac{F}{A}\right).$$

The force on the aluminum due to the steel is equal and opposite to the force on the steel due to the aluminum. Since the length (L) and area (A) are the same for both materials, we have

$$FL/A = Y_s(\Delta L - \Delta L_s) = Y_{Al}(\Delta L_{Al} - \Delta L).$$

We can solve for ΔL giving:

$$\Delta L = \left(\frac{Y_{Al}\Delta L_{Al} + Y_s\Delta L_s}{Y_{Al} + Y_s}\right)$$

Dividing by L and writing $\Delta L_{Al}/L = \alpha_{Al}\Delta T$ and $\Delta L_s/L = \alpha_s\Delta T$, we have:

$$\frac{\Delta L}{L} = \left(\frac{Y_{Al}\alpha_{Al} + Y_s\alpha_s}{Y_{Al} + Y_s}\right)\Delta T = 1.51 \times 10^{-5}\ \Delta T.$$

The strain in the aluminum for a 50 C° temperature change = 4.4 x 10⁻⁴. The stress in the aluminum for a 50 C° temperature change = 3.1 x 10⁷ N/m². The force is F = 2.8 x 10⁴ N with the sign (compression or extension) depending on the sign (+ or -) of the temperature change. The treatment used in Example 6 amounted to assuming that Young's modulus for steel was much larger than that for aluminum, which is a good approximation here.

Example 8

Derive a formula for the conversion from Btu's to calories.

Solution:

The respective units Btu and calorie were chosen so that c_w (heat capacity per unit weight) or c_m (heat capacity per unit mass) will be equal to unity for water. Thus since c_w = 1 Btu/lb·F° and C_m = 1 cal/g·C° are the specific heat capacities of the same substance, we can find the conversion formula by writing:

1 Btu/lb·F° = 1 cal/g·C°.

Now we use the conversions, 1 lb = 454 g (i.e. a quantity of water with a weight of 1 lb has a mass of 454 g.) and 1 F° = 5/9 C° so that

1 Btu /(454 g)(5/9 °C)= 1 cal/g·C°.

Thus we have:

1 Btu = 252 cal.

Example 9

Water flowing at a speed of 5 m/s falls over a 50 m high waterfall (approximate height of Niagara Falls) into a still pool below. Calculate the approximate rise in water temperature due to the conversion of mechanical energy into thermal energy.

Solution:

Let m be the mass of water and equate the loss of mechanical energy to the increase in thermal energy.

$$\frac{1}{2}\,mv^2 + mgh = mc\Delta T$$

The mass cancels out so we can solve for T to obtain:

$$\Delta T = \frac{1}{c}\left(\frac{1}{2}v^2 + gh\right)$$

Numerically:

$$\Delta T = \frac{(0.5)\left(25 \text{ m}^2/\text{s}^2\right) + \left(9.8 \text{ m}/\text{s}^2\right)(50 \text{ m})}{(4186 \text{ J}/\text{kgC})} \quad \text{in S.I. units}$$

Since 1 J = 1 kg·m2/s2, we have ΔT = 0.12°C.

As a historical note, Joule actually tried to measure this temperature increase but since it is so small, he was unsuccessful.

Example 10

A silver bullet with speed of 500 m/s initially at a temperature of 20 °C stops suddenly and all its mechanical energy is converted into thermal energy. What is its temperature rise?

Solution:

From Table 17-3 of the text we find c = 234 J/kg·C°. Equating the loss of mechanical energy to the increase in thermal energy, we have:

$$\frac{1}{2} mv^2 = mc\Delta T$$

The mass cancels out so $\Delta T = v^2/2c$.

Numerically, we have

$$\Delta T = \frac{(500 \text{ m}/\text{s})^2}{2(234 \text{ J}/\text{kgC})} = 534 \frac{(\text{m}/\text{s})^2}{(\text{m}/\text{s})^2} C^\circ = 534 \, C^\circ$$

This is a very large temperature increase but the melting point of silver is 960°C, so the only error we have made is in assuming that the specific heat is constant over this very large temperature interval. Had the bullet been a lead bullet (lead melts at 327°C), it would have melted. This might be the reason that vampires can only be taken out with silver bullets.

Example 11

Suppose in Example 9 the temperature of the flowing water and the still pool below the falls is about 0°C. If a block of ice flowing with the water goes over the falls, what fraction of the ice melts?

Solution:

Again we equate the mechanical energy loss to the increase in thermal energy. In this case however no temperature increase occurs but some fraction, f, of the ice melts. Let L be the latent heat for ice. (Take the value of L = 3.35 x 10⁵ J/kg.)

$$\frac{1}{2}\, mv^2 + mgh = fmL$$

Again m cancels out on both sides of the equation and the fraction, f, is:

$$f = \frac{\frac{1}{2}v^2 + gh}{L}$$

Numerically

$$f = \frac{(0.5)(25)(m/s)^2 + (9.8\ m/s^2)(50\ m)}{3.35\ x\ 10^5\ J/kg}$$

f = 1.5 x 10⁻³   or 0.15% melts.

What would happen if the change in mechanical energy was more than enough to melt all the ice? See the problem-solving strategy on calorimetry in the main text.

Example 12

A copper calorimeter of mass 2 kg initially contains 1.5 kg of ice at -10° C. How much heat energy must be added to convert all of the ice to water and then half of the water into steam?

Solution:

Refer to the problem-solving strategy on calorimetry in the main text. The heat added increases the temperature of the ice until the melting point of ice is reached, melts the ice, raises the temperature of the water until it reaches its vaporization point, and then vaporizes half the water. For the calorimeter, no phase change occurs so heat added only results in a temperature increase. We will calculate the heat required in five steps.

(a) Heat the ice to 0°C, c_{ice} = 2300 J/kg·C°.

$$\Delta Q_a = mc_{ice}\Delta T = (1.5\ kg)(2300\ J/kg·C°)(10°C)$$

$$\Delta Q_a = 3.45\ x\ 10^4\ J.$$

(b) Melt the ice at fixed temperature (0°). Take $L_F = 3.33 \times 10^5$ J/kg.

$$\Delta Q_b = mL_F = (1.5 \text{ kg})(3.33 \times 10^5 \text{ J/kg})$$

$$\Delta Q_b = 5.00 \times 10^5 \text{ J}$$

(c) Heat the water to 100°C. Take $c_w = 4190$ J/kg·C°.

$$\Delta Q_c = mc_w\Delta T = (1.5 \text{ kg})(4190 \text{ J/kgC°})(100° \text{ C})$$

$$\Delta Q_c = 6.28 \times 10^5 \text{ J.}$$

(d) Vaporize half the water at fixed temperature (100°C). Take $L_v = 2.256 \times 10^6$ J/kg.

$$\Delta Q_d = (0.5)mL_v \text{ (since only 1/2 of the total mass is being vaporized)}$$

$$\Delta Q_d = (0.5)(1.5 \text{ kg})(2.256 \times 10^6 \text{ J/kg})$$

$$\Delta Q_d = 1.69 \times 10^6 \text{ J}$$

(e) Heat the calorimeter from -10°C to 100°C; Take $C_c = 390$ J/kg·C°.

$$\Delta Q_e = M_c \, C_c\Delta T = (2 \text{ kg})(390 \text{ J/kg·C°})(110° \text{ C})$$

$$\Delta Q_e = 8.59 \times 10^4 \text{ J.}$$

The total heat energy required is the sum of these five contributions.

$$\Delta Q_T = 2.94 \times 10^6 \text{ J.}$$

If we had ignored the contribution of the copper calorimeter altogether, we would have made an error of 3%.

Example 13

A copper calorimeter of mass 2 kg contains 1.5 kg of ice at -10°C.
    (a) What will the final temperature be if the heat added, ΔQ, is equal to 5×10^5 J?
    (b) If $\Delta Q = 10^6$ J what will the final temperature be?

Solution:

Refer to the problem-solving strategy on calorimetry in the main text.
(a) The answer to the first part is trivial if we look at the numerical solution to Example 12. There we found that 5×10^5 J were required to melt all the ice but 3.45×10^4 J would be needed to heat the ice to 0°C. *Thus the final temperature is 0°C with nearly all the ice melted.*

Without reference to the previous problem we could make an error if we assumed that the heat added was more than enough to cause a phase change and that an additional temperature rise occurred. To illustrate this difficulty we write:

$$\Delta Q = m_{ice}c_{ice}(0°C + 10°C) + m_{ice}L_F + m_w c_w(T_f - 0°C) + m_c c_c(T_f - 10°C)$$

If we now substitute numerically (note $m_{ice} = m_w$) we find that $T_f = -5.99°C$! This nonphysical result came from introducing too many processes for the amount of heat energy available.

(b) If $\Delta Q = 10^6$ J, then we have enough heat energy to melt the ice so an additional temperature rise occurs.

$$\Delta Q = m_{ice}c_{ice}(10°C) + m_{ice}L_F + m_w c_w(T_f - 0°C) + m_c c_c(T_f + 10°C).$$

Numerically, we break this into five terms, two of which contain the unknown (T_f):

$$10^6 \text{ J} = 3.45 \times 10^4 \text{ J} + 5 \times 10^5 \text{ J} + (1.5 \text{ kg})(4190 \text{ J/kg·C°})T_f$$
$$+ (2 \text{ kg})(390 \text{ J/kg·C°})T_f + 7.8 \times 10^3 \text{ J}.$$

Solving for the unknown T_f;

$$T_f = 64.8°C.$$

If this temperature rise exceeds 100°C, we will have to reformulate the problem to include vaporization of the water as well.

Example 14

A brass rod of length 0.15 m, and cross-sectional area 10^{-4} m² has one end at 0° C and the other end at +5° C. How large is the heat current through the rod ?

Solution:

The heat current flows from the hot end to the cold end. We use the defining equation for the conduction heat current, H, and take the thermal conductivity to be $\kappa = 109$ J/(s·m·C°). Also we know $A = 10^{-4}$ m², L = 0.15 m, and $\Delta T = 5$ C°.

$$H = \kappa A \left(\frac{\Delta T}{L} \right)$$

$$H = (109 \text{ J/s·m·C°})(10^{-4} \text{ m}^2) \left(\frac{[5-0] \text{ C°}}{0.15 \text{ m}} \right)$$

$$= 0.363 \text{ J/s} = 0.363 \text{ Watts}$$

Example   15

Two rods of different materials but each of uniform cross-section area, A, are joined at the interface of area A, as shown in Fig. 17-3. The lengths of the rods are $\mathcal{L}_1$ and $\mathcal{L}_2$ and the respective thermal conductivities are κ_1 and κ_2. The ends are maintained at temperatures T_1 and T_2 where $T_1 > T_2$. Derive a general expression for the interface temperature T assuming all the heat conducted through rod 1 passes through rod 2. See the problem-solving strategy on heat conduction in the main text.

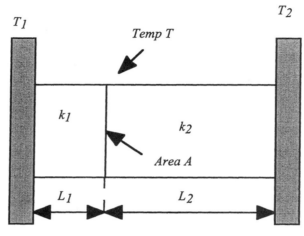

Figure   17-3

Solution:

The heat current through rod 1, $H_1 = \kappa_1 A(T_1 - T)/L_1$, is equal to the heat current through rod 2, $H_2 = \kappa_2 A(T - T_2)/L_2$. Equating the two expressions, we have:

$$\left(\frac{\kappa_2 A}{L_2}\right)(T - T_2) = \left(\frac{\kappa_1 A}{L_1}\right)(T_1 - T)$$

We solve this equation for the unknown interface temperature T to obtain:

$$T = \frac{\left(\dfrac{\kappa_1 T_1}{L_1}\right) + \left(\dfrac{\kappa_2 T_2}{L_2}\right)}{\left(\dfrac{\kappa_1}{L_1}\right) + \left(\dfrac{\kappa_2}{L_2}\right)}$$

(The common area A cancels out.)

To illustrate this formula, let rod 1 be the brass rod of the previous example and rod 2 be a steel rod [κ = 50.2 J/(s·m·C°)] of length 0.2 m and the same area as 1. If $T_1 - T_2 = 5°$ C with $T_2 = 0°$ C, then the interface temperature is equal to 3.72° C. For a steady heat current, the temperature gradient (dT/dx) will be equal to $-H/\kappa_1 A$ in rod 1 and equal to $-H/\kappa_2 A$ in rod 2. Thus the temperature will decrease linearly with x in both rods but with different slopes.

Example 16

The average radiated power from the sun reaching the earth's surface is given by the "solar constant", S = 1.395 kw/m². Assuming the sun radiates isotropically as an ideal blackbody, calculate the absolute temperature of the sun's surface. Refer to Fig. 17-4.

Figure 17-4

Solution:

Take for the distance between the earth and sun r = 1.5 x 10¹¹ m and the radius of the sun, R_S = 7 x 10⁸ m. If we draw a series of spheres concentric with the sun, we see that the energy per unit area per unit time (the solar constant for that sphere) multiplied by the surface area of the sphere $(4\pi r^2)$ is a constant and equal to the power radiated by the sun. Equating the power absorbed from the sun (with e = 1 for the sun) to the power radiated away, we obtain:

$$(4\pi R_s{}^2)\sigma T_s{}^4 = (4\pi r^2)S$$

Numerically:

$$T_s = \left[\left(\frac{r}{R_s}\right)^2 \frac{S}{\sigma}\right]^{1/4} = 5.8 \times 10^3 \text{ K} = 5800 \text{ K}.$$

In this problem we have used a measured value of the radiated power that falls on a unit area on the earth's surface (S) and two known lengths to deduce the temperature at the Sun's surface from Stefan's law.

Example 17

Assuming the moon is a perfect blackbody and absorbs all the radiation falling on it from the sun, estimate the temperature of the moon's surface.

Solution:

Since only an estimate is called for, we assume the distance between the moon and the sun is on average the same as the average distance between the earth and the sun. This means we can use the same solar constant, S, for the moon as we used for the earth. Let r_m be the radius of the moon. The heat current from the sun intercepted by the moon is:

$$H_{in} = \pi r_m^2 \, S.$$

In this expression, πr_m^2 is the area of a circle with radius equal to the moon's radius and is the effective area for absorption of energy from the sun. The heat current radiated by the moon is:

$$H_{out} = 4\pi r_m^2 \, \sigma T_m^4 \quad \text{(the whole area radiates and e = 1).}$$

If the moon is in thermal equilibrium, then $H_{in} = H_{out}$, giving:

$$T_m = \left(\frac{S}{4\sigma}\right)^{1/4} = \left[\frac{1.4 \times 10^3 \text{ W/m}^2}{4 \times 5.67 \times 10^{-8} \text{ W/m}^2 \cdot \text{K}^4}\right]^{1/4} = 280 \text{ K.}$$

Because of the assumptions made, this would apply to the earth as well. The model used is vastly oversimplified. The observed mean temperature is 287 K presumably due to the core and mantle being at elevated temperatures and cooling, so the assumption of thermal equilibrium was not justified for the present.

Example 18

A piece of copper with constant thickness, t, has a rectangular cross-section of variable area as shown in Fig. 17-5. At x = 0, y = y_0 and at x = L, y = $y_0/2$, changing linearly in between. The temperature is equal to T_1 at the origin and equal to T_2 at x = L with $T_1 > T_2$. Find the temperature as a function of x.

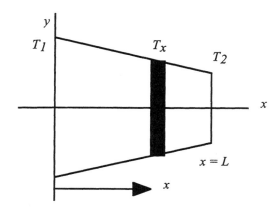

Figure 17-5

Solution:

The equation for the straight line y = y(x) is y = $y_0[1 - (x/2L)]$. Defining $A_0 = 2y_0 t$, the area, A(x) = 2yt is given by:

$$A(x) = A_0\left(1 - \frac{x}{2L}\right)$$

Here it is essential to recognize that the heat is constant through the sample and given by:

$$H = -\kappa \frac{A(x)}{x}\left[T(x) - T_1\right]$$

In particular at x = L, the heat current is:

$$H = \kappa \frac{A_0}{2L}\left[T_1 - T_2\right]$$

Equating these two expressions for H and solving for T(x), we obtain:

$$T(x) = T_1 - \frac{(x/2L)\left[T_1 - T_2\right]}{\left(1 - \frac{x}{2L}\right)}$$

371

QUIZ

1. Suppose an absolute temperature scale was created with the single fixed point (except absolute zero), the boiling point of water (100 °C), being assigned the value 1000 °A. Find the values of the following temperatures on this scale: (a) the freezing point of water, 0° C, and (b) normal body temperature, 37° C.

Answer: (a) 732° A, (b) 831.2° A

2. Sections for a railroad track are laid on a day when the temperature is 41° F and their length is 10.000 m. If the expansion coefficient of the metal is $\alpha = 1.2 \times 10^{-5}$ /(C°), what must be the spacing between adjacent sections so that they will just touch on a day when the temperature is 104° F?

Answer: 4.2×10^{-3} m.

3. A metal calorimeter of mass 0.05 kg contains 0.300 kg of water at 20° C. A second piece of the *same* metal of mass 0.20 kg is heated to 75° C and then dropped into the calorimeter plus water. At equilibrium, the final temperature is 35° C. Calculate the specific heat of the unknown metal from this data.

Answer: c = 2598 J/kg.

4. An aluminum pail (c_{Al} = 910 J/kg, m_{Al} = 1.5 kg) contains 1 kg of water and 2 kg of ice at 0 °C. If 3 kg of water at 70 °C is poured into this pail, calculate the final temperature of the pail plus water.

Answer: T_f = 7.92 °C.

5. Two pieces of metal with the same cross-sectional area of 0.02 m² are welded together. One piece is stainless steel and of length 0.3 m (κ_{ss} = 50.2 W/m·C°). The second piece is of length 0.2 m but its thermal conductivity is unknown. When the stainless steel end is maintained at a temperature of 50 °C and the other end maintained at 20 °C, the temperature of their common junction is 27 °C. (a) Calculate the heat current in the stainless steel piece. (b) Calculate the thermal conductivity of the unknown piece.

Answer: (a) H = 77.0 Watts, (b) κ_u = 110 W/m·C°.

6. Taking the mean distance from the earth to the sun to be 1.49×10^{11} m and that from the sun to Venus to be 1.08×10^{11} m, (a) use the known solar constant on earth, S = 1.4 kW/m², to calculate the solar constant on Venus. (b) Assuming Venus absorbs and radiates like a blackbody, calculate the equilibrium temperature on that planet.

Answer: (a) S = 2.66 kW/m², (b) T = 329 K (or 56 °C).

18
THERMAL PROPERTIES OF MATTER

OBJECTIVES

In this chapter, the concept of an equation of state is introduced. The ideal gas is used to illustrate this concept. A molecular microscopic model, the kinetic theory of gases, is used to derive the some of the observed macroscopic properties of ideal gases. Your objectives are to:

Apply the ideal gas equation of state to calculations of the pressure, volume, temperature, and number of moles.

Recognize the differences in behavior between the ideal gas and real substances.

Construct p-T diagrams of real substances showing the triple point and critical point and the solid, liquid, and vapor regions.

Recognize that ordinary matter is composed of enormous number of molecules and that the observed macroscopic properties are average properties.

Calculate the mean square speed of gas molecules as a function of temperature and the mass of the molecule.

Calculate the observable properties such as molar heat capacity at constant volume and the ratio C_p/C_V for monatomic (and other) gases using kinetic theory.

REVIEW

An equation of state connects the variables that characterize the equilibrium states of the system. For the next few chapters we will be considering gases so that the variables most frequently used are the volume (V), pressure (p), temperature (T) and the mass m. Since the above variables are not independent, we can express one of them as a function of the remaining variables:

$$V = f(p,T,m)$$

Other possibilities will be discussed later.

One useful idealization is the ideal gas in which p,V,T, and m are connected by the equation:

$$pV = nRT$$

where the mass, m, is contained in the factor n that gives the number of moles of gas in the volume V. The constant R is called the universal (same for all) gas constant. The various numerical values and different units are examined in Example 1.

If the number of moles of an ideal gas is constant, then the equation of state predicts that pV/T is a constant. This equation in three variables (only two of which are independent) defines a surface. Projection of this three dimensional surface onto two dimensions yields p vs T, V vs T, and p vs V diagrams. We will have use for such diagrams in the next few chapters.

Real substances have substantially different p,V,T surfaces from that of an ideal gas. This is because the real substances can undergo phase transformations such as liquefaction and solidification. For a real substance, although some functional relationships give good indications of much of the behavior (see Example 4), we should regard the p,V,T surfaces as graphical displays of experimental data.

Isobaric processes are those occurring at constant pressure while isothermal processes occur at constant temperature.

The two dimensional projection p vs T is called the phase diagram. As a rule, a point on this diagram gives the coordinates of a single phase but there are lines on this diagram where two phases (such as vapor and liquid) coexist. A positive slope of p versus T (dp/dT > 0) indicates that the volume (for fixed mass) contracts as you pass from vapor to liquid and again when you pass from liquid to solid. A negative slope of the fusion curve, (dp/dT < 0), usually is a signal that the substance expands upon freezing (as water does) but the phase boundary between liquid and solid helium is an exception that will not be treated here.

The fusion curve and the vaporization curve intersect at a point called the triple point where liquid, solid, and vapor can all coexist at a unique pressure and temperature. The vaporization curve terminates in a point called the critical point. If, at a temperature below the critical temperature, the pressure is increased from a point in the vapor phase to a point in the liquid phase, a discontinuous change in specific volume (as well as a latent heat) is found in crossing the vaporization curve. You can tell that a phase transformation has occurred since the two phases had different specific volumes at the same p and T. If, on the other hand, T is greater than the critical temperature and the pressure is increased, no discontinuous change in specific volume occurs so we can't tell the difference between vapor and liquid.

Along the vaporization curve, the pressure (called the vapor pressure) is a function of the temperature alone and does not depend on the volume of the container. For this reason it is easy to reduce the temperature of a liquid-vapor system in equilibrium simply by using a vacuum pump to reduce the pressure. The relationship between p and T along the vaporization curve is so reproducible that a measurement of the pressure accurately yields the temperature. The helium vapor pressure-temperature curve forms the basis of the provisional international temperature scale below 4.2 K.

The atomic nature of matter and the formation of molecules from atoms are well established. The attractive forces between electrically neutral molecules lead to the clustering of molecules in the liquid state and then in the solid state as the mean separation between particles gets smaller. These attractive forces are electrical in their origin and arise from the fact that while the average charge at a point in a molecule is zero, its instantaneous value is non-zero. This leads to distortions of the charge clouds of molecules and non-zero forces. In Example 6, the average spacing between molecules in a gas, liquid, and solid is calculated.

One very important observation concerning the atomic nature of matter is that one mole of a pure substance contains a fixed number of identical molecules. This number is called Avogadro's number, N_A, and is equal to 6.022×10^{23} molecules/mol. This fact coupled with the knowledge that 1 mole of an ideal gas occupies 22.4 liters at standard temperature and pressure (S.T.P.) enables you to calculate the number of molecules present in an ideal gas under any conditions of temperature and pressure. See Example 7.

The kinetic theory of an ideal gas was an early attempt to calculate the observed macroscopic properties of gases from a simple microscopic model. The usual approach (as in the text) is to find the pressure in a dilute (non-interacting) gas due to the molecules colliding with the walls. This pressure is equal to the time average force per unit area on the selected wall due to one molecule multiplied by the number of molecules in the volume of interest. In Example 8 the time average force is shown to be equal to the average momentum change per collision with the wall divided by the mean time between collisions (with the same wall). If the wall (of area L_yL_z) lies in the yz plane, the change in momentum per collision is $2mv_x$ as the y and z components are unaffected. Furthermore the mean time between collisions (with the same wall) is $2L_x/v_x$. Since the pressure, p_1, equals force per unit area,

$$p_1 = \frac{F}{L_xL_z}$$

$$p_1 = \frac{2\,mv_x}{(L_xL_z)(2\,L_x/v_x)} = \frac{m(v_x)^2}{V}$$

where $L_xL_yL_z = V$. This is the partial pressure due to one molecule. Since the effects due to the other molecules (N in number) are additive, the total pressure p is:

$$p = N\,\frac{m(v_x)^2}{V}$$

If we denote the average of a quantity by $(\ )_{av}$, then on average $(v_x^2)_{av} = (v_y^2)_{av} = (v_z^2)_{av}$, so $(v_x^2)_{av}$ is replaced by

$$\left(v_x^2\right)_{av} = \left(v_y^2\right)_{av} = \left(v_z^2\right)_{av} = \tfrac{1}{3}\left(v^2\right)_{av}$$

Since the average kinetic energy per molecule is $(1/2)mv^2$, this result can be written:

$$pV = \frac{2}{3}\,N\left(\frac{1}{2}\,mv^2\right)_{av}$$

This is the end of the road for this model. All other results are obtained by appealing to the ideal gas law. In particular these key results are:

(a) The translational kinetic energy of n moles of this monatomic gas is,

$$K_{tr} = \frac{3}{2} nRT$$

(b) The average translational kinetic energy per molecule depends only on T,

$$\left(\frac{1}{2} mv^2 \right)_{av} = \frac{3}{2} T \frac{R}{N_A} = \frac{3}{2} kT$$

This last result is used then to calculate the root-mean-square velocity (v_{rms}) for an ideal gas.

$$v_{rms} = \sqrt{(v^2)_{av}} = \sqrt{\frac{3\,kT}{m}} = \sqrt{\frac{3\,RT}{M}}$$

In the above expressions, m is the mass of an individual molecule while M is the mass per mole of the gas. See Examples 9, 10, and 11.

The first of the above results can be used to calculate the molar heat capacity of an ideal gas. Note that K_{tr} depends only on the temperature and $\Delta K_{tr} = (3/2) nR \Delta T$. At constant volume $\Delta Q = \Delta K_{tr} = nC_V \Delta T$ so for one mole, $C_V = 3/2\ R$. Furthermore, since $C_p = C_V + R$, we have

$$C_p = \frac{5}{2} R \quad \text{and} \ \gamma = \frac{C_p}{C_V} = \frac{5}{3} = 1.67.$$

This is the high point for the simple kinetic theory presented here, as all these results agree well with experimental values. The extension of this approach to more complicated molecules is more complex and is done by appealing to the principle known as 'the equipartition of energy'.

In its most general form, this principle assigns a 'degree of freedom', f, to each term in the total energy (sum of kinetic plus potential) which is a pure squared term in a coordinate or momentum. Since both the coordinates and momenta are as often negative as positive, the average over a long time of either one will be zero but the average of the square will be non-zero. For each such degree of freedom there is associated a thermal energy of kT/2. Based on this approach, a gas of N molecules, each with f degrees of freedom, has total kinetic energy of

$$K = \frac{f}{2} NkT$$

and a molar heat capacity at constant volume of

$$C_v = \frac{f}{2} N_A k = \frac{f}{2} R$$

(since $N_A k = R$). While this approach is clear enough, it requires some help from a more advanced theory, quantum mechanics, in order to give satisfactory agreement with experiment. For a monatomic gas, we obtain f = 3 so K = 3nRT/2 as required. To agree with room temperature measurements on *diatomic gases* (H_2, N_2, O_2, etc.) a value of f = 5 is needed. Diatomic gases actually have 8 degrees of freedom, not 5. This problem was resolved with the discovery of Quantum Mechanics, which modifies the equipartition principle.

PROBLEM-SOLVING STRATEGIES

The main text includes Problem-Solving Strategies for (1) Ideal Gases and (2) Kinetic-Molecular Theory. Now is an appropriate time to review those strategies. Express all temperatures as absolute temperatures, usually in Kelvin degrees. Frequently numerical substitution for R can be avoided by remembering that

$$R = \frac{p_s V_s}{T_s}$$

where p_s, V_s, and T_s are the standard pressure, volume, and temperature for one mole. This relationship is useful when two of the known quantities are easily expressed in terms of these standard values yielding the ratio of the unknown quantity to its standard value. Watch the units more carefully in this chapter than you may have in others.

QUESTIONS AND ANSWERS

QUESTION. Why does the earth have an atmosphere while the moon has no atmosphere?

ANSWER. The escape velocity of a molecule on earth is about 11 km/s, much higher than the rms (root-mean-square) speed of molecules in the atmosphere. Thus molecules stay in the earth's atmosphere. On the moon's surface, the gravitational potential is 20 times weaker that at the earth's surface so the escape speed is 11 km/s divided by $\sqrt{20}$ or 2400 m/s. This makes it more probable for a molecule to escape the moon's atmosphere.

QUESTION. To appreciate how large Avogadro's number is, imagine a 12 oz can containing an ideal gas under conditions of STP (standard temperature and pressure) that has a leak that permits 10 billion molecules per second to escape. How long would it take for the can to empty? (a) about an hour   (b) a day   (c) a year   (d) about 30000 years.

ANSWER. The correct estimate is (d). There are about 10^{22} molecules in the can and 3.15 x 10^7 seconds in a year. Thus at 10^{10} per second, it takes about 3 x 10^4 years.

EXAMPLES AND SOLUTIONS

Example 1

Show that the values of (a) R = 8.314 J/mol·K; (b) R = 8.314 x 10⁷ ergs/mol·K;
(c) R = 1.99 cal/mol·K; and (d) R = 8.20 x 10⁻² atm/mol; are all equivalent.

Solution:

On occasion it is difficult to remember the value of R, the molar gas constant, or to recall the value of R in the units most appropriate to the problem under consideration. If you can remember that 1 mole of any ideal gas occupies 22.4 liters at standard temperature and pressure, you can overcome either difficulty and calculate R as seen below.

(a) Using the ideal gas law

$$pV = nRT \quad or \quad R = pv/nT,$$

then standard pressure is 1 atm (1.013×10^5 Pa) and standard temperature is 0°C (273 K) so for n = 1, and using 22.4 L = 22.4×10^{-3} m³, we find for R:

$$R = \frac{\left(1.013 \times 10^5 \ N/m^2\right)\left(22.4 \times 10^{-3} \ m^3\right)}{\left(273 \ K\right)\left(1 \ mole\right)}$$

$$= 8.31 \ J/mol \cdot K$$

(b) Since 1 J = 10⁷ ergs, the conversion is trivial.

(c) Similarly, since 4.19 J = 1 cal, dividing the result (a) by this factor produces

$$R = 1.99 \ cal/mol \cdot K$$

(d) This result merely summarizes the starting data. Since standard pressure is 1 atm and standard temperature is 273 K we have:

$$R = \frac{\left(1 \ atm\right)\left(22.4 \ L\right)}{\left(273 \ K\right)\left(1 \ mole\right)}$$

$$R = 8.20 \times 10^{-2} \ atm \cdot l/mol \cdot K$$

This problem illustrates the importance of units in this chapter.

Example 2

Helium gas is admitted to a volume of 200 cm³ at a temperature of 77°K until the pressure is equal to 1 atm.
   (a) If the temperature of this container is raised to 20°C, what will the pressure be?
   (b) If the system has a pressure relief valve that will not permit the pressure to exceed 1 atm, what fraction of the gas remains at 20°C?

Solution:

This problem involves comparison of two states of an ideal gas.

(a) For this closed system, the number of moles, n, and the volume are both constant so:

$$\frac{P_i}{T_i} = \frac{P_f}{T_f}$$

Thus

$$\frac{P_f}{(1 \text{ atm})} = \frac{T_f}{T_i} = \frac{(273 \text{ K} + 20 \text{ K})}{(77 \text{ K})}$$

$P_f = 3.8$ atm.

(b) In this situation, both the pressure and volume are constant but the number of moles is not constant since gas can leak out the pressure relief valve. Therefore equating the pV products:

$$n_f(RT_f) = n_i(RT_i)$$

The fraction of gas left is:

$$n_f/n_i = T_i/T_f = 77/293 = 0.263.$$

Example 3

The molar volume of liquid helium is approximately 28 cm³/mol. A high pressure (100 atm) helium cylinder has a volume of 0.05 m³. How many liters of liquid helium can be produced from the gas in the cylinder?

Solution:

Since one liter contains 1000 cm3, one liter of liquid helium represents 35.7 moles. To calculate the number of moles stored in the high pressure cylinder we use the ideal gas law:

$$pV = nRT$$

Let

$$p = 100 \text{ atm} = 10^7 \text{ Pa}$$

$$V = 5 \times 10^{-2} \text{ m3}$$

$$R = 8.31 \text{ J/mol·K.}$$

and take room temperature for T(293°K). Then n is:

$$n = \frac{(10^7 \text{ Pa})(5 \times 10^{-2} \text{ m}^3)}{(8.31 \text{ J/K} \cdot \text{mol})(293 \text{ K})} = 2.05 \times 10^2 \text{ moles}$$

Thus dividing this by 35.7 moles per liter, we find that we can make 5.8 liters of liquid from one high pressure cylinder of gas.

Example 4

Treating N_2 as though it is an ideal gas, (a) calculate the volume occupied by a mole of this gas at the critical temperature (T_c = 126.2 K) and critical pressure (p_c = 33.9 x 10^5 N/m2). Express your answer as a ratio of the volume to the known critical volume (V_c) where V_c = 90.1 x 10-6 m3. (b) Calculate the pressure at V_c, T_c.

Solution:

(a) Using pV = nRT

$$V = \frac{nRT}{p} = \frac{(1 \text{ mole})(8.31 \text{ J/K} \cdot \text{mol})(126.2 \text{ K})}{(33.9 \times 10^5 \text{ N/m}^2)} = 3.095 \times 10^{-4} \text{m}^3$$

Since V_c = 90.1 x 10-6 m3, then V = 3.44 V_c .

(b) Solving the ideal gas law for p,

$$p = \frac{(1 \text{ mole})(8.31 \text{ J/K} \cdot \text{mol})(126.2 \text{ K})}{(90.1 \times 10^{-6} \text{m}^3)} = 1.16 \times 10^7 \text{ Pa.}$$

The critical pressure is p_c = 33.9 x 10^5 Pa so for the ratio we have p = 3.44 p_c.

Example 5

Assume the temperature and pressure in the lower part of the atmosphere are given by

$$T = T_0 - \alpha y$$

$$\ln\left(\frac{p_0}{p}\right) = \frac{Mg}{R\alpha} \ln\left(\frac{T_0}{T_0 - \alpha y}\right)$$

where T_0 and p_0 are the temperature and pressure at the surface, M the molecular mass, and $\alpha = 6 \times 10^{-3}$ C°/m. y is the height above the surface. Calculate the temperature and pressure at a height of 1 mi(1.6 km). Take M = 29 g as the mean molecular mass of air.

Solution:

We will assume T_0 = 293 K (or 20 °C). The temperature 1 mi above the earth's surface will be

$$T = T_0 - \alpha y = 293 \text{ K} - (6 \times 10^{-3} \text{ K/m})(1.6 \times 10^3 \text{ m}) = 293 \text{ K} - 9.6 \text{ K} = 283.4 \text{ K}.$$

To calculate the pressure, first evaluate the factor

$$\frac{Mg}{R\alpha} = \frac{\left(29 \times 10^{-3} \text{ kg/mol}\right)\left(9.8 \text{ m/s}^2\right)}{\left(8.31 \text{ J/K} \cdot \text{mol}\right)\left(6 \times 10^{-3} \text{ K/m}\right)} = 5.70$$

This yields

$$\ln (p_0/p) = (5.70) \ln (1.034)$$

or

$$p = 0.827 \, p_0.$$

Example 6

Find the average spacing between H_2 molecules, assuming the molecules are at the vertices of a fictitious cubic structure, in:
   (a) gaseous H_2 at S.T.P.
   (b) liquid H_2 at 20 K where the density is 41060 mole/m^3
   (c) solid H_2 at 4.2 K where the molar volume is 22.91 x 10^{-6} m^3/mol.

Solution:

(a) We assign a volume of a^3 to each molecule. This is extremely crude but useful in producing order-of-magnitude estimates. Since one mole, N_A molecules, occupies 22.4 L at S.T.P., we write:

$$N_A a^3 = 22.4 \text{ L/mol} = 22.4 \times 10^{-3} \text{ m}^3/\text{mol}$$

Substituting N_A, we have

$$a^3 = 3.72 \times 10^{-26} \text{ m}^3$$

$$a = 3.34 \times 10^{-9} \text{ m}$$

(b) The density is specified in moles per cubic meter so one cubic meter contains 41060 moles of liquid H_2. Letting a^3 be the volume per molecule, we have

$$1 \text{ m}^3 = 41060 \, N_A a^3$$

$$N_A a^3 = 2.44 \times 10^{-5} \text{ m}^3/\text{mol}$$

$$a = 3.43 \times 10^{-10} \text{ m}$$

(c) Here the molar volume is specified, yielding

$$N_A a^3 = 22.91 \times 10^{-6} \text{ m}^3/\text{mol}$$

$$a^3 = 3.80 \times 10^{-29} \text{ m}^3$$

$$a = 3.36 \times 10^{-10} \text{ m}$$

Thus the average spacing between molecules in the liquid and solid is about the same but this spacing in the condensed states is about a factor of ten smaller than the spacing in the *gas* at S.T.P.

Example 7

A common type of laboratory vacuum pump produces an ultimate (lowest) pressure of 10 microns. How many molecules (how many moles) of gas would be present in a volume of 0.15 m³ reduced to that pressure at 300 K?

Solution:

Here the pressure unit is tricky as it is based on the fact that atmospheric pressure is 76 cm of mercury (Hg). Thus 1 atm = 0.76m (of Hg). The unit, micron (μm), is one millionth of a meter of mercury so:

$$10 \ \mu m = 10 \times 10^{-6} \ m \ (1 \ atm/0.76 \ m) = 1.32 \times 10^{-5} \ atm$$

The volume 0.15 m³ is equal to 1.5×10^2 L. In these units we can write the molar gas constant, R, as:

$$R = \frac{(1 \ atm)(22.4 \ L)}{(273 \ K)(1 \ mole)}$$

Since the number of moles, n, is given by n = pV/RT we have:

$$n = \frac{(1.32 \times 10^{-5} \ atm)(1.5 \times 10^2 \ L)(273 \ K)(1 \ mole)}{(1 \ atm)(22.4 \ L)(300 \ K)}$$

$$= 8.04 \times 10^{-5} \ mole$$

There are 6.02×10^{23} molecules per mole so there are still 4.84×10^{19} molecules left in this "evacuated" volume.

Example 8

Using the impulse-momentum theorem, show that the time average force is equal to the ratio of the average momentum change per collision to the mean time between collisions.

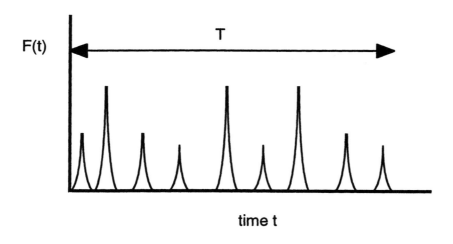

Figure 18-1

Solution:

The time average force, F_{av}, is given by:

$$F_{av} = \frac{1}{T} \int_0^T F(t)\, dt$$

where the integral gives us the total area under the F(t) spikes (see Fig. 18-1) during the time interval T without looking at the details of each collision. The impulse, I, is equal to the total change in momentum, and also given by the above integral,

$$I = \int_0^T F(t)\, dt = \Delta p_T$$

Thus $F_{av}T = \Delta p_T$. The total change in momentum is equal to the average change in momentum in one collision (Δp) multiplied by the number of collisions, N. Thus

$$F_{av} = \Delta p \left(\frac{N}{T}\right) = \frac{\Delta p}{T/N} = \frac{\Delta p}{\tau}$$

where $\tau = T/N$ is the mean time between collisions.

384

Example 9

In a gas at 300 K that is a mixture of the diatomic molecules H_2 ($M = 2 \times 10^{-3}$ kg/mol) and D_2 ($M = 4 \times 10^{-3}$ kg/mol) calculate the root-mean-square velocities of the molecules.

Solution:

The root-mean-square velocity is given by:

$$v_{rms} = \sqrt{\frac{3\,RT}{M}}$$

For H_2 (regular hydrogen)

$$v_{rms} = \sqrt{\frac{3\,(8.314\,\text{J/mol}\cdot\text{K})\,(300\,\text{K})}{(2 \times 10^{-3}\,\text{kg/mol})}}$$

$$= 1.93 \times 10^3 \text{ m/s.}$$

For D_2 (heavy hydrogen or deuterium)

$$v_{rms} = \sqrt{\frac{3\,(8.314\,\text{J/mol}\cdot\text{K})\,(300\,\text{K})}{(4 \times 10^{-3}\,\text{kg/mol})}}$$

$$= 1.37 \times 10^3 \text{ m/s.}$$

Because of the large relative difference in these speeds, it is relatively easy to separate the isotopes H_2 and D_2 by a process called effusion.

Example 10

A one liter flask at 293 K contains a mixture of 3 g of N_2 gas and 3 g of H_2 gas. Assuming the gases behave as ideal gases,
   (a) calculate the partial pressures exerted by N_2 and H_2,
   (b) calculate the root-mean-square speeds of N_2 and H_2.

Solution:

The mass of a mole of N_2 is 28 g while that of H_2 is 2 g.

(a) In the flask there are

$n(N_2) = (3/28)$ moles of N_2 and

$n(H_2) = (3/2)$ moles of H_2.

Applying the ideal gas law to each component *separately*, the partial pressures $p(N_2)$ and $p(H_2)$ are:

nitrogen

$$p(N_2) = \frac{n(N_2)RT}{V}$$

$$p(N_2) = \frac{3}{28} \frac{(8.314 \text{ J/mol} \cdot \text{K})(293 \text{ K})}{(10^{-3} \text{ m}^3)}$$

$p(N_2) = 2.61 \times 10^5$ Pa   (about 2 atm)

hydrogen

$$p(H_2) = \frac{n(H_2)RT}{V}$$

$$p(H_2) = \frac{3}{2} \frac{(8.314 \text{ J/mol} \cdot \text{K})(293 \text{ K})}{(10^{-3} \text{ m}^3)}$$

$p(H_2) = 36.5 \times 10^5$ Pa   (about 36.5 atm)

(b) The root-mean-square speeds are calculated from the expression

$$v_{rms} = \sqrt{\frac{3RT}{M}}$$

For N_2,

$$v_{rms} = \sqrt{\frac{3(8.314 \text{ J/mol} \cdot \text{K})(293 \text{ K})}{(28 \times 10^{-3} \text{ kg/mol})}} = 5.11 \times 10^2 \text{ m/s.}$$

For H_2,

$$v_{rms} = \sqrt{\frac{3(8.314 \text{ J/mol} \cdot \text{K})(293 \text{ K})}{(2 \times 10^{-3} \text{ kg/mol})}} = 1.91 \times 10^3 \text{ m/s.}$$

Example 11

The element carbon has atomic masses of 9, 10, 11, 12, 13, 14, 15 and 16. Only three of these isotopes live long enough to study by thermal means in the laboratory: 12, 13 and 14. Calculate the root-mean-square velocity of carbon 12 at 300 K and compare it with the same quantity for carbon 9 and carbon 16.

Solution:

The expression for the speed, v_{rms}, is:

$$v_{rms} = \sqrt{\frac{3kT}{m}}$$

For carbon 12, the mass of an atom is 12 times that of a hydrogen atom so:

$$m = 12(1.673 \times 10^{-27} \text{ kg}) = 2.01 \times 10^{-26} \text{ kg}.$$

Thus we have

$$v_{rms} = \sqrt{\frac{3\left(1.38 \times 10^{-23} \text{ J/K}\right)\left(300 \text{ K}\right)}{\left(2.01 \times 10^{-26} \text{ kg}\right)}}$$

$$v_{rms} = 7.87 \times 10^2 \text{ m/s}$$

For carbon 16, this speed is lower by the factor $(12/16)^{1/2}$ or

$$v_{rms} = 6.81 \times 10^2 \text{ m/s}$$

For the isotope with mass 9, the speed is higher than that of carbon 12 by the factor $(12/9)^{1/2}$ giving:

$$v_{rms} = 9.08 \times 10^2 \text{ m/s}$$

Example 12

Calculate the "escape" velocities at the surfaces of the earth and sun and determine whether any molecules have root-mean-square thermal velocities comparable to these escape velocities at 300 K.

Solution:

To just escape, we want the total energy at infinite separation to be zero. This condition determines the kinetic energy at the surface (of the earth or sun).

$$0 = \tfrac{1}{2}\,mv^2 - \frac{GMm}{R}$$

The formula for v is then:

$$v = \sqrt{\frac{2GM}{R}} \; .$$

For the sun, $M = 2 \times 10^{30}$ kg and $R = 7 \times 10^8$ m. $G = 6.67 \times 10^{-11}$ N·m^2/kg^2. Numerically, for the sun, we have

$$v_{sun} = 6.18 \times 10^5 \text{ m/s}$$

For the earth, we can use $GM = gR_E^2$ where $g = 9.8$ m/s^2 and $R_E = 6.4 \times 10^6$ m. Substituting these values into the general formula, we obtain:

$$v_{earth} = 1.12 \times 10^4 \text{ m/s}$$

The thermal velocity, at 300 K, for an H_2 molecule is 1.927×10^3 m/s. The thermal velocity of a hydrogen atom would be $(2)^{1/2}$ larger than this or 2.7×10^3 m/s. This is less than the escape velocity on the earth and much less than the escape velocity on the sun. The sun's temperature is more like 6000 K so the thermal velocity of a hydrogen atom at 6000 K would be 1.2×10^4 m/s which is still well below the escape velocity. However on the moon's surface, the gravitational potential is 20 times smaller than at the earth's surface so the escape velocity is 4.5 times smaller. This is just enough to let molecules leave the moon.

Example 13

Along the phase equilibrium line between liquid and gas, the slope is given by:

$$\frac{dp}{dT} = \frac{L_v}{T\left(V_g - V_L\right)}$$

where L_v is the heat of vaporization, V_g the volume of a mole of gas, and V_L the volume of a mole of liquid. Assuming the gas obeys the ideal gas law and that $V_g \gg V_L$, obtain the vapor pressure as a function of temperature.

Solution:

Making the approximation $V_g \gg V_L$, we have

$$\frac{dp}{dT} \cong \frac{L_v}{TV_g}$$

Replacing the gas volume, $V_g = RT/p$ in the denominator and rearranging,

$$\frac{dp}{p} = \frac{L_v}{R}\left(\frac{dT}{T^2}\right).$$

Integrating this expression yields

$$\ln\left(\frac{p}{p_0}\right) = -\frac{L_v}{RT}$$

where p_0 is the pressure at infinitely high temperature. Solving for p,

$$p = p_0 \exp\left(-\frac{L_v}{RT}\right).$$

This shows that the equilibrium vapor pressure decreases (but not linearly) with decreasing temperature.

___

QUIZ

1. A 50 liter tank at 293 K contains N_2 gas at a pressure of 0.6 atm.
   (a) How many moles of O_2 must be added to make a mixture that is 80% N_2 and 20% O_2?
   (b) What will the final pressure be?

Answer: (a) 0.312 moles of O_2, (b) p_f = 0.75 atm.

2. A 50 liter tank contains 3 moles of gas, the composition of the gas being 20 % oxygen and 80 % nitrogen. If the temperature of the tank is 293 K, (a) calculate the pressure in the tank and (b) the pressure that would be obtained if all the oxygen molecules were removed leaving only nitrogen.

Answer: (a) $p = 1.46 \times 10^5$ Pa, (b) $p_N = 1.17 \times 10^5$ Pa.

3. A cylinder of volume 0.5 liter fitted with a piston contains an ideal gas at 400 K with a pressure of 6 atm. The piston is pushed in at constant pressure until the volume is halved, the temperature increasing to 450 K. (a) Calculate the number of moles in the system initially. (b) How many moles remain at the end?

Answer: (a) $n_i = 9.14 \times 10^{-2}$ moles, (b) $n_f = 4.06 \times 10^{-2}$ moles.

4. One mole of helium (M = 4 g per mole) at S.T.P. is in a cubic container. Calculate the mean time between collisions of a given molecule with the same wall.

Answer: $t = 7.48 \times 10^{-4}$ s

5. Assuming conduction electrons in a metal behave as ideal gas particles, calculate the mean thermal velocity of such an electron at 300 K and 4 K.

Answer: At 300 K, $v_{rms} = 1.17 \times 10^5$ m/s
        At 4 K, $v_{rms} = 1.35 \times 10^4$ m/s.

6. A particle of mass 6×10^{-27} kg, traveling with a constant velocity of 1600 $m \cdot s^{-1}$ in the x direction (between collisions with the walls) strikes walls of cross-sectional area A = 0.04 m^2 at x = 0 and x = 0.2 m reversing its velocity without loss of energy. Ignoring gravity, calculate the following quantities: (a) the magnitude of the change in linear momentum per collision with the wall, (b) the mean time between collisions of one particle with the same wall, (c) the net average force exerted by this particle on one wall, and (d) the number of such particles required to produce a pressure of 10^5 $N \cdot m^{-2}$ on one wall.

Answer: (a) $\Delta p = 1.92 \times 10^{-23}$ kg·m/s, (b) $\Delta t = 2.5 \times 10^{-4}$ s, (c) $F_1 = 7.68 \times 10^{-20}$ N,
        (d) $N = 5.21 \times 10^{22}$ molecules.

19
THE FIRST LAW
OF THERMODYNAMICS

OBJECTIVES

In this chapter the first law of thermodynamics is stated. It is basically a restatement of energy conservation. Your objectives are to:

Calculate the heat taken in by the system of interest for various processes (constant volume, constant pressure, or constant temperature).

Calculate the work done by the system for the same processes.

Apply the first law of thermodynamics to the ideal gas.

REVIEW

Thermodynamics enables you to analyze physical systems without reference to their microscopic properties. Usually the universe is divided into two parts, a system of interest and the rest of the universe, which makes up the environment for the 'system'. The energy of the universe is taken as constant, but the system can change its energy by transfer of heat, the performance of work, or both, by interacting with the rest of the universe.

The work done *by* a fluid, W, as it changes from a state labeled 1 to a state labeled 2 is equal to the area under a pressure (p) versus volume (V) curve connecting the points (p_1, V_1) and (p_2, V_2).

$$W = \int_1^2 p \, dV$$

Since an infinite number of curves (leading to an infinite number of values of W) can connect the points 1 and 2, the work done *by* the system in general *depends on the path taken* between points 1 and 2. This is illustrated in Example 1. Since the pressure is always greater than or equal to zero, the work done by the system in an expansion (the final volume is larger than the initial volume) is positive and the work done by the system in a contraction (final volume less than initial volume) is negative.

Like work, the heat taken in or rejected by a system depends on the path taken between its initial and final thermal equilibrium states. Thus the "heat content" of a body is not a well defined concept, although heat transfer is. Heat taken in by the system will be considered positive while heat rejected will be negative.

While neither the heat taken in nor the work done by a system are functions of just the coordinates of the initial and final equilibrium states and depend on the path taken between these states, their difference $\Delta Q - \Delta W$ is independent of the path and depends only on the coordinates of the initial and final states. This "state function", $U_2 - U_1 = \Delta U$, is called the *internal energy.* The first law of thermodynamics can be stated in the following way: there exists a function, ΔU, of the thermodynamic coordinates (p,V, and T for an ideal gas) of the initial and final equilibrium states of the system that is equal to the difference of the heat taken in by the system and the work done by the system.

$$\Delta U = \Delta Q - \Delta W$$

It might be easier to remember this equation if it is rewritten in the form:

$$\Delta Q = \Delta U + \Delta W$$

Here the heat taken in is equal to the increase in internal energy plus the work done by the system. The first law of thermodynamics recognizes heat as a form of energy and basically is a restatement of energy conservation.

Five specific thermodynamic processes are given special names:

(1) Adiabatic Process: no heat energy enters or leaves the system so $\Delta Q = 0$. If $\Delta Q = 0$, the first law implies that $\Delta U = -\Delta W$ so that if one quantity, ΔU or ΔW, is known, the other can be found.

(2) Isochoric Process: there is no volume change, $\Delta V = 0$, so no work is done and $\Delta Q = \Delta U$.

(3) Isothermal Process: there is no temperature change, $\Delta T = 0$.

(4) Isobaric Process: there is no pressure change, $\Delta p = 0$.

(5) Throttling Process: this is a specific adiabatic process in which the sum of the internal energy (U) and the product of pressure and volume is held constant (i.e. pV + U = constant).

The first law of thermodynamics can also be stated in terms of infinitesimal processes so that it appears in differential form:

$$dQ = dU + dW$$

Again, for emphasis, we note that to calculate ΔQ or ΔW from dQ or dW, the path must be specified but to obtain ΔU we only need to know the initial and final equilibrium states.

Many of these general (and very powerful) ideas can be applied profitably to the ideal gas. For the ideal gas, it is found that the internal energy depends *only* on the temperature (T) and not on the pressure or volume. This is a very important conclusion and we will use it frequently although in this chapter it is simply asserted without proof. (See Example 2).

The two specific heat capacities that are important for the ideal gas are c_p, the specific heat at constant pressure and the specific heat at constant volume, c_v. C_p and C_v are the heat capacities of 1 mole--in this case 1 mole of ideal gas.

When the heat taken in (dQ) by the system (containing n moles), at constant volume, results in a temperature increase (dT), the heat capacity per mole, C_v, is equal to:

$$nC_v = \frac{dQ}{dT} \quad \text{at constant volume (V).}$$

Since $dV = 0$ here (isochoric process), no work is done and the first law gives $dQ = dU$. We can conclude that $dU = nC_v\, dT$. The change in U is always independent of the path and since U is a function of T only for an ideal gas, we can now calculate ΔU for *any* process for an ideal gas:

$$\Delta U_{ideal} = nC_v\Delta T$$

If heat is added at constant pressure, then the molar heat capacity, C_p, is defined by:

$$nC_p = \frac{dQ}{dT} \quad \text{at constant pressure (p).}$$

For dQ we can use the first law and write $dQ = dU + pdV$. We know that $dU = nC_vdT$ and since p is constant, $pdV = nRdT$. Substituting these results into the equation for C_p yields:

$$C_p = C_v + R.$$

The ratio of C_p/C_v is denoted γ and will be regarded as an experimental quantity for our purposes although it can be calculated by appealing to microscopic models of gases. See Examples 3 and 4.

When an ideal gas undergoes an adiabatic ($dQ = 0$) expansion or compression, the values of the thermodynamic variables are constrained by the two equations:

$$TV^{\gamma-1} = \text{constant;} \quad \text{and } pV^{\gamma} = \text{constant.}$$

Since $dQ = 0$, the work done by the gas is most simply obtained by first calculating the internal energy change from the temperature change and then writing that

$$\Delta W = -\Delta U = -nC_v\Delta T.$$

This procedure is illustrated in Example 5.

PROBLEM-SOLVING STRATEGIES

Nearly all calculations in this chapter involve the ideal gas. The internal energy of an ideal gas depends *only* on the temperature and is usually the easiest quantity to calculate in a problem as $\Delta U = nC_v\Delta T$. For many of the processes considered here, either ΔW or ΔQ can be calculated without undo difficulty. By using the first law of thermodynamics, the third or unknown quantity can be found.

Sign conventions will be important here. ΔQ is positive if heat is taken in by the system, negative otherwise. The work done *by* the system is positive when the final volume is larger than the initial volume.

QUESTIONS AND ANSWERS

QUESTION. Suppose a massless container of helium gas is dropped from a height of 50 meters and suffers a completely inelastic collision with the ground, transferring all the original potential energy of the gas into internal energy. Will there be a change in the gas's temperature? Will it increase or decrease? Would the temperature change be the same for nitrogen gas as for the helium gas? Treat both as ideal gases.

ANSWER. The initial potential energy at height 50 m is converted into kinetic energy during the fall. By increasing the kinetic energy, ignoring the container, the internal energy will increase and so will the temperature. If the gas is N_2 rather than He, the individual molecule masses will be greater by a factor of 7 so the potential energy change will be 7 times larger and the temperature change will be larger (but not by a factor of 7 because the specific heat of the diatomic gas is different from that of a monatomic gas).

QUESTION. Is it possible to have a complete thermodynamic cycle composed of (a) two adiabatic processes and one isothermal process or (b) two isothermal processes and one adiabatic process?

ANSWER. No, in both cases. Either case would require that one point of the cycle would have, passing through it, either two adiabatics or two isothermals. This is impossible because the slope of either curve is uniquely defined by the p, V coordinates of a point on a pV diagram.

EXAMPLES AND SOLUTIONS

Example 1

Consider n moles of an ideal gas that undergo the constant volume and constant pressure processes along the path shown in the Fig. 19-1a from the initial state a to b (a → b), then b to c (b → c), c to d (c → d), and finally back to a, d to a (d → a). These are two constant volume processes and two constant pressure process. Calculate (a) the work done by the system; and (b) the heat taken in by the system for each of these processes. (c) Show that the net heat taken in by the gas is equal to the net work done by the gas.

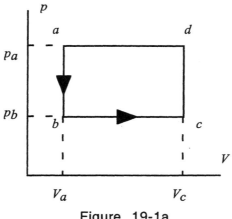

p

a d

p_a

p_b

b c

V

V_a V_c

Figure 19-1a

Solution:

(a) The work done by the ideal gas:

Process (a → b). No work is done because dV = 0 along the vertical line (a → b).

Process (b → c). $W_{b-c} = \int_b^c p\, dV = p_b \int_b^c dV = p_b (V_c - V_b) > 0$

Work is done <u>by</u> the gas.

Process (c → d). No work is done because dV = 0 along the vertical line (c → d).

Process (d → a). $W_{d-a} = \int_d^a p\, dV = p_a \int_d^a dV = p_a (V_a - V_c) < 0$

Work is done <u>on</u> the gas.

(b) Heat taken in by the ideal gas.

Process (a → b). $\Delta Q = nC_V \Delta T = nC_V(T_b - T_a) < 0$ [Constant volume process.]

Process (b → c). $\Delta Q = nC_p \Delta T = nC_p(T_c - T_b) > 0$ [Constant pressure process.]

Process (c → d). $\Delta Q = nC_V \Delta T = nC_V(T_d - T_c) < 0$ [Constant volume process.]

Process (d → a). $\Delta Q = nC_p \Delta T = nC_p(T_a - T_d) < 0$ [Constant pressure process.]

(c) If we sum all the contributions, we find that the net heat taken in is equal to the net work done by the gas.

$$\Delta Q_{net} = nC_V(T_b - T_a) + nC_p(T_c - T_b) + nC_V(T_d - T_c) + nC_p(T_a - T_d).$$

$$= nC_p(T_c - T_b + T_a - T_d) + nC_v(T_b - T_a + T_d - T_c)$$
$$= (nC_p - nC_v)(T_c - T_b + T_a - T_d) = nR(T_c - T_b + T_a - T_d)$$

Since $C_p = C_v + R$.

$$W_{net} = p_b(V_c - V_b) + p_a(V_a - V_c) = nRT_c - nRT_b + nRT_a - nRT_d = nR(T_c - T_b + T_a - T_d)$$

where we have used the ideal gas law in the last step to convert the products of p and V into the absolute temperature of the state in question.

Example 2

Referring to Fig. 19-1b, calculate the heat taken in by n moles of an ideal gas in going from a to c by the straight line process shown. Note that this process is neither a constant volume process nor a constant pressure process so we can't calculate the heat taken in directly as we did for the constant volume and constant pressure processes of Example 1.

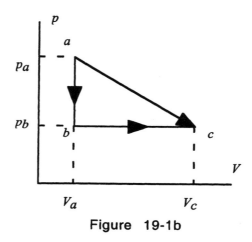

Figure 19-1b

Solution:

The change in internal energy, between points a and c is the same whether we go from a to c directly or go from a to b and then b to c. Even though the process from a to c is not a constant volume process, the change in internal energy is still given by:

$$\Delta U = nC_v(T_c - T_a)$$

The work done is the area of the triangle (area = base X height/2).

$$\Delta W = p_b(V_c - V_a) + (p_a - p_b)(V_c - V_a)/2 = (p_a + p_b)(V_c - V_a)/2$$

The net heat taken in is equal to

$$\Delta Q_T = \Delta W + \Delta U$$

$$= (p_a + p_b)(V_c - V_a)/2 + nC_v(T_c - T_a)$$

The heat taken in along this path and the work done by the gas are both different than computed in Example 1 using the processes (a → b) and (b → c) but the _difference_ between the heat taken in and the work done by the gas _is_ the same (independent of path).

Example 3

An ideal monatomic gas (n moles) is taken from point 1 on the T_1 isotherm to point 3 on the T_2 isotherm along the path 1→2→3 shown in Fig. 19-2.

    (a) In terms of the parameters shown, calculate the change in internal energy of the gas and the heat that must be added to it in this process.

    (b) Suppose the path 1→4→3 is followed instead. Calculate the heat added along this path.

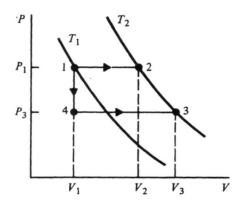

Figure 19-2

Solution:

For this gas assume C_v and C_p are known and constant.

(a) In going from 1 to 3, the internal energy change depends on the temperature change (but not on the path taken).

$$\Delta U = nC_v(T_3 - T_1)$$

The heat taken in along 1 to 2 can be written as:

$$\Delta Q = n\, C_p(T_2 - T_1)$$

or the first law can be used to give:

$$\Delta Q = \Delta U + \Delta W$$

$$\Delta Q = nC_v(T_2 - T_1) + p_1(V_2 - V_1)$$

Along the path 2 to 3, the internal energy change is zero (because the temperature is constant) so the heat taken in can be calculated from the work done:

$$\Delta Q = \Delta W = \int_{V_2}^{V_3} p\,dV = nRT_2 \int_{V_2}^{V_3} \frac{dV}{V}$$

$$\Delta Q = nRT_2 \ln\left(\frac{V_3}{V_2}\right)$$

Therefore along the path $1 \rightarrow 2 \rightarrow 3$,

$$\Delta Q = nC_p\left(T_2 - T_1\right) + nRT_2 \ln\left(\frac{V_3}{V_2}\right)$$

(b) Since the work done in the process from 1 to 4 is zero, the total work done is:

$$\Delta W = P_3(V_3 - V_1).$$

The change in internal energy is the same, $nC_v(T_2 - T_1)$, so:

$$\Delta Q = nC_v(T_2 - T_1) + P_3(V_3 - V_1)$$

If we subtract the heat taken in along $1 \rightarrow 2 \rightarrow 3$ from the heat taken in along $1 \rightarrow 4 \rightarrow 3$, the difference is equal to:

$$(P_1 - P_3)V_1 - nRT_2 \ln (V_3/V_2)$$

We can conclude that for different paths, ΔQ is different.

Example 4

One mole of an ideal monatomic gas starts at point A in Fig. 19-3 ($T_a = 273$ K, $P_a = 1$ atm.) and undergoes an adiabatic expansion to point B where $V_b = 2V_a$. This is followed by an isothermal compression to the original volume at C ($V_c = V_a$) and a pressure increase at constant volume back to the original pressure. Take $\gamma = 5/3$.
    (a) Compute the temperature at point B.
    (b) Compute the pressure at point C.
    (c) Compute the total work done for this cycle.

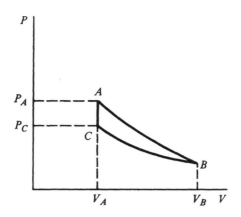

Figure 19-3

Solution:

(a) $T_b V_b^{\gamma-1} = T_a V_a^{\gamma-1}$

so ~

$$T_b = \frac{T_a V_a^{\gamma-1}}{V_b^{\gamma-1}} = (273 \text{ K})\left(\frac{V_a}{V_b}\right)^{\gamma-1}$$

$$T_b = (273 \text{ K})\left(\frac{V_a}{V_b}\right)^{(5/3)-1} = \frac{(273 \text{ K})}{(2)^{2/3}} = 172 \text{ K}.$$

(b) The temperature at C is the same as at B and $V_c = V_a$, so the ratio

$$\frac{p_a V_a}{T_a} = \frac{p_c V_c}{T_c}$$

simplifies to:

$$p_c = p_a\left(\frac{T_c}{T_a}\right)$$

$$p_c = (1 \text{ atm})\left(\frac{173 \text{ K}}{273 \text{ K}}\right) = 0.630 \text{ atm}.$$

(c) The total work done by the gas in the cycle is equal to the area ABC enclosed by the cycle in the P-V plane.

$$\Delta W_{AB} = -\Delta U_{AB} = -C_v(T_b - T_a) \quad \text{(since } \Delta Q = 0)$$

$$\Delta W_{BC} = RT_b[\ln(V_c/V_b)] \quad \text{(since T is constant)}$$

and the work done is zero from C to A since no volume change occurs. Summing these contributions:

$$W_{NET} = C_v(T_a - T_b) - RT_b \, (\ln 2)$$

Using $C_v = 3R/2$ for an ideal monatomic gas, we have

$$W_{NET} = 268 \text{ J.}$$

Example 5

For the thermodynamic cycle shown in Fig. 19-4 use the equation
$$TV^{\gamma - 1} = \text{constant}$$
and the ideal gas law to show that:

$$\frac{V_2}{V_1} = \frac{V_3}{V_4}$$

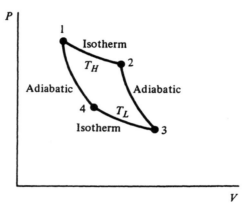

Figure 19-4

Solution:

For the adiabatic connecting points 1 and 4,

$$T_H V_1^{\gamma - 1} = T_L V_4^{\gamma - 1}$$

and for the adiabatic connecting 2 and 3,
$$T_H V_2^{\gamma - 1} = T_L V_3^{\gamma - 1}$$
Dividing the first equation by the second to eliminate the temperatures,
$$\left(\frac{V_1}{V_2}\right)^{\gamma - 1} = \left(\frac{V_4}{V_3}\right)^{\gamma - 1}$$
Since both sides are raised to the same power, we can conclude that

$$V_2/V_1 = V_3/V_4.$$

Example 6

An ideal diatomic gas initially at a pressure of 1 atm, a temperature of 300 K, and a volume of 0.05 m³ is compressed adiabatically to a final volume of 0.03 m³. Calculate the final temperature and pressure and the work done on the gas.

Solution:

For an adiabatic process, $TV^{\gamma-1}$ is constant, as is pV^{γ}. First we find the pressure, p_f,

$$p_f V_f^{\gamma} = p_i V_i^{\gamma}$$

or by re-arranging

$$\frac{p_f}{p_i} = \left(\frac{V_i}{V_f}\right)^{\gamma}$$

Substituting, using $\gamma = 1.40$ for a diatomic gas, we have

$$\frac{p_f}{(1\ \text{atm})} = \left(\frac{0.05\ \text{m}^3}{0.03\ \text{m}^3}\right)^{1.40} = 2.04$$

$p_f = 2.04$ atm.

We can find T_f by using either the ideal gas law, since p_f and V_f are now known, or the adiabatic relations

$$T_f V_f^{\gamma-1} = T_i V_i^{\gamma-1}$$

$$\frac{T_f}{T_i} = \left(\frac{V_i}{V_f}\right)^{\gamma-1}$$

Using the last formula we have

$$\frac{T_f}{(300\ \text{K})} = \left(\frac{0.05\ \text{m}^3}{0.03\ \text{m}^3}\right)^{1.40-1} = 1.23$$

Thus

$T_f = 368$ K.

To calculate the work done by the gas, we calculate the integral

$$\Delta W = \int_{V_i}^{V_f} p\ dV$$

where for p we substitute

$$p = \frac{K}{V^\gamma} \quad \text{where the constant } K = pV^\gamma \text{ at any particular point.}$$

Completing this integral gives:

$$\Delta W = \frac{K}{1-\gamma}\left[V_f^{1-\gamma} - V_i^{1-\gamma}\right]$$

This can be put in a more recognizable form by noting that

(a) $\quad KV_f^{1-\gamma} = p_f V_f^\gamma V_f^{1-\gamma} = p_f V_f = nRT_f$

(b) $\quad KV_i^{1-\gamma} = p_i V_i^\gamma V_i^{1-\gamma} = p_i V_i = nRT_i$

and

(c) $\quad 1 - \gamma = 1 - (C_p/C_v) = (C_v - C_p)/C_v = -R/C_v$

Making all these substitutions, we get for the work done by the gas,

$$\Delta W = -nC_v(T_f - T_i)$$

We note that this equals $-\Delta U$, a result dictated by the first law of thermodynamics and the fact that $\Delta Q = 0$ (so that $\Delta W = -\Delta U$) for an adiabatic process. In order to avoid the above integration and tricky substitutions, we can obtain the work done on the gas (the negative of the work done by the gas) by calculating ΔU directly. To finish this calculation we use the specific heat value for H_2 from Table 17-1,

$$C_v = 20.42 \text{ J/mol·K}$$

and calculate the number of moles present. At standard temperature and pressure one mole of an ideal gas occupies 0.0224 m3. Thus the number of moles is

$$n = \frac{(1 \text{ mole})(1 \text{ atm})(0.05 \text{ m}^3)(273 \text{ K})}{(1 \text{ atm})(0.0224 \text{ m}^3)(300 \text{ K})} = 2.03 \text{ moles}$$

Therefore the work done on the gas:

$$\Delta W_{on} = -\Delta W_{by} = (2.03)(20.42 \text{ J})(68) = 2821 \text{ J}$$

QUIZ

1. Two moles of argon gas (C_v = 12.47 J/mol·K) are compressed at a constant pressure of 2 atm from a volume of 5 liters to a volume of 2 liters. During this process, the absolute temperature increases by 20%. How much heat is taken in by the system?

Answer: Approximately 300 J of heat is rejected by the system in this process.

2. Calculate the work done by 1.5 moles of an ideal gas (C_v = 20.76 J/mol·K) in going along a straight line path in the p-V plane from the point (p_1, V_1) to the point (p_2, V_2) where p_1 = 3 atm, V_1 = 1 liter, p_2 = 1 atm, and V_2 = 2 liters. This problem can be done without integration.

Answer: 2 atm·liter or about 200 J.

3. An ideal gas with C_v = (5/2)R and γ = 1.4 expands adiabatically from an initial pressure and volume (p_i = 20 atm, V_i = 2 liters) to a final pressure of 1 atm. Assuming there are 1.5 moles of gas present, calculate (a) the initial temperature, (b) the final volume, (c) the final temperature, and (d) the work done by the gas.

Answer: (a) T_i = 325.6 K, (b) V_f = 17.0 liters, (c) T_f = 138.3 K, (d) W = 5838 J.

4. An ideal gas with γ = 1.50 is initially at a pressure of 1 atm and a temperature of T_a = 300 K (point a). Its volume is 0.60 liters. At *constant volume*, the gas absorbs heat until its temperature reaches T_b where the pressure is 2.5 atm (point b). The gas is then expanded adiabatically until its temperature is 300 K again (point c). (a) Calculate the number of moles in the system. (b) Calculate the temperature T_b. (c) Calculate the volume after the adiabatic expansion to point c. (d) Calculate the heat taken in for the constant volume process connecting points a and b.

Answer: (a) 2.44 x 10^{-2} , (b) 750 K, (c) 3.75 liters, (d) 183 J.

5. The internal energy of one mole of a "ficticious" ideal gas is given by

$$U = (5/2)RT + 3RT_0$$

where T_0 is a constant. Calculate (a) the specific heat at constant volume for this gas and (b) the value of $\gamma = C_p/C_v$.

Answer: (a) C_v = (5/2)R, (b) γ = (7/5) = 1.4.

20

THE SECOND LAW OF THERMODYNAMICS

OBJECTIVES

In this chapter the second law of thermodynamics is stated and the limitations that it places on physical processes are given. Heat engines and refrigerators are used to illustrate the concepts on which the second law is based. Your objectives are to:

Calculate the heat taken in and work done per cycle for several heat engines.

Calculate the thermal efficiency of heat engines and the coefficient of performance for refrigerators or heat pumps from the fundamental definitions.

Understand the relationship between the efficiency of a Carnot engine and the absolute temperatures of the hot and cold reservoirs.

Calculate the entropy changes that accompany both reversible and irreversible processes.

REVIEW

The conversion of heat energy into mechanical work is a topic of great practical importance. The sweeping general nature of thermodynamics allows us to know the fundamental limitations of proposed systems without examining each detail of the operation.

The heat engines of greatest importance are those that take the working substance, freon, gasoline, steam, ideal gas, etc., through a complete cycle. The internal energy, U, is a definite function of the thermodynamic coordinates, no matter what the working substance is, so that in a complete cycle we must have $\Delta U = 0$ as the system is returned to its initial state. Applied to this very same cycle, the first law of thermodynamics gives:

$Q = W$

Thus the net heat taken in during the entire cycle is equal to the net work done by the engine in the cycle. The quantity Q, the net heat, is actually the sum of the heat absorbed per cycle, Q_H, and the heat rejected per cycle, Q_C or $Q = Q_H + Q_C = W$. Since Q_C is a negative number, Q_H will be larger than W. The heat rejected per cycle is usually wasted so the thermal efficiency, e, is defined to be the ratio of the work done by the engine, W, to the heat absorbed, Q_H.

$$e = \frac{W}{Q_H} = \frac{Q_H + Q_C}{Q_H} = 1 - \left| \frac{Q_C}{Q_H} \right|$$

We should note here that since Q_C is negative (and less than Q_H in absolute value or W couldn't be positive) that $e \leq 1$. See Example 1.

Various types of heat engines can be analyzed in order to gain facility in calculating the quantities needed to obtain the efficiency. Some notable examples are: the internal combustion engine, the Diesel engine, and the steam engine (the cycle for this engine was given in Example 2 in Chapter 17). It is interesting to note that the efficiencies of the internal combustion engine and the Diesel engine depend only on the expansion and compression ratios (see Example 3) and not directly on the temperatures.

If the heat engine is run backwards (only reversible engines can be run backwards), it becomes a refrigerator (or a heat pump) in that mechanical work is supplied, heat is extracted from some "cold" reservoir and delivered, along with the work, to the "hot" reservoir. The coefficient of performance, K, of a refrigerator is the (negative) ratio of the heat extracted from the cold source to the mechanical work needed to do this.

$$K = \frac{Q_C}{W} = - \frac{Q_C}{Q_H + Q_C}$$

The coefficient of performance for a heat pump is

$$K' = \frac{Q_H}{W}$$

These coefficients are always positive (as Q_H and W are negative for the refrigerator) but can easily be greater than unity. From the K and e definitions, it is clear that they are related. In fact

$$K = \frac{1 - e}{e}$$

also

$$K' = K + 1$$

so that an inefficient Carnot heat engine, run backwards as a heat pump or refrigerator, has a high performance coefficient.

The second law of thermodynamics is stated in the following way: "It is impossible for any system to undergo a process in which it absorbs heat from a reservoir at a single temperature and converts it completely into mechanical work, while ending in the same state in which it began." There are several alternative statements but it has been shown that they are all equivalent. The second law places stringent limitations on physical processes and tells us that there are some things that we cannot do even though they are allowed by the conservation of energy, momentum, angular momentum, etc.

Closely connected with any discussion of the second law is the Carnot cycle in which one uses an ideal gas as a working substance and considers two adiabatic processes and two isothermal processes. We used this cycle in Chapter 19, Example 5. Since the irreversible process of heat flow due to a temperature gradient is avoided by letting heat enter or leave the system only during isothermal processes, this cycle has the highest efficiency of any thermodynamic cycle. We can use this fact to set limits of performance on real heat engines (which can never be as efficient as the idealized Carnot engines). See Examples 3 and 4.

It is important to remember that

$$e_{carnot} = 1 - \frac{T_C}{T_H}$$

is the efficiency of a Carnot engine where T_C and T_H are the absolute temperatures of the cold and hot reservoirs respectively. Clearly

$$K = \frac{T_C}{T_H - T_C} \; .$$

The above expression for the Carnot efficiency (e_{Carnot}) forms the basis of the absolute temperature scale. This scale is identical, for all practical purposes, to the ideal gas temperature scale but in fact is independent of the nature of the working substance.

Numerous examples of irreversible processes, processes that can proceed only in one direction in nature, are given in the text. All processes involving friction are irreversible (but there are many others as well). The reverse process to a sliding block coming to rest on a rough surface would be that of a block initially at rest on this surface to gain kinetic energy (and speed) by extracting heat energy from the surface thereby cooling it. The opposite of an irreversible process is a reversible one (the direction being unimportant) and such a process can be thought of as the sum of a series of infinitesimal processes that take the system from one equilibrium state to another nearby equilibrium state. Irreversible processes start and end at equilibrium states, but in between, the system is out of equilibrium.

It has been pointed out that work and heat are not simply functions of the thermodynamic coordinates, as internal energy is, but depend on the actual path taken between the initial and final states. The *entropy*, S, of a system is like the internal energy in that the entropy change in a given process depends only on the coordinates of the initial and final states and not on the path. For a reversible process the infinitesimal change in entropy can be defined as:

TdS = dQ

For a reversible process in which the system goes from state 1 to state 2, the entropy change, ΔS, is given by:

$$\Delta S = \int_1^2 \frac{dQ}{T}$$

Any path connecting 1 and 2 may be used to evaluate the integral. For an ideal gas the entropy is obtained using this definition in Example 5 and the entropy change occurring during a reversible isothermal expansion is calculated in Example 6.

As many of the processes we wish to study are irreversible ones, we need a technique for finding the entropy changes occurring there as our previous formula allows us to find ΔS only for a reversible process. If a system goes from state 1 to a state 2 by means of an irreversible process, we can still find the entropy change (since it depends only on the coordinates of the initial and final states) by inventing a reversible process that connects states 1 and 2 and then calculating the previous integral. This is illustrated in Example 7.

The critical statement in the text regarding entropy and its connection with the second law of thermodynamics is: "When all systems taking part in a process are included, the entropy either remains constant or increases." No process is possible in which the entropy of the universe decreases. The arrow representing the direction of time (from past to present to future) is oriented by the existence of irreversible processes.

PROBLEM-SOLVING STRATEGIES

For cycles, the change in internal energy is zero. Nevertheless it is easier to calculate the work done in an adiabatic process by calculating the change in internal energy *for that process* and then using the first law of thermodynamics. This approach will also work for constant volume processes where ΔQ can be obtained from a calculation of ΔU.

For irreversible processes the entropy change ΔS can be calculated by replacing the actual process by a reversible process connecting the initial and final states.

QUESTIONS AND ANSWERS

QUESTION. The temperature of the ocean's surface can reach approximately 30 C while at a kilometer below the surface its temperature is about 5 C. Is it possible to run a heat engine and extract useful energy under these conditions?

ANSWER. Yes. This process was suggested long ago and now is known by the acronym OTEC (Ocean Thermal Energy Conversion). It has worked in small scale demonstration projects but its Carnot efficiency is low, about .08 or 8%.

QUESTION. On a hot day, why not open the refrigerator door to cool the house?

ANSWER. If the refrigerator exhausts its heat into the house (as most do), more heat would enter the house than removed by the refrigerator so the house would actually warm up.

EXAMPLES AND SOLUTIONS

Example 1

Calculate the efficiency of a heat engine using an ideal monatomic gas as the working substance, following the cycle shown in Fig. 20-1. Take $\gamma = 1.4$ and let the ratio (r) of V_b to V_a be 2.5.

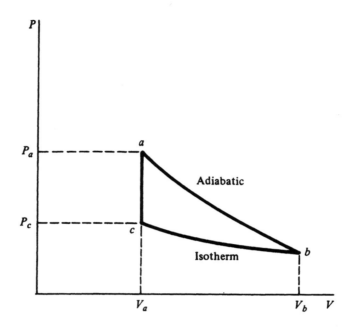

Figure 20-1

Solution:

No heat is taken in or rejected along ab. Since $\Delta U = 0$ along bc, we have

$$Q_C = W_{bc} = nRT_c \ln(V_a/V_b)$$

Since $V_a < V_b$ both Q_c and W_{bc} are negative. Along ca, the heat taken in is:

$$Q_H = nC_V(T_a - T_c).$$

Q_H is positive here since $T_a > T_c$.

There is no net change in U for this cyclic process so the net work done by the gas is equal to the sum of the above two terms. Along ab, we have

$$T_a V_a^{\gamma-1} = T_c V_b^{\gamma-1}$$

so

$$T_a = T_c \left(\frac{V_b}{V_a}\right)^{\gamma-1}$$

If the ratio of the two volumes is called $r = V_b/V_a$, the efficiency can be written in terms of r.

$$e = 1 + \frac{Q_C}{Q_H}$$

$$e = 1 - \frac{\left[(\gamma-1)\ln r\right]}{\left(r^{\gamma-1} - 1\right)} = 0.172$$

Increasing the ratio r will increase the efficiency.

Example 2

Calculate the efficiency of the Diesel cycle shown in Fig. 20-2 below, in terms of the temperatures T_a, T_b, T_c, and T_d.

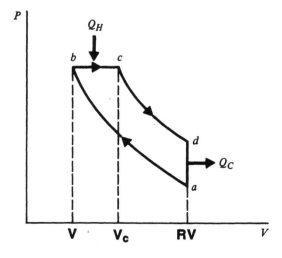

Figure 20-2

Solution:

R is the compression ratio and the expansion ratio is given by: $E = (RV/V_c)$.

Although these two ratios are the same in the internal combustion cycle, they are not in the Diesel cycle. Processes ab and cd are adiabatics so we have

$$Q_H = nC_p(T_c - T_b) > 0$$

and

$$Q_C = nC_V(T_a - T_d) < 0.$$

As the two processes where heat enters or leaves are constant pressure and constant volume processes respectively, the efficiency, e, is given by

$$e = \frac{Q_H + Q_C}{Q_H} = 1 + \frac{Q_C}{Q_H}$$

$$e = 1 - \frac{T_d - T_a}{\gamma \left(T_c - T_b \right)}$$

since $C_p/C_V = \gamma$. This is the complete answer but it can also be written in terms of the expansion and compression ratios, as will be seen in the next example.

Example 3

Evaluate the efficiency of the Diesel cycle of the previous example using R = 18, E = 6, and γ = 1.4.

Solution:

(a) Since points b and c are at the same pressure, we have

$$\frac{V}{T_b} = \frac{V_c}{T_c} \quad \text{and } E = \frac{RV}{V_c} \quad \text{so}$$

$$T_c = T_b \frac{R}{E}$$

(b) Along the adiabatic,

$$T_b V^{\gamma - 1} = T_a (RV)^{\gamma - 1}$$

Thus

$$T_b = T_a (R)^{\gamma - 1} \quad \text{and } T_c = T_a (R)^{\gamma - 1} \frac{R}{E}$$

(c) Finally along the adiabatic cd, we have

$$T_c V_c^{\gamma - 1} = T_d (RV)^{\gamma - 1}$$

We obtain

$$T_c = T_d (E)^{\gamma - 1}$$

and we can write

$$T_d = (E)^{1 - \gamma} R E^{-1} T_a = \left(\frac{R}{E}\right)^{\gamma} T_a$$

Now all temperatures are expressed in terms of T_a so the efficiency is:

$$e = 1 - \frac{\left[(R/E)^{\gamma} - 1\right]}{\gamma \left(R^{\gamma}/E - R^{\gamma}/R\right)}$$

This can also be written as:

$$e = 1 - \frac{\left[(1/E)^{\gamma} - (1/R)^{\gamma}\right]}{\gamma (1/E - 1/R)}$$

$$e = 1 - \frac{\left[(1/6)^{1.4} - (1/18)^{1.4}\right]}{(1.40)(1/6 - 1/18)} = 0.59.$$

Example 4

The mean temperature of deep ground water is 5°C whereas the mean house temperature is about 20°C.

    (a) What is the efficiency of a Carnot engine operating between these two temperatures?

    (b) What is the coefficient of performance of a Carnot refrigerator operating between these two temperatures?

    (c) What is the coefficient of performance of a Carnot heat pump here?

Solution:

(a) The efficiency is given in terms of Kelvin temperatures so the above Celsius temperatures must be converted.

$$e = 1 - \frac{T_C}{T_H} = 1 - \frac{278 \text{ K}}{293 \text{ K}}$$

$$e = 0.051 \quad \text{or} \quad 5.1\%$$

(b) The coefficient of performance, K, is given by:

$$K = \frac{T_C}{T_H - T_C} = \frac{278 \text{ K}}{293 \text{ K} - 278 \text{ K}} = 18.5$$

(c) The coefficient of performance for the heat pump is:

$$K' = K + 1 = 18.5 + 1 = 19.5$$

This device is practical for heating a house in winter.

Example 5

Calculate the entropy, S, for a system that obeys the ideal gas law.

Solution:

For an ideal gas, the internal energy depends on the temperature alone, so:

$$dU = nC_V dT$$

$$dW = p\, dV = \frac{nRT}{V}\, dV$$

Since the entropy is related to dQ through $TdS = dQ$ (for a reversible process only) the first law yields:

$$dQ = dU + dW$$

$$T\, dS = nC_V\, dT + \frac{nRT}{V}\, dV$$

Dividing by T and treating n, C_V, and R as constants,

$$dS = d(nC_V \ln T + nR \ln V + \text{const.})$$

We can regard the entropy as the quantity in the bracket above. It can be put in a more useful form by noting that

$$\frac{R}{C_V} = \gamma - 1$$

Then we have

$$S = nC_V \left[\ln T + (\gamma - 1) \ln V \right] + \text{constant.}$$

$$S = nC_V \ln \left(TV^{\gamma - 1} \right) + \text{constant.}$$

In an adiabatic process, where the entropy remains constant, it is apparent from this form that $TV^{\gamma - 1}$ must remain constant.

Example 6

Calculate the entropy change of an ideal gas during an isothermal expansion from an initial volume of V to a final volume of 2V.

Solution:

For the ideal gas dU = 0 in an isothermal process, so we have

$$T\, dS = dQ = 0 + dW$$

$$dW = p\, dV = \frac{nRT}{V}\, dV$$

which yields

$$dS = \frac{nR}{V}\, dV$$

Integrating from the initial volume to the final volume, we have

$$\Delta S = nR \ln 2V - nR \ln V = nR \ln 2$$

If we use the expression for the ideal gas entropy,

$$S = nC_V \ln \left(TV^{\gamma - 1} \right) + K$$

then,

$$S_f - S_i = \Delta S = nC_v \ln \left(T(2V)^{\gamma - 1} \right) - nC_v \ln \left(T(V)^{\gamma - 1} \right)$$

$$\Delta S = nC_v \ln \left(\frac{T(2V)^{\gamma - 1}}{T(V)^{\gamma - 1}} \right) = nC_v \ln (2)^{\gamma - 1} = nC_v \left(\gamma - 1 \right) \ln 2$$

$$= n R \ln 2 \qquad \text{[Note that } R = (\gamma - 1)C_V\text{]}$$

Example 7

Equal volumes of water at 80° C and 20° C are mixed together. Calculate the increase in entropy for a total volume of 1 m³.

Solution:

Assuming the heat capacities are all constant and independent of temperature, the final temperature is *midway* (i.e. 50° C) between the above two temperatures: the heat lost by one volume is gained by the other. To calculate the entropy change in this irreversible process, we must replace the actual process by a fictitious reversible one that still connects the initial and final states. In this case, let the volume remain fixed and find the heat taken in (or rejected) in a process that would take the system from the initial temperature to the final temperature. Since $dQ = C_V dT = TdS$ for this process,

$$\Delta S = C_v \int_{T_i}^{T_f} \frac{dT}{T} = C_v \ln \left(\frac{T_f}{T_i} \right)$$

For the "hot" water, $T_f = 323$ K while $T_i = 353$ K whereas for the "cold" water, $T_f = 323$ K and $T_i = 293$ K. Numerically

$$\frac{\Delta S}{C_v} = 2 \ln (323) - [\ln (353) + \ln (293)] = 8.66 \times 10^{-3}$$

For C_V we can take 4.18×10^6 J/m³·(C°). Since the volume is 1 m³, the final result for ΔS is

$$\Delta S = 3.62 \times 10^4 \text{ J/C}° = 3.62 \times 10^4 \text{ J/K}.$$

QUIZ

1. For a Carnot engine with operating temperatures of 800 K and 300 K, (a) calculate the efficiency. (b) To improve the efficiency, it is decided to either raise the higher temperature by 10 K or to lower the lower temperature by 10 K. Which would result in the largest improvement?

Answer: (a) e = 0.625 (b) Reducing the lower temperature by 10 K gives e = 0.6375 whereas raising the higher temperature by 10 K only gives e = 0.6296.

2. In a thermally isolated container, 60 g of water at 80 C is mixed with 40 g of water at 20 C. Assume the specific heat of water is constant. (a) Calculate the final temperature of the mixture and (b) the entropy change of the system.

Answer: (a) T_f = 56 C or 329 K, (b) ΔS = 1.72 J/K.

3. An ideal gas with γ = 5/3 is used for a thermodynamic cycle that starts at point $\underline{a}$ where p,V, and T are p_0, V_0, and T_0. The gas expands to point $\underline{b}$ along a straight line path until both the pressure and volume are double that at point $\underline{a}$. Next the gas goes to point $\underline{c}$, where the pressure is p_0 along a constant volume path. Finally the gas returns to point $\underline{a}$ along a constant pressure path. The figure is a triangle in the pV plane. Calculate the thermodynamic efficiency of this cycle.

Answer: e = (1/12).

4. An ideal gas with γ = 5/3 is used for a thermodynamic cycle that starts at point $\underline{a}$ where p,V, and T are p_0, V_0, and T_0. Heat added to the gas in going from $\underline{a}$ to $\underline{b}$, at constant volume, causes the pressure to double. Next the gas expands at the constant pressure of $2p_0$ until the volume at point $\underline{c}$ is double the original volume. In going from $\underline{c}$ to $\underline{d}$, heat is removed, at constant volume until the pressure equals the original pressure (p_0). Finally, the gas returns to point $\underline{a}$ at constant pressure. This cycle is a rectangle in the pV plane with twice the area of the triangle in the previous question. (a) Calculate the thermodynamic efficiency of this cycle and (b) that of a Carnot engine operating between the highest and lowest temperatures in the cycle.

Answer: (a) e = (1/5); (b) e_C = (3/4).

5. The pressures, volumes, and temperatures of three points in the pV plane are: p_a = 3 atm, V_a = 6 L, T_a = 500 K, p_b = 2 atm, V_b = 9 L, $T_b = T_a$, p_c = 1.53 atm, $V_c = V_b$, T_c = 381.6 K. A thermodynamic cycle of an ideal gas (γ = 5/3) starts at point $\underline{a}$ and proceeds to $\underline{b}$ by an isothermal process. From $\underline{b}$ to $\underline{c}$, a constant volume process is followed. From $\underline{c}$ to $\underline{a}$, the process is adiabatic. (a) Calculate the efficiency of this cycle and (b) that of a Carnot engine operating between 500 K and 381.6 K.

Answer: (a) e = 0.124, (b) e_C = 0.237.